P9-CDH-178

# FOUNDATIONS OF HIGHER MATHEMATICS

# ABOUT THE AUTHORS

## PETER FLETCHER

(Ph.D., University of North Carolina at Chapel Hill) is professor emeritus at Virginia Polytechnic Institute and State University where he taught for twenty-six years. He is a coauthor of a research monograph on quasi-uniform spaces (Marcel-Dekker, 1982) and continues to do research in this area and in number theory.

## C. WAYNE PATTY

(Ph.D., University of Georgia) has been a member of the Mathematics Department at Virginia Polytechnic Institute and State University since 1967 and was Department Head for twenty-five years. Prior to joining the faculty at Virginia Polytechnic Institute and State University he was Associate Professor at the University of North Carolina at Chapel Hill. Professor Patty has written research articles in topology as well as undergraduate texts in this and other fields.

3rd EDITION

# FOUNDATIONS OF HIGHER MATHEMATICS

**PETER FLETCHER**
**C. WAYNE PATTY**
*VPI & State University*

Australia • Brazil • Japan • Korea • Mexico • Singapore • Spain • United Kingdom • United States

**Foundations of Higher Mathematics, Third Edition**
**Peter Fletcher, C. Wayne Patty**

Sponsoring Editor: Steve Quigley

Production Coordinator: Monique A. Calello

Production Services: York Production Services

Editorial Assistant: Anna Aleksandrowicz

Manufacturing Coordinator: Wendy Kilborn

Marketing Manager: Marianne Rutter

Cover/Text Designer: Kathleen Wilson

Art Studio: York Production Services

Compositor: The Charlesworth Group

© 1996, 1992, 1988 Brooks/Cole, Cengage Learning

ALL RIGHTS RESERVED. No part of this work covered by the copyright herein may be reproduced, transmitted, stored, or used in any form or by any means graphic, electronic, or mechanical, including but not limited to photocopying, recording, scanning, digitizing, taping, Web distribution, information networks, or information storage and retrieval systems, except as permitted under Section 107 or 108 of the 1976 United States Copyright Act, without the prior written permission of the publisher.

For product information and technology assistance, contact us at
**Cengage Learning Customer & Sales Support, 1-800-354-9706**

For permission to use material from this text or product,
submit all requests online at **cengage.com/permissions**
Further permissions questions can be emailed to
**permissionrequest@cengage.com**

**Library of Congress Cataloging-in-Publication Data**
Fletcher, Peter
Foundations of higher mathematics/Peter Fletcher & C. Wayne Patty. — 3rd ed.
p. cm.
Includes bibliographical references and index.
ISBN-13: 978-0-534-95166-5 (hardcover)
ISBN-10: 0-534-95166-X (hardcover)
1. Mathematics. I. Patty, C. Wayne. II. Title.
QA39.2.F59 2000 95-35544
511.3—dc20 CIP

**Brooks/Cole**
10 Davis Drive
Belmont, CA 94002-3098
USA

Cengage Learning is a leading provider of customized learning solutions with office locations around the globe, including Singapore, the United Kingdom, Australia, Mexico, Brazil, and Japan. Locate your local office at: **international.cengage.com/region**

Cengage Learning products are represented in Canada by Nelson Education, Ltd.

For your course and learning solutions, visit **academic.cengage.com**

Purchase any of our products at your local college store or at our preferred online store **www.ichapters.com**

Printed in the United States of America
17 18 19 20 21 15 14 13 12

# CONTENTS

# PREFACE

*Foundations of Higher Mathematics*, Third Edition, introduces students to basic techniques of writing proofs and acquaints them with some fundamental ideas that are used throughout mathematics. We have in mind such topics as the relationship between equivalence relations and partitions, techniques such as mathematical induction and the pigeonhole principle, theorems such as the Euclidean algorithm and the Schroeder-Bernstein Theorem, and definitions of such fundamental terms as countable set, function, and indexed family. As even this abbreviated list indicates, there is much mathematics left in the no-man's-land between freshman calculus and advanced undergraduate mathematics. But our principal concern is not content, it is what the artist Ben Shahn has called "the shape of content." Results and even their demonstration are merely the context within which students learn to discover mathematical truth. The students have no doubt already taken courses in which the objective is to develop the skill to use results and arguments that have been thought up by others. The purpose of our text is to develop the students' ability to think mathematically and to distinguish mathematical thinking from wishful thinking. For this reason, the answer section is much abbreviated, and the hints given there, while valid, do not necessarily lead to an easy or straightforward proof. Similarly, the text's examples are only general examples, not models of some routine. The students' understanding is our paramount concern. We have chosen not to prove the binomial theorem by induction, because the combinatorial argument makes clear what makes the theorem work. Many texts present the axiom of induction in terms of a sequence $P(n)$ of statements. We think students gain a better understanding of the axiom if it, like the Least-Natural-Number Principle, is stated as an axiom about subsets of natural numbers. Twice we have glossed over subtleties. In the chapter on sets, we have taken $A=B$ to mean by definition that $A \subseteq B$ and $B \subseteq A$. That isn't quite right. In the chapter on induction we have confessed the need for the Theorem of Inductive Definition, but we have not given this theorem or its proof.

Curiously, in both cases what proves difficult is not solving the problem, but understanding that there is a problem that needs to be solved.

Although our starting points, logic and set theory, are each vast subjects of mathematics, we have taken from them only what we need to get going. Thus, set theory is presented as intuitive set theory, with a level of naïveté far exceeding what Paul Halmos has labeled "Naive Set Theory." Our presentation of logic is also restrained. We introduce only enough formal logic to allow students to determine whether or not some given piece of reasoning is spurious. Thus many of the most important laws of logic never appear in our text. But we have taken care to distinguish between proof by contrapositive and proof by contradiction, and to indicate when an indirect proof may be more likely to succeed than a direct argument. Our experience has been that students, after starting out uneasy about indirect proofs, soon become so enamored with proofs by contradiction that they must be prodded to use a direct proof, whenever such a proof is available.

Our coverage of induction is somewhat more extensive than is customary. Induction is often left to high school courses or to brief discussions tucked away as appendixes to freshman calculus texts. We believe, however, that the axiom is too important to be left outside the college curriculum or to be wrestled with alongside the concepts of limit and derivative. Most of the problems at the end of Section 3.3 are accessible to those who have covered only the first two sections, so Section 3.3 can be treated as a well-stocked larder of interesting problems involving the Axiom of Induction.

Although our presentation of relations and functions follows the usual set-theoretic approach based on Kasmir Kuratowski's definition of ordered pair, we have made the concepts of both function and relation concrete by giving many examples in the plane and by stressing that the study of relations and the study of directed graphs are one and the same. Directed graphs clarify the relationship between partitions and equivalence relations (in that the components of the directed graph of an equivalence relation are the members of the associated partition). Directed graphs also aid in the study of permutations and of linear and partial orders. Many texts do not cover order relations, but we believe that their inclusion is justified by the importance of orders in algebra, analysis, and computer science.

The study of equivalence relations and partitions is both geometric and algebraic. For example, from geometry and topology we have the notion of identification, which allows vivid geometric interpretations, and from algebra there is Karl Gauss's construction of congruence classes and the construction of a field (group) of quotients. These latter constructions lead to the fundamentally algebraic issue of determining whether certain natural operations are well defined. A historical presentation would begin with Gauss's congruence relations, but we have found it helpful to start with identifications, which at first seem more concrete and "real" to most students.

This third edition implements changes suggested by those who have taught from the previous editions. We have made few structural changes to the text. The chapter on combinatorial proofs has been repositioned so that it follows the chapter on functions; much of the material on graphs that was in the chapter on relations is now in the chapter on combinatorics. A chapter introducing the careful use of limits and epsilon-delta proofs replaces a chapter on the development of the real-number system.

Although the structure of the text has not been altered greatly, the third edition represents a major revision. At the authors' university, the administration recently initiated a program, "Writing across the curriculum." This program calls for written projects from students in every discipline. The course for which *Foundations of Higher Mathematics* is intended is a natural place to emphasize written work, and we have incorporated many features in this third edition to encourage more writing. Early on, we have addressed problems students are likely to encounter in writing proofs and we have expanded the bibliography to include articles that students can use as the basis for short essays as well as articles by well-known mathematicians about mathematical writing. Throughout the text, we have added writing exercises sufficient to form the basis for a semester-long project in which students build a portfolio of written work. Finally, we have included anecdotes about some of the mathematicians whose mathematics we encounter. These anecdotes are not intended as substitutes for biographies, but we hope that they may pique students' interest in the history of mathematics by showing that those who have created our subject had personalities, as well as personae and feelings, as well as thoughts.

There are several different one-semester courses that can be taught from this text; consequently, what constitutes the core material depends to a large extent on the discretion of the instructor. The instructor who is using the text for the first time should plan to cover the following material from the first five chapters: Sections 1.1 through 1.5, 2.1, 2.2, 3.1 through 3.3 (and 3.4 and 3.5 if the instructor plans to cover Chapter 8), 4.1 through 4.5, 5.1 and 5.5.

Because the concept of function is fundamental to the study of freshman and sophomore calculus, the average student will come to Chapter 5 already knowing a great deal. (It is our experience, however, that every class should work through Sections 5.1 and 5.5.) Sections 5.2 and 5.6 are needed only in Chapters 7 and 8. As the content of Section 5.4 is taken straight from freshman calculus, this section may be treated as review or omitted entirely.

The remaining chapters are included to challenge the student's understanding of what has come before, and to demonstrate the value of the theorems and methods of proof that have been previously mastered. Any chapter may be selected and it is also possible to hop about selecting a few topics from several chapters. Chapter 6 provides an introduction to combinatorial arguments. This chapter is especially important to those students who plan to pursue discrete mathematics or computer science; for other students we consider it an antidote to the chapter on induction, because it presents core results that are better established by counting than by induction. The chapter comprises a proof of the Binomial Theorem and the Principle of Inclusion-Exclusion and ends with a brief introduction to graph theory.

Chapter 7 is the most abstract and challenging of the remaining work. Here is a recipe for taking some of the sting from this chapter. The proof of the Schroeder-Bernstein Theorem may be omitted. Indeed, it is a valuable to exercise to disallow the theorem, on the eminently reasonable grounds that it has not been proved, and actually construct the needed one-to-one functions (see Exercises 13(b) and 21, for example). Section 7.4 presents a proof that the countable union of countable sets is countable. We have taken some care in this proof to show exactly where it needs the (countable) Axiom of Choice. Section 7.4 can be replaced by Cantor's famous zig-zag argument that the countable union of countably infinite sets is countably infinite.

After completing the first five chapters of this text, many instructors may prefer to move directly to algebra or analysis; the last two chapters give an introduction to these subjects. The pace of these last chapters is more rapid than in previous chapters, and most of the proofs are left to the student. These two chapters are completely independent, but between them they make use of virtually all that has come before.

Previously, we have been emphasizing how to select material for a one-semester course. We believe that in a full year's course an instructor will be able to cover all the material and end with the thorough review that Chapters 8 and 9 provide.

## CORE

| Chapter | Sections |
|---|---|
| 1 | 1.1 through 1.5* |
| 2 | 2.1 and 2.2 |
| 3 | 3.1 and 3.2† |
| 4 | 4.1 through 4.5 |
| 5 | 5.1 and 5.5 |

* The discussion of the element-chasing method from Section 1.6 is also recommended.

† Even if the instructor does not cover Section 3.3, some of the problems at the end of Section 3.3 should be assigned.

## ACKNOWLEDGMENTS

We have incorporated the ideas and insights of our reviewers and thank them for sharing their experiences with us: Edward A. Azoff, *The University of Georgia*; Sherry Ettlich, *South Oregon State College*; Paul Fairbanks, *Bridgewater State College*; Holly B. Puterbaugh, *The University of Vermont*; Lt. Col. John S. Robertson, *U.S. Military Academy at West Point*; James K. Strayer, *Lock Haven University*; and Mark E. Watkins, *Syracuse University*.

We also thank Steve Quigley, our editor at PWS Publishing Company, for his professional support and encouragement. Thanks also to Monique Calello, our production coordinator, Wendy Kilborn, our manufacturing coordinator, and Anna Aleksandrowicz, our editorial assistant. Finally we wish to thank Tamra Winters, of York Production Services, for her first-rate work in the production of this text.

***Peter Fletcher***
***C. Wayne Patty***

# FOUNDATIONS OF
# HIGHER MATHEMATICS

CHAPTER

# 1 The Logic and Language of Proofs

In higher-level mathematics courses, students are asked to construct proofs. The students must first "see in their minds" why the theorem they are trying to prove is correct. That is, they must discover a method of proof. Once this discovery has been made, they must be able to communicate their discovery to others who "speak the language." Since a proof is a communication of an argument, the standards by which a proof is deemed acceptable are to some degree a matter of judgment. One colleague, teacher, or referee might accept an argument that another would not. But there is a set of rules, the laws of logic, and the purpose of this chapter is to study the rules that govern the structure and presentation of mathematical proofs. Happily, the rules we need are nowhere near so complicated as those that govern, say, baseball or cricket.

## 1.1 *Propositions*

Vast progress has been made in understanding the foundations of mathematics. During the nineteenth century, investigation revealed that the lack of rigor in calculus was due largely to a lack of understanding of the real number system. During the latter part of this period, Richard Dedekind (pronounced Day-de-kin(d), 1831–1916) succeeded in constructing the real numbers from the integers. With the set theory developed primarily by Georg Cantor (1845–1918) at the end of the nineteenth century, mathematicians were able to define the integers directly in terms of the concept of a set. By the beginning of the twentieth century, it became clear that many basic mathematical concepts could be created entirely from logical concepts. Soon, however, intuitive reasoning about sets led to paradoxes; that is, to statements that seemed to be both true and false. A new type of scientist, the mathematical logician, came into being.

We do not attempt to present a systematic development of logic, but we do discuss informally some ideas that are useful in proving theorems.

**Definition**

A **proposition** (or **statement**) is a sentence that is either true or false.

Some examples of propositions are: "$2+3=6$," "May 9, 1986, was a Friday," and "Skunks have a pair of glands from which a secretion with a pleasant odor is ejected."

Some sentences that are not propositions are "He is handsome," "What time is it?", and "This sentence is false." Imagine the last sentence as a true–false question. The question would be unfair exactly because the sentence is not a proposition. If it is true, then it is false; and if it is false, then it is true.

Since a proposition must be a sentence that is either true or false, no question can be a proposition. The sentence, "He is handsome," is not a proposition for two reasons. First, we do not know to whom "he" refers, and even if we did, there might be disagreement over whether he is handsome. Hereafter, all sentences we use in this text are presumed to be statements. Thus if we later consider the sentence "Sally Lou has blue eyes," we expect that you will not worry who Sally Lou is or what it means to say that her eyes are truly blue.

In ordinary English, we are all adept at building new propositions from old ones. There are three basic ways of building a new proposition. We can connect two propositions with "and," we can connect two propositions with "or," or we can negate a proposition. In this text, we use capital letters $P$, $Q$, $R$, and so on, to represent propositions, $P \wedge Q$, to represent $P$ *and* $Q$, $P \vee Q$ to represent $P$ *or* $Q$, and $\neg P$ to represent the denial of a proposition $P$.

Suppose we are given boxes, $\Box$ and $\Diamond$, into which we can place propositions. Since, by definition, any proposition placed in the box can have only two truth values, we can tabulate the behavior of $\wedge$, $\vee$, and $\neg$ and thereby define these signs by listing all possible inputs and the corresponding possible outputs in tabular form.The resulting tables are called truth tables for $\wedge$, $\vee$, and $\neg$. The definitions of $\wedge$, $\vee$, and $\neg$ are given by the following truth tables.

AND

| $\Box$ | $\Diamond$ | $\Box \wedge \Diamond$ |
|---|---|---|
| T | T | T |
| T | F | F |
| F | T | F |
| F | F | F |

OR

| $\Box$ | $\Diamond$ | $\Box \vee \Diamond$ |
|---|---|---|
| T | T | T |
| T | F | T |
| F | T | T |
| F | F | F |

NOT

| $\Box$ | $\neg\Box$ |
|---|---|
| T | F |
| F | T |

The truth table for $\wedge$ gives meaning to this symbol by specifying that $P \wedge Q$ is true whenever $P$ and $Q$ are both true propositions and false when either $P$ or $Q$ is false. You can quickly verify that the table exhausts all combinations of truth values for any two propositions $P$ and $Q$ that might be placed in the boxes □ and ◇. Thus, given any two propositions $P$ and $Q$, the truth table for $\wedge$ determines the truth value of $P \wedge Q$ in terms of the truth values of $P$ and $Q$.

**Definition**

The proposition "$P \wedge Q$" is referred to as the **conjunction** of $P$ and $Q$.

We can think of "$P \wedge Q$" as "$P$ and $Q$"; that is, $P \wedge Q$ is true if and only if

$P$ is true and also $Q$ is true.

Each of $P$ and $Q$ is true.

$P$ and $Q$ are both true.

The conjunction of $P$ and $Q$ is false if and only if at least one of $P$ or $Q$ is false.

We illustrate the truth-value combinations for conjunctions with the following four compound propositions.

$P$: Broccoli is a vegetable and $3 < \sqrt{17}$.

$Q$: Mercury is the closest planet to the sun and $3 > \sqrt{17}$.

$R$: Shakespeare was a Roman emperor and $|3| = |-3|$.

$S$: Zimbabwe is a country in South America and 8 is a prime number.

Proposition $P$ is true because it is composed of two simple propositions, each of which is true. Propositions $Q$ and $R$ are both false because each is composed of one proposition that is true and one that is false. Proposition $S$ is false because it is composed of two simple propositions, each of which is false.

**Definition**

If $P$ is a proposition, the assertion that $P$ is false is commonly known as the **denial**, or **negation**, of $P$.

The proposition "$\neg P$" means "not $P$," and it is used to denote the negation of $P$. For example, let $P$ be the proposition "5 is an odd integer." Then $P$ is true and hence the negation ($\neg P$), "5 is not an odd integer," is false. If $Q$ is the proposition "8 is prime," then $Q$ is false. Thus the negation ($\neg Q$), "8 is not prime," is true.

**Definition**

The proposition "$P \vee Q$" is referred to as the **disjunction** of $P$ and $Q$.

We can think of "$P \vee Q$" as "$P$ or $Q$"; that is, $P \vee Q$ is true if and only if at least one of $P$ or $Q$ is true. It is important that we understand the use of the word *or*. Suppose you walk into a fast-food restaurant and place an order. The person behind the counter says, "Is this to eat here or to take out?" and you respond, "Yes." Obviously there is a lack of communication. The person behind the counter is using the word *or* in its strongest sense (that is, as an **exclusive** disjunction meaning $P$ or $Q$ but not both), and you are using the word *or* in the mathematical sense (that is, as an **inclusive** disjunction meaning $P$ or $Q$ or both). Remember, to the mathematician, "$P$ or $Q$" means $P$ or $Q$ or both. Incidentally, the Latin word *vel* means *or* in the inclusive sense, and one way to remember the meaning of $\vee$ is to think of the v in vel. Another way is to remember that the *d*isjunction symbol points *d*own.

The disjunction of $P$ and $Q$ is false if and only if $P$ and $Q$ are false.

We illustrate the truth-value combinations for disjunctions with the following four compound propositions.

$P$: Gauss was a mathematician or $\pi > 3$.

$Q$: Fermat was a mathematician or the earth is flat.

$R$: The earth is flat or 16 is an even number.

$S$: Pasteur invented the airplane or the moon is made of pasteurized cheese.

Proposition $P$ is true because it is composed of two simple propositions, each of which is true. Propositions $Q$ and $R$ are both true because each is composed of one proposition that is true and one that is false. Proposition $S$ is false because it is composed of two propositions, each of which is false.

**Definition**

The two connectives, $\rightarrow$ and $\leftrightarrow$, called the **conditional** and the **biconditional**, are defined by the following truth tables.

CONDITIONAL
(implies)

| $\square$ | $\diamond$ | $\square \rightarrow \diamond$ |
|---|---|---|
| T | T | T |
| T | F | F |
| F | T | T |
| F | F | T |

BICONDITIONAL
(if and only if)

| $\square$ | $\diamond$ | $\square \leftrightarrow \diamond$ |
|---|---|---|
| T | T | T |
| T | F | F |
| F | T | F |
| F | F | T |

**Definition**

The proposition "$P \rightarrow Q$" is referred to as a **conditional** proposition.

The conditional proposition $P \rightarrow Q$ occurs frequently in our daily lives, and the English language has many ways in which to express this proposition. Here are some of the ways in which we can express that $P \rightarrow Q$ is true.

If $P$, then $Q$.

If $P$ is true, then $Q$ is also true.

$P$ is true only if $Q$ is true.

$P$ implies $Q$.

$Q$ is true whenever $P$ is true.

For $P$ to be true, it is necessary for $Q$ to be true.

For $Q$ to be true, it is sufficient for $P$ to be true.

Let's examine the truth values of $P \rightarrow Q$. Suppose Peter makes the statement "If the sun is shining, then Wayne is playing golf." In order to determine when the statement "$P$ implies $Q$" is true, let's ask ourselves when Peter is telling the truth. In the first case (that is, when the sun is shining and Wayne is playing golf), Peter has told the truth. In the second case (when the sun is shining, but Wayne is not playing golf), Peter has not told the truth. In the last two cases, we would not want to say that Peter is a liar because the sun is not shining: Peter said only that something would happen if the sun were shining. Therefore, the statement "$P$ implies $Q$" is true in three of the four possibilities.

**Definition**

In the conditional proposition "If $P$, then $Q$," $P$ is called the **hypothesis**, or **condition**, and $Q$ is called the **conclusion**. (Other commonly used terms are **antecedent** in place of hypothesis and **consequent** in place of conclusion.)

We illustrate the truth value combinations for the conditional with the following four compound propositions.

$P$: If $\pi > 3$, then 7 is a positive integer.

$Q$: If a robin is a bird, then 4 is a prime number.

$R$: If a cauliflower is a bird, then a robin is a bird.

$S$: If 5 is an even integer, then no one will pass this course.

Proposition $P$ is true because it is composed of two propositions, each of which is true. Proposition $Q$ is false because it is composed of a true hypothesis and a false conclusion. Propositions $R$ and $S$ are both true because the hypothesis of each is false. Note that although $S$ is true, $S$ does not say that no one will pass this course. The truth table of the conditional is easy to remember because there is only one case in which the implication is false—namely, when the hypothesis (antecedent) is true and the conditional (consequent) is false.

Consider again the proposition "If the sun is shining, then Wayne is playing golf." This is, of course, a proposition of the form $P \rightarrow Q$, and there is another proposition, $Q \rightarrow P$, that states that "If Wayne is playing golf, then the sun is shining." This proposition, $Q \to P$, is called the **converse** of $P \to Q$. Note that for an avid golfer like Wayne, the proposition "If the sun is shining, then Wayne is playing golf" may be true, but its converse may not be true. The formal definition of converse is as follows.

**Definition**

If $P$ and $Q$ are propositions, then the **converse** of $P \to Q$ is $Q \to P$.

If $P$ is the conditional proposition "If Au is the symbol for gold, then Be is the symbol for lead," then the converse of $P$ is the conditional proposition "If Be is the symbol for lead, then Au is the symbol for gold." Since $P$ is false and the converse of $P$ is true, we have an example that clearly shows that a conditional proposition and its converse can have opposite truth values. This is illustrated in the following table.

| $P$ | $Q$ | $P \to Q$ | $Q \to P$ |
|---|---|---|---|
| T | T | T | T |
| T | F | F | T |
| F | T | T | F |
| F | F | T | T |

There are two cases in which $P \to Q$ and $Q \to P$ have the same truth value, but there are also two cases in which they do not.

**Definition**

If $P$ and $Q$ are propositions, then the **contrapositive** of $P \to Q$ is $\neg Q \to \neg P$.

The contrapositive of "If $4 < 3$, then the moon is made of cheese" is "If the moon is not made of cheese, then $4 \geqslant 3$." Note that when the contrapositive of a conditional proposition is written, the denial of the conclusion becomes the hypothesis, and the denial of the hypothesis becomes the conclusion. In the preceding example, both the conditional proposition and the contrapositive of this conditional proposition are true. As we will see in the next section, this is not merely a coincidence.

**Definition**

The proposition "$P \leftrightarrow Q$" is referred to as a **biconditional** proposition.

We can think of "$P \leftrightarrow Q$" as follows:

$P$ if and only if $Q$.

For $P$ to be true, it is necessary and sufficient for $Q$ to be true.

$P$ implies $Q$ and $Q$ implies $P$.

$P$ is equivalent to $Q$.

The abbreviation "iff" is sometimes used for "if and only if." An example of a biconditional proposition is "The sun is shining if and only if Wayne is playing golf." In order for this biconditional proposition to be true, the proposition "If the sun is shining, then Wayne is playing golf" and its converse must be true.

A biconditional proposition "$P$ if and only if $Q$" involves two conditional propositions, "If $P$, then $Q$" and "If $Q$, then $P$." Note that "$P$ if and only if $Q$" is true whenever "If $P$, then $Q$" and "If $Q$, then $P$" are both true; and "$P$ if and only if $Q$" is false whenever either "If $P$, then $Q$" or "If $Q$, then $P$" is false.

---

## EXERCISES 1.1

**1.** Which of the following are propositions?

**a)** $2^2 + 3^2 = 17$

**b)** $8x^3 + 6x^2 - 4x + 2$

**c)** If $n$ is a positive integer, then the sum of the first $n$ positive integers is given by $n(n+1)/2$.

**d)** Will you marry me?

**e)** No, I am already married to Leslie!

**2.** Use the definition of $\rightarrow$ to determine the conditions under which each of the following compound propositions is true.

**a)** If $8 > 5$, then $3 < 1$.

**b)** If $a = 2$, then $1 < 3$.

**c)** If $5 > 8$, then $3 < 1$.

**d)** If $1 < 3$, then $a = 2$.

**e)** If $1 > 3$, then $a = 2$.

**3.** Identify the hypothesis and the conclusion in each of the following compound propositions.

**a)** If Mary is 24 years old, then I am a monkey's uncle.

**b)** $n^2$ is odd whenever $n$ is an odd integer.

**c)** That $r$ is a rational number implies that $r^2$ is rational.

**d)** When $a$ is irrational, $a^2 + a$ is irrational.

**e)** When $a$ is rational and $b$ is irrational, $a + b$ is irrational.

**f)** If $a$ is irrational, then $a^2$ and $2a$ are irrational.

**g)** In order to pass the driver's test, the candidate must be able to parallel park.

**h)** In order to pass the vision test, it is sufficient for the candidate to read the line Q S Z P W M 4.

**4.** Assume that "Mary is a girl" is a true statement and that "Mary is ten years old" is a true statement. Which of the following are true?

**a)** If Mary is ten years old, then Mary is a girl.

**b)** Mary is ten years old if and only if Mary is a girl.

**5.** Assume that "Joe is a girl" is a false statement and that "Mary is ten years old" is a true statement. Which of the following are true?

**a)** If Mary is ten years old, then Joe is a girl.

**b)** If Joe is a girl, then Mary is ten years old.

**6.** Assume that "Joe is a girl" is a false statement and that "Joe is ten years old" is a false statement. Which of the following are true?

**a)** Joe is ten years old or Joe is a girl.

**b)** If Joe is ten years old, then Joe is a girl.

**c)** Joe is ten years old if and only if Joe is a girl.

**d)** Joe is not a ten year-old girl.

**7.** Write the converse and the contrapositive of each of the following propositions.

**a)** If $\sqrt{2} < \sqrt{5}$, then $2 < 5$. **b)** If $2 \geqslant 5$, then $\sqrt{2} \geqslant \sqrt{5}$.

**8.** On a certain island (Manhattan), the inhabitants are divided into two types, those who always tell the truth and those who always lie. One day a visitor stops three inhabitants of the island to ask directions to a well-known museum (The Guggenheim). "All three of us are liars," warns the first inhabitant, "Not so; only two of us are liars," says the second. "Not so," says the third, "the other two guys are lying." Which, if any, of the three islanders can the visitor trust to give honest directions?

**9.** A visitor to an island whose inhabitants *always* tell the truth or *always* lie encounters an islander who makes the following two statements.

**1.** I love Bertha.

**2.** If I love Bertha, then I love Sally Lou.

Does the islander love Bertha? Does he love Sally Lou? Argue persuasively in a sentence or two that your assessment is correct.

**10.** Jimmy the Greek has been asked to give odds on two basketball games to be played by an obscure college we will call "Tech." Tech is to play State U and Southern U. Although Jimmy is quite knowledgeable about the Boston Celtics, the truth is that he has not scouted the Tech team in years. It is suggested that Jimmy should consult the oracles, which he does. The oracle at Delphi says, "Tech will beat State U or Tech will beat Southern U." Jimmy is quite pleased with this information until someone mentions to him that the oracles are known for their equivocal sayings and that he should seek a second opinion just to be on the safe side. He does. The oracle at Xanthi says, "If Tech doesn't beat State U, then Tech will beat Southern U." Now Jimmy is distraught. He has the vague feeling that one of the oracles is saying more than the other, but he's not sure. Straighten Jimmy out. [*Authors' remark*: We hope that the word ORacle reminds you of the word OR.]

**11.** William Sessions, Director of the F.B.I., addressed the National Press Club in September 1988 and was asked a final question, which he promised to answer: "As a Texan, would you like to see a Texan President or a Texan Vice-President?" Bearing in mind that the Republicans were running a Texan for President (George Bush) and the Democrats were running a Texan for Vice-President (Lloyd Bentsen) and bearing in mind that William Sessions is an intelligent man, guess correctly the answer Sessions gave.

**12.** Explain why the connective $\oplus$ defined by

| $\square$ | $\diamond$ | $\oplus$ |
|---|---|---|
| T | T | F |
| T | F | T |
| F | T | T |
| F | F | F |

is sometimes called *exclusive or*.

**13. a)** Write the converse of the contrapositive of the following statement: If $3 > 1$, then $5 > 1$.

**b)** Write the contrapositive of the converse of the following statement: If $3 > 1$, then $5 > 1$.

**14.** Do there exist propositions $p$ and $q$ such that both $p \to q$ and its converse are true? Do there exist propositions $p$ and $q$ such that both $p \to q$ and its converse are false. Justify each answer with an example or argument.

## 1.2 *Expressions and Tautologies*

In the preceding section, we considered several fundamental ways in which to combine propositions to form compound propositions and we used boxes (squares and diamonds) as placeholders in writing down the associated truth tables. There is an obvious disadvantage to our placeholder notation. What other figures can we use? We could use pentagons, hexagons, and so on, but if we are considering a compound proposition made up of many simple propositions, it is annoying to count up sides, and although the human eye can readily distinguish between a square and a pentagon, the distinction between a small nonagon and a small decagon is not so easily apparent. We therefore introduce the idea of a **propositional expression**. We are given an alphabet of **variables**, and we have $\vee$, $\wedge$, $\neg$, and parentheses. We add the signs $\to$ and $\leftrightarrow$. Using variables, these signs, and parentheses, we form expressions. We rely on the reader's intuition rather than attempt a formal definition of a propositional expression. Each variable is a propositional expression, and if the preceding symbols are put together in a way that appears meaningful, the result is an expression. For example, $\neg(X \vee Y) \wedge (\neg X \to (Y \leftrightarrow Z))$ is an expression, as is $(X \vee Y) \leftrightarrow (X \wedge Y)$, but we hope that you do not believe that $XX \to)) \wedge (Y \leftrightarrow$ is also an expression. Note that it makes no sense to ask if the expression $(X \vee Y) \leftrightarrow (X \wedge Y)$ is true. As it is written, the expression has no meaning whatsoever, but when we replace the variables $X$ and $Y$ with propositions $P$ and $Q$, we obtain a proposition $P \vee Q \leftrightarrow P \wedge Q$, which like any other proposition is either true or false.

**Remark** A proposition is either true or false. An expression has no meaning until its variables are replaced by propositions; thus an expression cannot be true or false. Each expression has a truth table. A proposition has no truth table, because by definition a proposition has only one truth value.

**Definition**

Two expressions are **equivalent** provided they have the same truth values for all possible values of true or false for all variables appearing in either expression. When two expressions $X$ and $Y$ are equivalent, we write $X \Leftrightarrow Y$ and say that $X$ and $Y$ are **logically equivalent**.

The signs $\leftrightarrow$ (biconditional) and $\Leftrightarrow$ (logical equivalence) should not be confused. The first is one of the signs we can use to **form** expressions, but the second sign cannot appear in any propositional expression. In other words, an expression may contain the biconditional ($\leftrightarrow$), but logical equivalence ($\Leftrightarrow$) tells something about the relationship of two expressions. We consider some examples to illustrate the notion of equivalent expressions.

**EXAMPLE 1**

Are the expressions $\neg X \vee Y$ and $X \to Y$ equivalent?

**ANALYSIS** Since we have only two variables, the truth table of each expression has four rows.

| $X$ | $Y$ | $X \to Y$ |
|---|---|---|
| T | T | T |
| T | F | F |
| F | T | T |
| F | F | T |

| $X$ | $Y$ | $\neg X$ | $\neg X \vee Y$ |
|---|---|---|---|
| T | T | F | T |
| T | F | F | F |
| F | T | T | T |
| F | F | T | T |

We have checked all possible truth values, and in each of the four cases the resulting values for $X \to Y$ and $\neg X \vee Y$ are the same. Therefore, the expressions are equivalent. ❑

**EXAMPLE 2**

Are the expressions $W \vee X$ and $Y \vee X$ equivalent?

**ANALYSIS** Substituting a false proposition for $W$, a false proposition for $X$, and a true proposition for $Y$ yields a false proposition for the first expression and a true proposition for the second expression. Therefore, the two expressions are not equivalent. ❑

It is instructive to compare the truth tables of the two expressions in Example 2.

So far we have been using an alphabet of variables consisting of $W$, $X$, and $Y$ and have reserved the letters $P$, $Q$, $R$, and $S$ to name propositions. The laws of logic, however, are generally stated using the letters $P$, $Q$, $R$, and $S$; thus, from now on we will use the letters both for the names of propositions and as variables of our expressions. The result, unfortunately, is instantaneous ambiguity. If we write $P \to Q$, we may be referring to the expression $P \to Q$, in which case $P$ and $Q$ are variables, or we may have in mind two propositions such as

$P =$ George Washington slept at the Waldorf Astoria, and

$Q =$ The Waldorf Astoria is a historic site,

in which case $P \to Q$ is a proposition.

Since you are required to determine from context whether you are considering a proposition rather than an expression, it is worthwhile to consider an example that illustrates how to make this determination.

**EXAMPLE 3** Which of the following statements are propositions and which are expressions?

**a)** If $2 = 3$, then $\sqrt{2}$ is a rational number.

**b)** $P$ and $Q$ are solutions of the equation $x^2 - 5x + 6 = 0$.

**ANALYSIS** The statement in part (a) is true; therefore, it is a proposition. The statement in part (b) has no meaning until $P$ and $Q$ are replaced by propositions. Thus this statement is an expression. If $P$ is replaced by the proposition $x = 2$ and $Q$ is replaced by the proposition $x = 3$, then the resulting proposition is true. If $P$ is replaced by $x = 4$ and $Q$ is replaced by $x = 5$, then the resulting proposition is false. ❑

Some compound expressions always yield a true proposition no matter what propositions replace their variables, and some compound expressions always yield a false proposition. Consider, for example, the truth tables for the expressions $P \vee \neg P$ and $P \wedge \neg P$, which follow.

| $P$ | $\neg P$ | $P \vee \neg P$ |
|---|---|---|
| T | F | T |
| F | T | T |

| $P$ | $\neg P$ | $P \wedge \neg P$ |
|---|---|---|
| T | F | F |
| F | T | F |

Note that $P \vee \neg P$ is always true and that $P \wedge \neg P$ is always false.

**Definition** A propositional expression is called a **tautology** if it yields a true proposition regardless of what propositions replace its variables. A propositional expression is called a **contradiction** if it yields a false proposition regardless of what propositions replace its variables.

An important tautology is $(P \to Q) \leftrightarrow (\neg Q \to \neg P)$; that is, a conditional proposition is equivalent to its contrapositive. As usual, we consider all possible truth values of $P$ and $Q$ to see that this propositional expression is a tautology.

| $P$ | $Q$ | $P \to Q$ | $\neg Q$ | $\neg P$ | $\neg Q \to \neg P$ | $(P \to Q) \leftrightarrow (\neg Q \to \neg P)$ |
|---|---|---|---|---|---|---|
| T | T | T | F | F | T | T |
| T | F | F | T | F | F | T |
| F | T | T | F | T | T | T |
| F | F | T | T | T | T | T |

Five more tautologies that are useful from time to time are given by the following proposition. Parts (c) and (d) are called **de Morgan's laws** in honor of the logician Augustus de Morgan (______*–1871).

Personal Note

## Augustus de Morgan (______*–1871)

De Morgan was born in India, but his family returned to England when he was less than a year old. After obtaining an undergraduate degree at Trinity College, Cambridge, de Morgan refused to sign that he was a member of the Church of England (even though his family were prominent members of the Church); with this refusal he became ineligible for a graduate degree or position at Trinity College. He accepted the position of Professor of Mathematics at London University, a university founded on the principle of religious (and sectarian) freedom. There he became a colleague of another famous mathematician who had been denied access to Trinity College on account of religious prejudice, J. J. Sylvester. After 30 years service to London University, de Morgan resigned his position when he became convinced that the University's refusal to hire a Unitarian minister to chair its department of philosophy was based on religious prejudice.

**Proposition 1.1**

Each of the following propositional expressions is a tautology.

**a)** $P \vee Q \leftrightarrow (\neg P \to Q)$
**b)** $\neg(P \to Q) \leftrightarrow P \wedge \neg Q$
**c)** $\neg(P \vee Q) \leftrightarrow \neg P \wedge \neg Q$
**d)** $\neg(P \wedge Q) \leftrightarrow \neg P \vee \neg Q$
**e)** $\neg(\neg P) \leftrightarrow P$

**PROOF** We prove part (d) by considering all possible combinations of truth values of $P$ and $Q$. We leave the proof of the remaining parts of Proposition 1.1 as an exercise.

---

*We have not given the year of de Morgan's birth, since it reveals the solution to de Morgan's own mathematical problem:

"I was $x$ years of age in the year $x^2$."

| $P$ | $Q$ | $\neg P$ | $\neg Q$ | $P \wedge Q$ | $\neg(P \wedge Q)$ | $\neg P \vee \neg Q$ | $\neg(P \wedge Q) \leftrightarrow \neg P \vee \neg Q$ |
|---|---|---|---|---|---|---|---|
| T | T | F | F | T | F | F | T |
| T | F | F | T | F | T | T | T |
| F | T | T | F | F | T | T | T |
| F | F | T | T | F | T | T | T |

As we can see, $\neg(P \wedge Q) \leftrightarrow \neg P \vee \neg Q$ is always true, so it is a tautology. (You are asked in Exercise 29 to give an alternative proof of part (d) of Proposition 1.1.) □

A few comments concerning the meaning of the tautologies of Proposition 1.1 are appropriate at this point. Consider the following problem: Prove that if $x$ and $y$ are real numbers such that $xy = 0$, then $x = 0$ or $y = 0$. We are considering an implication of the form $H \rightarrow P \vee Q$, where $H$ is the hypothesis "$x$ and $y$ are real numbers such that $xy = 0$," $P$ is "$x = 0$," and $Q$ is "$y = 0$." According to Proposition 1.1(a), to establish the conclusion $P \vee Q$, it suffices to show that $\neg P \rightarrow Q$. The proof proceeds along this path. Suppose that $x \neq 0$ ($\neg P$). Then $1/x$ is a real number, and $y = (1/x)(xy) = (1/x)(0) = 0$. (1) Are we finished? (2) What about the case $y \neq 0$? (3) Must we consider the case $y \neq 0$ and prove that in this case $x = 0$? The correct answers to these questions are (1) yes, (2) irrelevant, and (3) no. If you did not answer these questions correctly, you have not yet realized the full power of Proposition 1.1(a).

Proposition 1.1(b) tells us how to deny an "if, then" statement. Consider the statement "If the sun is shining, then Wayne is playing golf." We are considering a statement of the form $P \rightarrow Q$, where $P$ is the statement "The sun is shining," and $Q$ is the statement "Wayne is playing golf." Is the denial "If the sun is not shining, Wayne is playing golf" ($\neg P \rightarrow Q$)? How about, "If the sun is not shining, Wayne is not playing golf" ($\neg P \rightarrow \neg Q$) or "If the sun is shining, Wayne is not playing golf" ($P \rightarrow \neg Q$)? According to part (b), the denial is $P \wedge \neg Q$; that is, the denial is "The sun is shining, and Wayne is not playing golf." A comparison of the following table with the table to be constructed in Exercise 28(b) shows that none of the other three statements constitutes a correct denial.

| | | | | FIRST TRY | SECOND TRY | THIRD TRY |
|---|---|---|---|---|---|---|
| $P$ | $Q$ | $\neg P$ | $\neg Q$ | $\neg P \rightarrow Q$ | $\neg P \rightarrow \neg Q$ | $P \rightarrow \neg Q$ |
| T | T | F | F | T | T | F |
| T | F | F | T | T | T | T |
| F | T | T | F | T | F | T |
| F | F | T | T | F | T | T |

Part (c) says that we deny a compound proposition involving *or* by replacing the *or* with *and* and denying the two simple propositions. Consider the statement "It is raining, or Wayne is playing golf." According to part (c), the denial is $\neg P \wedge \neg Q$; that is, the denial is "It is not raining, and Wayne is not playing golf."

Part (d) is the analogue of part (c). It says that we deny a compound proposition involving *and* by replacing the *and* with *or* and denying the two simple propositions. Consider the statement "The sun is shining, and Wayne is playing golf." This is the compound statement $P \wedge Q$, where $P$ is the statement "The sun is shining," and $Q$ is the statement "Wayne is playing golf." According to part (d), the denial is $\neg P \vee \neg Q$; that is, the denial is "The sun is not shining or Wayne is not playing golf." If "The sun is shining and Wayne is playing golf" is not true, then is it possible that "The sun is not shining, and Wayne is not playing golf"?

We could list many tautologies, but this is not a textbook for a logic course. In fact, in this book, we are interested in tautologies only as aids for adjudicating any disagreements about the logic used to write a proof. Therefore, we have chosen to list only some of the tautologies that you may find helpful in proving theorems. In particular, many obvious tautologies are not listed.

**Proposition 1.2**

Each of the following propositional expressions is a tautology.

**a)** $P \rightarrow P$

**b)** $(P \leftrightarrow Q) \rightarrow (Q \leftrightarrow P)$

**c)** $(P \leftrightarrow Q) \leftrightarrow (\neg P \leftrightarrow \neg Q)$

**d)** $[(P \rightarrow Q) \wedge (Q \rightarrow P)] \leftrightarrow (P \leftrightarrow Q)$

**e)** $[(P \rightarrow Q) \wedge (Q \rightarrow R)] \rightarrow (P \rightarrow R)$ (Transitivity)

**f)** $[(P \leftrightarrow Q) \wedge (Q \leftrightarrow R)] \rightarrow (P \leftrightarrow R)$ (Transitivity)

**g)** $[P \vee (Q \wedge R)] \leftrightarrow [(P \vee Q) \wedge (P \vee R)]$ (Distributivity)

**h)** $[P \wedge (Q \vee R)] \leftrightarrow [(P \wedge Q) \vee (P \wedge R)]$ (Distributivity)

**i)** $[(P \vee Q) \vee R] \leftrightarrow [P \vee (Q \vee R)]$ (Associativity)

**j)** $[(P \wedge Q) \wedge R] \leftrightarrow [P \wedge (Q \wedge R)]$ (Associativity)

**k)** $\{(\neg P) \rightarrow [Q \wedge (\neg Q)]\} \rightarrow P$

**PROOF** The proof of Proposition 1.2 is left as an exercise. □

**Proposition 1.3**

Each of the following propositional expressions is a tautology.

**a)** $(P \leftrightarrow Q) \rightarrow (R \wedge P \leftrightarrow R \wedge Q)$

**b)** $(P \leftrightarrow Q) \rightarrow (R \vee P \leftrightarrow R \vee Q)$

**Proposition 1.4**

Each of the following expressions is a contradiction.

**a)** $(P \to Q) \wedge (P \wedge \neg Q)$

**b)** $[(P \vee Q) \wedge \neg P] \wedge (\neg Q)$

**c)** $(P \wedge Q) \wedge (\neg P)$

**PROOF** We prove part (a) and leave the proofs of parts (b) and (c) as exercises.

By Proposition 1.1(b), $P \wedge \neg Q \leftrightarrow \neg(P \to Q)$. By definition, $\neg(P \to Q) \wedge (P \to Q)$ is a contradiction. Therefore, $(P \to Q) \wedge (P \wedge \neg Q)$ is a contradiction by Proposition 1.3(a). (Of course, it can also be proved by constructing the appropriate table.) □

As another example of a proof of a logical proposition that does not rely on the construction of a table, we deduce Proposition 1.1(c) from Propositions 1.1(a), 1.2(c), 1.1(b), and 1.2(f).

By Proposition 1.1(a), $P \vee Q \leftrightarrow \neg P \to Q$. Therefore, by Proposition 1.2(c), $\neg(P \vee Q) \leftrightarrow \neg(\neg P \to Q)$. By Proposition 1.1(b), $\neg(\neg P \to Q) \leftrightarrow \neg P \wedge \neg Q$. Hence, by Proposition 1.2(f), $\neg(P \vee Q) \leftrightarrow \neg P \wedge \neg Q$.

## Exercises 1.2

**15.** Prepare a truth table for each of the following expressions.

**a)** $P \wedge Q \to P \vee Q$ **b)** $P \to (Q \to R)$ **c)** $(P \to Q) \to R$

**16.** Which of the following pairs of expressions are equivalent?

**a)** $\neg(P \vee Q), \neg P \vee \neg Q$ **b)** $\neg P \wedge \neg Q, \neg(P \wedge \neg Q)$ **c)** $\neg(P \leftrightarrow Q), \neg P \leftrightarrow Q$

**17. a)** Using only negation and disjunction, find an expression in terms of variables $P$ and $Q$ that is equivalent to $P \to Q$.

**b)** Verify your answer to part (a) using an appropriate table.

**18.** Construct a truth table for each of the following expressions.

**a)** $\neg(P \wedge Q)$ **b)** $\neg P \wedge \neg Q$

**c)** $(P \to Q) \leftrightarrow (\neg Q \to \neg P)$ **d)** $P \to (Q \to R)$

**e)** $(P \to Q) \to R$ **f)** $\neg(P \vee Q) \to R$

**g)** $\neg((P \vee Q) \to R)$

**19.** Let $P$ be the proposition $2 = 5$. If possible, find a proposition $Q$ such that $P \vee Q \leftrightarrow P \wedge Q$ is true. If possible, find a proposition $Q$ such that $P \vee Q \leftrightarrow P \wedge Q$ is false.

**20. a)** Write the truth tables for the expressions $(P \leftrightarrow Q) \to R$ and $P \leftrightarrow (Q \to R)$.

**b)** Find propositions $P$, $Q$, and $R$ such that one of the two compound propositions $(P \leftrightarrow Q) \to R$ and $P \leftrightarrow (Q \to R)$ is true and the other is false.

**21.** Under the assumption that $P \to Q$ is false, give truth values for $Q \to P$, $P \leftrightarrow Q$, $P \vee Q$, and $\neg P \wedge \neg Q$.

**22.** Which of the following statements are true?

**a)** $(P \leftrightarrow Q) \Leftrightarrow (Q \leftrightarrow P)$ **b)** $\neg(P \leftrightarrow Q) \Leftrightarrow (\neg P \to \neg Q)$

**c)** $(\neg P \leftrightarrow Q) \Leftrightarrow (P \leftrightarrow \neg Q)$ **d)** $(\neg P \to Q) \Leftrightarrow (\neg P \wedge \neg Q)$

**23.** Let $P$, $Q$, and $R$ be variables. Prove that each of the following expressions is a tautology.

**a)** $P \to P$ **b)** $\neg(\neg P) \leftrightarrow P$

**c)** $P \wedge (Q \vee R) \leftrightarrow (P \wedge Q) \vee (P \wedge R)$ **d)** $\neg(P \vee Q) \leftrightarrow \neg P \wedge \neg Q$

**24.** Prove that each of the following propositional expressions is a contradiction.

**a)** $((P \vee Q) \wedge \neg P) \wedge \neg Q$ **b)** $(P \wedge Q) \wedge \neg P$

**25.** Write, in symbols, the converse, the contrapositive, and the negation of each of the following propositional expressions.

**a)** $P \to (Q \vee R)$ **b)** $P \to (Q \wedge R)$ **c)** $(P \vee Q) \to R$ **d)** $(P \wedge Q) \to R$

**26.** Which of the following expressions is logically equivalent to $P \to Q$?

**a)** $P \to Q$ **b)** $\neg P \to Q$ **c)** $\neg Q \to P$ **d)** $\neg P \to \neg Q$

**e)** $\neg Q \to \neg P$ **f)** $P \vee \neg Q$ **g)** $P \wedge \neg Q$ **h)** $\neg P \vee Q$

**i)** $\neg P \wedge Q$ **j)** $P \vee Q$ **k)** $P \wedge Q$ **l)** $Q \to P$

**27.** Suppose each of the following three statements is true.

John is smart.

John or Mary is ten years old.

If Mary is ten years old, then John is not smart.

Which of the following statements are true?

**a)** Mary is ten years old.

**b)** John is ten years old.

**c)** Either John or Mary is not ten years old.

**28.** By constructing truth tables, prove each of the following.

**a)** Proposition 1.1(a) **b)** Proposition 1.1(b) **c)** Proposition 1.1(c)

**29.** Assuming Propositions 1.1(c) and (e) and Proposition 1.2(c), deduce Proposition 1.1(d) without constructing a table.

**30.** Prove each of the following.

**a)** Proposition 1.2(d) **b)** Proposition 1.2(e)

**31.** Use the results of Exercise 30 to show that if $P$, $Q$, and $R$ are propositions and $P \to Q$, $Q \to R$, and $R \to P$ are true, then either $P$, $Q$, and $R$ are all true or $P$, $Q$, and $R$ are all false. *Do not construct a truth table.*

**32.** Prove each of the following parts of Proposition 1.2.

**a)** Part (a) **b)** Part (b) **c)** Part (c) **d)** Part (f)

**e)** Part (g) **f)** Part (h) **g)** Part (i) **h)** Part (j)

**i)** Part (k)

**33.** Prove Proposition 1.3.

34. Prove each of the following.
 a) Proposition 1.4(b)  b) Proposition 1.4(c)

35. Write a useful negation of each of the following propositions.
 a) If $2 \neq 4$, then $f(2) \neq f(4)$.  b) If $4 > 2$, then $f(4) > f(2)$.
 c) If $a < b$, then $a^2 < b^2$.  d) If $ab = 0$, then $a = 0$ or $b = 0$.

36. Write the contrapositive of each of the following propositions.
 a) If $ab \neq 0$, then $a = 0$ or $b = 0$.  b) If $a \neq 0$ or $b \neq 0$, then $ab \neq 0$.

37. Write the converse of each of the propositions in Exercise 36.

38. Write the negation of the propositions in Exercise 36.

39. Write the negation and the contrapositive of the following statement: If $x$ is an even integer or $x > 17$, then $x$ is a multiple of 4 and $x \geqslant 5$.

*40. Give an example of two equivalent expressions that do not have the same truth table.

## 1.3 Quantifiers

Since our primary purpose is to teach you how to prove theorems, we have discussed propositional logic with this in mind. We now wish to discuss another topic in logic—namely, quantifiers. Be aware that our description of this topic is brief; we do not give the in-depth approach required for a rigorous presentation such as would be found in a course on logic. We hope, however, that enough is said here to help you write the negation of propositions and to prepare you to use quantifiers in the study of sets.

By a **sign**, we mean a mark that can be recognized, but we are interested only in mathematical signs. In this section, we discuss two types of signs, **constants** and **variables**. We are given a universe from which all constants are drawn, and so it is easy to say what a constant is; it is just a member of our given universe. A sentence such as "$x < 4$" is not a proposition. It contains a **variable**, which we could replace with a specific real number to make the sentence a proposition. Such a sentence is called a **propositional function** or **open sentence**. The set of objects that can replace the variable (in this case, the set of real numbers) is called the **universe** or **set of meanings**, and the word, or sign, in a propositional function to be replaced is called a **variable**. Thus, for example, in the open sentence "$x < 4$," the sign $x$ is being used as a variable. Any object in the set of meanings must be a **constant**; that is, it must belong to the given universe. The collection of objects in the set of meanings that can be substituted to make a propositional function a true proposition is called the **truth set** of the propositional function. Thus the truth set of "$x < 4$" , the set of all real numbers that are less than 4.

*An exercise requiring the use of noncore material or one that is particularly difficult is marked with an asterisk.

The expression "$x$ is the least positive even integer" is a *propositional function.* Because the variable $x$ is present, this is not a proposition. If we replace $x$ with 2, we obtain a true proposition. If $x$ is replaced by any other constant, we obtain a false proposition. Thus the truth set contains only 2, assuming, of course, that 2 is in our universe of constants.

Suppose that the set of meanings for each of the following propositional functions is the set of real numbers.

**1.** $x^2 + 2x + 16 = 0$

**2.** $\sin^2 x + \cos^2 x = 1$

**3.** $x^2 - 5x + 6 = 0$

The truth set of (1) is empty, the truth set of (2) is the set of real numbers, and the truth set of (3) consists of only the numbers 2 and 3.

Of course, a propositional function may have more than one variable. If the set of meanings for the propositional function "$x + y = 2$" is the plane, then the truth set is the line with slope $-1$ and $y$-intercept 2.

Note that the sentence "For all real numbers $x$, $x + 2 = 2 + x$" is not a propositional function, because this sentence is a true sentence. The $x$ occurs in this expression as an **apparent** or **bound variable**. If a variable occurs in an expression in such a way that in order to turn the expression into a proposition it is necessary to replace the variable with a constant, then the variable occurs in the expression as an **actual** or **free variable**. The variables in the following expressions all occur as actual variables:

The population of $x$ is over one million.

$x + 3 = 7$

$x$ is married to $y$.

Note that the set of meanings of each of these propositional functions is clear by context.

In the propositional function $\int_2^x y^2\,dy = 39$, $x$ occurs as an actual variable and $y$ occurs as an apparent variable (see Exercise 51).

Phrases such as

for some $x$

there exists an element $x$

there are $x$ and $y$

are called **existential quantifiers**, and the symbol $\exists$ is used to denote them. Each of the following propositions contains an existential quantifier:

**1.** There is a real number $x$ such that $x + 2 = 5$.

**2.** There are real numbers $x$ and $y$ such that $x + y = 7$ and $x - 1 = 1$.

**3.** There exists a real number $x$ such that $x^2 - 5x + 6 = 0$.

**4.** $x^3 + 1 = 0$ for some real number $x$.

Phrases such as

for each $x$

for any $x$

for all $x$

are called **universal quantifiers**, and the symbol $\forall$ is used to denote them. Each of the following propositions contains a universal quantifier:

**1.** For each real number $x$, $x^2 \geqslant x$.

**2.** Any even integer is a multiple of 5.

**3.** $x + 1 > x$ for all real numbers $x$.

If $P(x)$ is a propositional function, then "For all $x$, $P(x)$" is denoted by $(\forall x)P(x)$, and "There exists $x$ such that $P(x)$" is denoted by $(\exists x)P(x)$. The two quantifiers can be used together in a single statement. For example, "For every real number $x$, there is a real number $y$ such that $2^y = x$" is denoted by $(\forall x)(\exists y)(2^y = x)$. Note that the variables governed by the existential and universal quantifiers occur as bound variables, which is why $(\forall x)P(x)$ and $(\exists x)P(x)$ are propositions and not merely propositional functions.

Sometimes the universal and existential quantifiers can be tucked away in a statement so unobtrusively that it is hard to tell that they are there. We call such occurrences of these quantifiers *hidden quantifiers.* The statement "If $n$ is a positive integer, then $n$ is the sum of 4 perfect squares," has two hidden quantifiers, even though the telltale phrases "all," "any," "there are," "for each," and "for some" are not used in formulating the statement. We can find the hidden quantifiers by recasting the sentence: "If $n$ is any positive integer, then there are four perfect squares whose sum is $n$." The expression "If $n$ is any positive integer" is denoted by

$$(\forall n)(n \text{ is a positive integer})$$

and the expression "There are four perfect squares whose sum is $n$" is denoted by

$$(\exists p_1^2, p_2^2, p_3^2, p_4^2)(n = p_1^2 + p_2^2 + p_3^2 + p_4^2),$$

and we can write

$$(\forall n)(n \text{ is a positive integer})(\exists p_1^2, p_2^2, p_3^2, p_4^2)(n = p_1^2 + p_2^2 + p_3^2 + p_4^2)$$

It is particularly easy to hide a universal quantifier. Here is a list of some of the more common disguises of this quantifier.

| | |
|---|---|
| Whenever | $x^2 > 0$, whenever $x > 0$. |
| If | If $n$ is even, $n^2$ is even. |
| A | A rolling stone gathers no moss. |

For a — For a decimal to represent a fraction, it is sufficient that it be a terminating decimal.

(All) — (All) Repeating decimals can be written as fractions.

As we will see, for certain propositions, before we can prove anything by either the contrapositive method or the contradiction method, we must first write the negation. How do we write the negation of propositions that involve quantifiers? Let's consider some examples. If $P$ is the proposition "For each real number $x$, $x^2 \geqslant 0$," what is the negation of $P$? First, $P$ says that something (namely, $x^2 \geqslant 0$) is true for each real number $x$. So if $P$ is not true, then there must be at least one real number $x$ such that $x^2 < 0$. Therefore, the negation of $P$ is "There is a real number $x$ such that $x^2 < 0$." In symbols, $P$ is the proposition $(\forall x)(x^2 \geqslant 0)$ and $\urcorner P$ is the proposition $(\exists x)(x^2 < 0)$.

If $Q$ is the proposition "There is a real number $x$ such that $x^2 + 2 = 0$," what is the negation of $Q$? Since $Q$ says that we can find a real number $x$ such that $x^2 + 2 = 0$, the negation of $Q$ should say that we cannot find such a real number; that is, no matter which real number $x$ we have, it is not the case that $x^2 + 2 = 0$. Thus a useful denial of $Q$ is "For each real number $x$, $x^2 + 2 \neq 0$." In symbols, $Q$ is the proposition $(\exists x)(x^2 + 2 = 0)$ and $\urcorner Q$ is the proposition $(\forall x)(x^2 + 2 \neq 0)$.

Note that in the preceding examples, we have denied the propositions $(\forall x)P(x)$ and $(\exists x)Q(x)$ by replacing $\forall$ with $\exists$ and $\exists$ with $\forall$ and denying $P(x)$ and $Q(x)$. The following theorem tells us that this is precisely the method we use in order to deny propositions that involve quantifiers. In presenting and discussing this theorem, we say that two propositions are "equivalent" as shorthand for saying that both propositions are true or both propositions are false.

**Theorem 1.5**

If $P(x)$ is a propositional function with variable $x$, then

**a)** $\urcorner[(\forall x)P(x)]$ is equivalent to $(\exists x)\urcorner[P(x)]$.

**b)** $\urcorner[(\exists x)P(x)]$ is equivalent to $(\forall x)\urcorner[P(x)]$.

**PROOF**

**a)** Suppose the proposition $\urcorner[(\forall x)P(x)]$ is true. Then $(\forall x)P(x)$ is false, so the truth set of $P(x)$ is not the universe. Therefore, the truth set of $\urcorner[(P(x)]$ is nonempty, and the proposition $(\exists x)\urcorner[P(x)]$ is true.

Now suppose the proposition $(\exists x)\urcorner[P(x)]$ is true. Then the truth set of $\urcorner[P(x)]$ is nonempty, so the truth set of $P(x)$ is not the universe. Therefore, the proposition $(\forall x)P(x)$ is false, so $\urcorner[(\forall x)P(x)]$ is true.

**b)** The proof of part (b), which is similar to that of part (a) is left as Exercise 55. □

Although the theorem appears formidable, it is really quite simple. In order to deny "There exists $x$ such that $x < 4$," we replace "there exists" with "for each" and deny $x < 4$. In other words, the denial of "There exists $x$ such that $x < 4$" is "For each $x$, $x \geqslant 4$." Likewise, to deny "For each $x$, $x^2 \geqslant 0$," we replace "for each" with "there exists" and deny $x^2 \geqslant 0$. Thus the denial of "For each $x$, $x^2 \geqslant 0$" is "There exists $x$ such that $x^2 < 0$."

Consider the proposition "For every real number $x$, there is a real number $y$ such that $2^y = x$." In symbols, this proposition can be denoted by $(\forall x)(\exists y)(2^y = x)$. Then, by Theorem 1.5(a), $\neg[(\forall x)(\exists y)(2^y = x)]$ is equivalent to $(\exists x)\neg[(\exists y)(2^y = x)]$. By Theorem 1.5(b), the last statement is equivalent to $(\exists x)(\forall y)\neg(2^y = x)$. Then we note that this symbolic statement says "There is a real number $x$ with the property that if $y$ is any real number, then $2^y \neq x$."

Consider next the proposition "There is a real number $x$ such that if $y$ is any real number, then $x + y = 2$." Using symbols, we can denote this proposition by $(\exists x)(\forall y)(x + y = 2)$. Then, by Theorem 1.5(b), $\neg[(\exists x)(\forall y)(x + y = 2)]$ is equivalent to $(\forall x)\neg[(\forall y)(x + y = 2)]$. By Theorem 1.5(a), the last statement is equivalent to $(\forall x)(\exists y)\neg(x + y = 2)$. Then we note that this symbolic statement says, "For each real number $x$, there is a real number $y$ such that $x + y \neq 2$."

**Since the reader of any proof has the right to expect correct English, care should be taken to see that any denial is a complete sentence.**

Recall from elementary calculus that the limit of $f(x)$, as $x$ approaches $a$, is $L$, provided that for each $\varepsilon > 0$ there is a $\delta > 0$ such that if $|x - a| < \delta$ and $x \neq a$, then $|f(x) - L| < \varepsilon$. This definition involves several quantifiers. In fact, if you are not careful, you will overlook one quantifier in the definition. Let's see if we can find all the quantifiers. The first one occurs in the phrase "for each $\varepsilon > 0$," and the second one occurs in the phrase "there is a $\delta > 0$." Are there any more? These are the only ones that are given **explicitly** in the definition. However, there is one hidden quantifier: "if $|x - a| < \delta$" means "if $x$ is any real number such that $|x - a| < \delta$." Therefore, the third quantifier in the definition occurs in the phrase "for each real number $x$ such that...." In symbols, we say that the limit of $f(x)$, as $x$ approaches $a$, is $L$, provided that

$$(\forall \varepsilon > 0)(\exists \delta > 0)(\forall x)(0 < |x - a| < \delta \rightarrow |f(x) - L| < \varepsilon)$$

In Exercise 46, you will be asked to write the negation of this statement.

Consider the statement "If $x$ is any real number, then $x \geqslant 0$." This statement is obviously false because $-3$ is a real number and $-3 < 0$. We say that $x = -3$ is a **counterexample** to the statement. In general, if we have a propositional function $P(x)$ that involves the quantifier "for each," then a counterexample to the propositional function is an object $a$ under discussion such that $P(a)$ is not true. In the example above, we have $(\forall x)(x$ is a real number $\rightarrow x \geqslant 0)$. Since the statement "$-3$ is a real number" does not imply that $-3 \geqslant 0$, $x = -3$ is a counterexample.

**Definition**

If $P(x)$ is a propositional function with variable $x$, then a **counterexample** to $(\forall x)P(x)$ is an object $t$ in the set of meanings such that $P(t)$ is false.

Often we are faced with a statement such as "For each $x$ in some universe, if $P(x)$, then $Q(x)$." We may not know whether the statement is true, so we must either prove it or provide a counterexample. In order to prove it, we would have to show that "For each $x$ in the universe, $P(x)$ implies $Q(x)$." In order to provide a counterexample, we would have to find a $t$ in the universe such that $P(t)$ is true but $Q(t)$ is false.

The following definitions will be needed in Example 4 as well as in the exercises at the end of Section 1.4.

**Definition**

An integer $m$ **divides** an integer $n$ if there is an integer $q$ such that $n = m \times q$. If the integer $m$ divides the integer $n$, then we say that $m$ is a **divisor** of $n$, $n$ is **divisible by** $m$, and $n$ is a **multiple of** $m$.

**Definition**

Natural numbers greater than 1 whose only divisors among the natural numbers are themselves and 1 are called **primes**.

**Definition**

An integer $n$ is a **perfect square** provided there is an integer $k$ such that $n = k^2$.

**EXAMPLE 4**

**a)** Either prove or find a counterexample to the statement "If $n$ is prime, then $2^n - 1$ is prime."

**b)** Either prove or find a counterexample to the statement "If $n$ is a perfect square, then $2^n - 1$ is not prime."

**ANALYSIS**

**a)** In this case, we are dealing with the prime numbers, so if we are going to be able to provide a counterexample, then we must find a prime number $n$ such that $2^n - 1$ is not prime. Let's construct a table.

| $n$ | $2^n - 1$ |
|---|---|
| 2 | 3 |
| 3 | 7 |
| 5 | 31 |
| 7 | 127 |
| 11 | 2047 |

Since $2047 = 23 \times 89$, $n = 11$ provides a counterexample to our statement.

**b)** As in part (a), we construct a table.

| $n$ | $2^n - 1$ |
|---|---|
| 1 | 1 |
| 4 | $15 = 3 \times 5$ |
| 9 | $511 = 7 \times 73$ |
| 16 | $65{,}535 = 3 \times 5 \times 17 \times 257$ |

Although this brief search has not turned up a counterexample, it has not been fruitless because it suggests a proof (see Exercise 97 in Section 1.6). ❑

**Warning** The symbols ∀ and ∃ are symbols of logic. Although mathematicians often use these symbols informally as a sort of shorthand, their use in formal mathematics, outside of mathematical logic, is not standard. That is, the use of ∀ and ∃ in mathematical literature is uncommon.

## EXERCISES 1.3

**41.** Write a nontrivial negation of each of the following statements.

**a)** All cows eat grass.

**b)** There is a horse that does not eat grass.

**c)** There is a car that is blue and weighs less than 4000 pounds.

**d)** Every math book is either blue or hard to read.

**e)** Some cows are spotted.

**f)** No car has 15 cylinders.

**g)** Some cars are old but are still in good running condition.

**42.** An integer $x$ has **property *P*** provided that for all integers $a$ and $b$, whenever $x$ divides $ab$, $x$ divides $a$ or $x$ divides $b$. Explain what it means to say that $x$ does not have property $P$.

**43.** A group $G$ is **cyclic** provided that there is a member $a$ of $G$ such that for each member $g$ of $G$, there is an integer $n$ such that $a^n = g$. Explain in a useful way what it means to say that a group $G$ is not cyclic.

**44.** Write a useful denial of the statement "For each pair of real numbers $a$ and $b$ with $a < b$, there is a rational number $r$ such that $a < r < b$."

**45.** A real number $u$ is the least upper bound for a set $T$ of real numbers provided that $u$ is an upper bound for $T$ and for each $\varepsilon > 0$, there is a member $x$ of $T$ such that $x > u - \varepsilon$. Explain in a useful way what it means to say that $u$ is not the least upper bound for a set $T$ of real numbers.

**46.** Without using any negative words, state what it means to say that the limit of $f(x)$, as $x$ approaches $a$, is not $L$.

**47.** A sequence $\{x_n\}$ is a **Cauchy sequence** provided that for each $\varepsilon > 0$, there is a natural number $N$ such that if $m, n > N$, then $|x_n - x_m| < \varepsilon$. Without using any negative words, state what it means to say that $\{x_n\}$ is not a Cauchy sequence.

**48.** Write the negation and the contrapositive of the following statement without using any negative words: "If $x$ is a positive number, then there is an $\varepsilon > 0$ such that $x < \varepsilon$ and $1/\varepsilon < x$."

**49.** Prove part (b) of Theorem 1.5 without using part (a).

**50.** Find the truth set of the propositional function $(x^2+1)(x-3)(x^2-2)(2x-3) = 0$, when it is given that the set of meanings of this propositional function is each of the following.

**a)** $\mathbb{Z}$ **b)** $\mathbb{Q}$ **c)** $\mathbb{R}$ **d)** $\mathbb{C}$

**51.** Find the truth set of the propositional function $\int_2^x y^2\,dy = 39$, when it is given that the set of meanings of this propositional function is $\mathbb{R}$.

**52.** Find the hidden quantifiers in each of the following statements.

**a)** Tax returns must be postmarked no later than April 15th.

**b)** $x^2 + y^2 + z^2$ cannot be of the form $8k + 7$ when $x$, $y$, and $z$ are odd.

**c)** The sum of two integers is even if and only if they are both even or both odd.

**d)** A stitch in time saves nine.

**e)** $x^2 < x$, if $0 < x < 1$.

**f)** If a function is not continuous, it is not differentiable.

**53.** Write a useful negation of parts (a), (c), (e), and (f) of Exercise 52.

**54.** Both of the following statements have the set of positive real numbers as their set of meanings. Which statement is true?

**a)** $(\forall x)(\exists y)(x < y^2)$ **b)** $(\exists y)(\forall x)(x < y^2)$

**55.** Prove that $\neg[(\exists x)P(x)]$ is equivalent to $(\forall x)\neg[P(x)]$.

## 1.4 *Methods of Proof*

In this section, we discuss several types of proofs and give some examples of proofs. You are then provided the opportunity to do some rather simple proofs.

Let us examine a proposition of the form "If $P$, then $Q$" or "$P \to Q$." We know that there are three conditions under which this proposition is true:

1. Each of $P$ and $Q$ is true.
2. $P$ is false and $Q$ is true.
3. $P$ is false and $Q$ is false.

We first observe that if $P$ is false, then $P \to Q$ is true. Therefore, in order to prove that $P$ implies $Q$, it is sufficient to assume that $P$ is true, and, under this assumption, prove that $Q$ is true. The **direct method** of proof is precisely this method; that is, we assume that $P$ is true, and under this assumption, we proceed through a logical sequence of steps to arrive at the conclusion that $Q$ is true. Let's illustrate the direct method with a simple example. But first we recall the following definition.

**Definition**

An integer $n$ is said to be **odd** provided there is an integer $k$ such that $n = 2k + 1$, and $n$ is said to be **even** provided there is an integer $q$ such that $n = 2q$.

Thus, 17 is odd because there is an integer 8 such that $17 = 2(8) + 1$, and 24 is even because there is an integer 12 such that $24 = 2(12)$.

**Theorem 1.6**

If $n$ is an odd integer, then $n^2$ is odd.

**PROOF** The hypothesis is "$n$ is an odd integer," and the conclusion is "$n^2$ is odd." So we begin by assuming that $n$ is odd. Then, by definition, there is an integer $k$ such that $n = 2k + 1$. Thus,

$$n^2 = (2k + 1)(2k + 1) = 4k^2 + 4k + 1 = 2(2k^2 + 2k) + 1$$

Therefore, there is an integer $m$ (namely, $m = 2k^2 + 2k$) such that $n^2 = 2m + 1$. So, by definition, $n^2$ is odd. We have proved that if the hypothesis is true, then the conclusion is true; so the proof is complete. □

In Section 1.2, we saw that the propositional expression "$P \to Q$" is equivalent to the contrapositive "$\neg Q \to \neg P$." Thus Theorem 1.6 can be restated: If $n^2$ is an even integer, then $n$ is even. In general, another way of proving any proposition of the form "If $P$, then $Q$" is to assume that $Q$ is not true, and, under this assumption, proceed through a logical sequence of steps to arrive at the conclusion that $P$ is not true. This is the **contrapositive method** of proof. Let's illustrate the contrapositive method with another simple theorem.

**Theorem 1.7**

Suppose that $m$ and $b$ are real numbers with $m \neq 0$ and that $f$ is the linear function defined by $f(x) = mx + b$. If $x \neq y$, then $f(x) \neq f(y)$.

**PROOF** We want to prove that $x \neq y$ implies $f(x) \neq f(y)$. The contrapositive is $f(x) = f(y)$ implies $x = y$. Suppose that $f(x) = f(y)$. Then $mx + b = my + b$. If we subtract $b$ from both sides of this equation and then divide both sides of the equation by $m$, we conclude that $x = y$. Therefore, we have proved that $f(x) = f(y)$ implies $x = y$. Since the contrapositive is known to be equivalent to the original implication, we have proved the original implication. □

Another method of proof is the **contradiction method**. By Proposition 1.1(b), the propositional expression $\neg(P \to Q) \leftrightarrow P \wedge \neg Q$ is a tautology. Thus, $\neg(P \to Q)$ and $P \wedge \neg Q$ are logically equivalent. So if we want to show that $P$

implies $Q$, we **begin by assuming that $P$ is true and $Q$ is false**. The idea is to reach a contradiction.

Another way of regarding the contradiction method is to recall that $P \rightarrow Q$ is true except in the case that $P$ is true and $Q$ is false. In the contradiction method, we rule out this possibility by assuming that it does happen and then reaching a contradiction.

The most difficult question in the contradiction method is "What contradiction are we looking for?" There are no specific guidelines because each proof gives rise to its own contradictions, but *any* contradiction will do.

The advantage of the contradiction method over the contrapositive method is that we get two statements from which to reason rather than just one. The disadvantage is that we have no definite knowledge of where the contradiction will occur. In general, the best rule of thumb is to use contradiction when the statement "not $Q$" gives some useful information. We consider a simple example whose proof relies on Exercises 56 and 57 at the end of this section.

**Theorem 1.8**

Suppose $a$, $b$, and $c$ are integers. If $a$ and $b$ are even and $c$ is odd, then the equation $ax + by = c$ does not have an integral solution for $x$ and $y$.

**PROOF** Suppose there are even integers $a$ and $b$ and an odd integer $c$ such that $ax + by = c$ has an integral solution for $x$ and $y$. Then if $x_0$ and $y_0$ are integral solutions of this equation, we have $ax_0 + by_0 = c$. By Exercise 56, $ax_0$ and $by_0$ are even. Thus by Exercise 57, $ax_0 + by_0 = c$ is even. This is a contradiction. □

We give another example of the contradiction method of proof, but first we recall the following definitions.

**Definition**

A real number $r$ is **rational** provided there are integers $m$ and $n$, with $n \neq 0$, such that $r = m/n$, and $r$ is **irrational** provided it is not rational.

So 0.25 is rational because $0.25 = 1/4$, and Theorem 1.9 tells us that $\sqrt{2}$ is irrational.

**Theorem 1.9**

If $r$ is a real number such that $r^2 = 2$, then $r$ is irrational.

**PROOF** Suppose that $r^2 = 2$ and $r$ is not irrational. Then $r$ is rational, so there are integers $m$ and $n$ such that $r = m/n$. We can assume that $m$ and $n$

have no common divisors greater than 1 because if they did, we could divide both numerator and denominator by the greatest common divisor. So $r^2 = m^2/n^2$, and hence $m^2 = r^2n^2$. Since $r^2 = 2$, we have $m^2 = 2n^2$. Hence $m^2$ must be even. By Theorem 1.6, $m$ must be even. (Why?) Therefore, there is an integer $p$ such that $m = 2p$. So $2n^2 = m^2 = 4p^2$, and hence $n^2 = 2p^2$. It follows that $n^2$ is even, and again by Theorem 1.6, $n$ must be even. Since $m$ and $n$ are both even, they have a common divisor greater than 1. This is a contradiction to our assumption that $m$ and $n$ have no common divisors greater than 1. So we have established the theorem. □

Perhaps the three methods of proof we have discussed can be clarified by comparing them in a chart. In each case, we are concerned with proving that $P$ implies $Q$.

| Direct method | Contrapositive method | Contradiction method |
|---|---|---|
| Assume $P$. | Assume $\neg Q$. | Assume $P$ and $\neg Q$. |
| ⋮ | ⋮ | ⋮ |
| (logical sequence of steps) | (logical sequence of steps) | (logical sequence of steps) |
| ⋮ | ⋮ | ⋮ |
| Conclude $Q$. | Conclude $\neg P$. | Conclude $R$ and $\neg R$. |

The advantage and the difficulty of the contradiction method are illustrated by the chart. We have more information to work with because we are assuming both $P$ and $\neg Q$. On the other hand, $R$ suddenly appears (that is, we do not know $R$ before we begin the proof.)

We use Theorem 1.10 to illustrate the three basic methods of proof. Since we have no need of the theorem itself, we ask you for now to assume all sorts of results from calculus and to connive at any gaps in the arguments given. In particular, let us recall that $D_x(e^x) = e^x$ and if $u$ is any real number, then $\ln(e^u) = u$. Let us also recall Rolle's Theorem.

**Rolle's Theorem**

If the function $f$ is continuous on the closed interval $[a, b]$, differentiable on the open interval $(a, b)$, and $f(a) = f(b)$, then there exists $c$ in the open interval $(a, b)$ such that $f'(c) = 0$.

**Theorem 1.10**

Let $x$ and $y$ be real numbers. If $x \neq y$, then $e^x \neq e^y$.

**DIRECT PROOF** If $x \neq y$, either $x > y$ or $y > x$. We can assume that $x > y$ because $x$ and $y$ are real numbers, and we can always let $x$ be the larger of the two. Then there is a positive number $r$ such that $x = y + r$, and $e^x = e^{y+r} = e^y \times e^r$. Since $e > 1$ and $r > 0$, $e^r > 1$. Hence $e^y \times e^r > e^y$. Therefore, $e^x \neq e^y$. □

**CONTRAPOSITIVE PROOF** Suppose that $e^x = e^y$. Then

$$x = \ln(e^x) = \ln(e^y) = y \qquad \square$$

**CONTRADICTION PROOF** Suppose that $e^x = e^y$ and yet $x \neq y$. By Rolle's Theorem, there is a $z$ between $x$ and $y$ such that $e^z = 0$. This is a contradiction because $e > 0$ and hence $e^u > 0$ for each real number $u$. $\square$

In the next section, we discuss the contradiction method in more detail. The rest of this section is devoted to the direct method and the contrapositive method. The exercises at the end of this section should also be done by using one of these two methods.

Let us consider another theorem and prove it using the direct method.

**Theorem 1.11**

> If $a$, $b$, and $c$ are real numbers, then
> $$a^2 + b^2 + c^2 \geqslant ab + bc + ca$$

**PROOF** Suppose that $a$, $b$, and $c$ are real numbers. Because $x^2 \geqslant 0$ for each real number $x$,

$$(a-b)^2 + (b-c)^2 + (c-a)^2 \geqslant 0$$

By squaring the indicated terms, we conclude that

$$(a^2 - 2ab + b^2) + (b^2 - 2bc + c^2) + (c^2 - 2ca + a^2) \geqslant 0$$

Then by rearranging the terms, we have

$$2a^2 + 2b^2 + 2c^2 \geqslant 2ab + 2bc + 2ca$$

Dividing by 2, we obtain the conclusion. $\square$

You may wonder why we began with the formula

$$(a-b)^2 + (b-c)^2 + (c-a)^2 \geqslant 0$$

The answer is that we first worked backward. We know, for example, that $(a-b)^2$ is a nonnegative number and that $(a-b)^2 = a^2 - 2ab + b^2$. If we simply subtract $ab$ from both sides of the inequality we want to prove, then we have $a^2 - ab + b^2 + c^2 \geqslant bc + ca$. So in order to get a perfect square, we must multiply $-ab$ by 2. Therefore, working backward, we first multiply both sides of the inequality by 2.

$$2a^2 + 2b^2 + 2c^2 \geqslant 2ab + 2bc + 2ca$$

$$a^2 - 2ab + b^2 + b^2 - 2bc + c^2 + c^2 - 2ca + a^2 \geqslant 0$$

$$(a-b)^2 + (b-c)^2 + (c-a)^2 \geqslant 0$$

Recall that $Q \to P$ is not equivalent to $P \to Q$, so working backward does not constitute a proof. Working backward does, however, give us a clue as to where to begin our proof. But we must make sure that the steps are reversible before we have a proof.

Recall the discussion of the meaning of "if and only if" in Section 1.1. Suppose that $P$ and $Q$ are propositions and we have a theorem that states: "$P$ if and only if $Q$." Then we really have two theorems. That is, we must prove that if $P$ is true, then $Q$ is true, and we must prove that if $Q$ is true, then $P$ is true. Consider the following theorems.

**Theorem 1.12**

An integer $n$ is even if and only if $n^2$ is even.

**PROOF** The contrapositive of Theorem 1.6 tells us that if $n^2$ is even, then $n$ is even. So, in order to complete the proof of the theorem, it is sufficient to show that if $n$ is even, then $n^2$ is even.

Suppose that $n$ is even. Then, by definition, there is an integer $k$ such that $n = 2k$. Therefore,

$$n^2 = (2k)^2 = 4k^2 = 2(2k^2)$$

Hence, by definition, $n^2$ is even. □

**Theorem 1.13**

Let $x$ be a real number. Then $x = 1$ if and only if $x^3 - 3x^2 + 4x - 2 = 0$.

**PROOF** We know that $x$ is a real number, and we must prove two things: (1) If $x = 1$, then $x^3 - 3x^2 + 4x - 2 = 0$, and (2) if $x^3 - 3x^2 + 4x - 2 = 0$, then $x = 1$.

First suppose $x = 1$. Then

$$x^3 - 3x^2 + 4x - 2 = 1 - 3 + 4 - 2 = 0$$

This proves the first part of the theorem—namely that if $x = 1$, then $x^3 - 3x^2 + 4x - 2 = 0$.

Now let us prove the second part. We assume that $x^3 - 3x^2 + 4x - 2 = 0$. Since

$$x^3 - 3x^2 + 4x - 2 = (x - 1)(x^2 - 2x + 2)$$

either $x - 1 = 0$ or $x^2 - 2x + 2 = 0$. (The last step relies on the well-known property of real numbers—that whenever the product of two real numbers is 0, then at least one of the numbers is 0.) If $x^2 - 2x + 2 = 0$, the quadratic formula yields $x = 1 \pm i$. This contradicts the hypothesis that $x$ is real. Therefore, $x^2 - 2x + 2 \neq 0$. Hence $x - 1 = 0$, and thus, $x = 1$. □

In the second half of the preceding proof, we have propositions $x - 1 = 0$ and $x^2 - 2x + 2 = 0$ such that $(x - 1 = 0) \vee (x^2 - 2x + 2 = 0)$ is true. We show that $x^2 - 2x + 2 = 0$ implies that $x$ is not a real number. Therefore, by the contrapositive, since $x$ is a real number, $x^2 - 2x + 2 \neq 0$. Thus, in order for $(x - 1 = 0) \vee (x^2 - 2x + 2 = 0)$ to be true, $x - 1 = 0$ must be true.

## EXERCISES 1.4

**56.** Prove that any multiple of an even integer is even.

**57.** Prove that the sum of two even integers is even.

**58.** Let $A$, $B$, and $C$ be integers. Prove that if $A$ divides $B$ and $B$ divides $C$, then $A$ divides $C$.

**59.** Let $A$ and $B$ be integers and let $D$ be a positive integer.

**a)** Prove the following proposition: If $D$ divides $A$ and $D$ divides $B$, then $D$ divides both $A + B$ and $A - B$.

**b)** Is the converse of the proposition given in (a) true? If so, prove it; if not, give a counterexample.

**60.** Prove that if $D$ is an odd integer that divides both the sum and the difference of two integers $A$ and $B$, then $D$ divides both $A$ and $B$.

**61.** Prove that no integer is both even and odd.

**62.** Let $A$, $B$, $C$, $M$, and $N$ be integers. Prove that if $A$ divides each of $B$ and $C$, then $A$ divides $NB + MC$.

**63.** Let $A$ be an even integer and $B$ be an odd integer. Prove that $A + B$ is odd and $AB$ is even.

**64.** Prove that if $x$ and $y$ are real numbers such that $x > 2$ and $y > 3$, then the area of the rectangle with corners at $(x, y)$, $(x, -y)$, $(-x, y)$, and $(-x, -y)$ is greater than 24.

**65.** Prove that if $x$ and $y$ are real numbers such that $x < -2$ and $y < -3$, then the distance between $(x, y)$ and $(1, 4)$ is greater than 7.

**66.** Let $x$ be a real number. Prove that $x = -1$ if and only if $x^3 + x^2 + x + 1 = 0$.

**67.** Let $A$ and $B$ be integers. Prove that if $AB$ is odd, then $A + B$ is even.

**68.** If $x$ is any natural number that 37 does not divide, then it is known that there are natural numbers $m$ and $n$ such that $nx - 37m = 1$. Use this fact to show that if $x$ and $y$ are natural numbers and 37 divides $xy$, then 37 divides $x$ or 37 divides $y$.

**69.** Let $A$ and $B$ be integers with $B \neq 0$. Prove that if $A$ divides $B$, then $|A| \leqslant |B|$.

**70.** Let $n$ be an integer such that $n^2$ is even. Prove that $n^2$ is divisible by 4.

**71.** Prove that for any natural number $n$, either $n$ is a prime or a perfect square, or $n$ divides $(n - 1)!$

**72.** A student wishes to prove that if $p$ and $q$ are positive integers with $q > p$ such that $q - p$ divides $q - 1$, then $q - p$ also divides $p - 1$. The student writes: Because $q - p$ divides $q - 1$, there is an integer $a$ such that $a(q - p) = q - 1$. Suppose $b(q - p) = p - 1$. Then $a(q - p) - b(q - p) = q - p$ and because $q > p$, $a - b = 1$. Thus $b = a - 1$. What, if anything, is wrong with the student's proof?

**73.** Let $a$ and $b$ be positive integers such that $a^2 = b^3$.

**a)** Prove that if $a$ is even, 4 divides $a$.

**b)** Given that 4 divides $a$, prove that 4 divides $b$.

**c)** Given that 4 divides $b$, prove that 8 divides $a$.

**d)** Give an example of positive integers $a$ and $b$ such that $a$ is even and $a^2 = b^3$, but $b$ is not divisible by 8.

**74.** In this exercise, we give both a "proof" and a "counterexample" to the conjecture "If $x$, $y$, and $z$ are natural numbers such that $xz$ divides $yz$, then $x$ divides $y$." Is the "proof" correct? If not, why not? Is the "counterexample" correct? If not, why not?

"Proof": Suppose $x$, $y$, and $z$ are natural numbers such that $xz$ divides $yz$ but $x$ does not divide $y$. Then $y$ divides $x$, so there is an integer $p$ such that $x = py$. Thus $xz = (py)z = p(yz)$. Therefore $yz$ divides $xz$, and this is a contradiction because $xz$ divides $yz$.

"Counterexample": Let $x = 4$, $y = 6$, and $z = 16$. Then $xz$ divides $yz$, but $x$ does not divide $y$.

## 1.5 *The Contradiction Method of Proof*

In the preceding section, we discussed briefly the contradiction method of proof and proved two theorems using this method. Since this is an extremely important method of proof, we examine it in more detail in this section.

Once again, we are concerned with proving that $P$ implies $Q$, and we begin by assuming that $P$ is true, just as we would using the direct method of proof. However, in order to reach the desired conclusion that $Q$ is true, we proceed by trying to find some reason that $Q$ cannot be false. For example, suppose we assume that $P$ is true and $Q$ is false and somehow reach the conclusion that the number 17 is irrational. Then surely you would be convinced that it is not possible for $P$ to be true and $Q$ to be false. In this example, the contradiction we reach is $R \wedge \neg R$, where $R$ is the proposition "17 is rational."

Recall that $P \to Q$ is true except when $P$ is true and $Q$ is false. In a proof by contradiction, we begin by assuming that $P$ is true and $Q$ is false and then reach a contradiction. As examples, in the proof of Theorem 1.9, the contradiction was that $m$ and $n$ had no common divisors but they were both even. In the proof by contradiction of Theorem 1.10, we had $e^z = 0$ for some real number $z$, but you know that if $a$ is a positive real number and $b$ is any real number, then $a^b \neq 0$.

The logical basis for the contradiction method of proof is the tautology $[\neg Q \to (R \wedge \neg R)] \to Q$ (see Proposition 1.2k). That is, if the negation of $Q$ leads to a contradiction, $R \wedge \neg R$, then we can conclude that $Q$ is true.

When do you use the contradiction method of proof? Once again, there is no definite answer. As we have previously stated, more information is available in the contradiction method. If, when you use the direct method and the contrapositive method, it appears that not enough information is available, then a proof by contradiction may be in order. Another general guideline is that if we are proving that something does *not* happen, then it may be easier to assume that it does happen and reach a contradiction.

We now give some more examples of proof by contradiction.

**Theorem 1.14**

There do not exist prime numbers $a$, $b$, and $c$ such that $a^3 + b^3 = c^3$.

**PROOF** The proof is by contradiction. We want to show that if $a$, $b$, and $c$ are primes, then $a^3 + b^3 \neq c^3$.

Suppose there are primes $a$, $b$, and $c$ such that $a^3 + b^3 = c^3$. If $a$ and $b$ are odd, then $a^3$ and $b^3$ are odd, so $c^3$ is even. Therefore, $c$ is an even prime, so $c = 2$. But this is impossible because $a$ and $b$ are both greater than 2.

So, at least one of $a$ and $b$ is even, and hence we can assume that $b = 2$. Therefore,

$$8 = c^3 - a^3 = (c - a)(c^2 + ca + a^2)$$

Now $a$ and $c$ are primes, so each of $c^2$, $ca$, and $a^2$ is greater than or equal to 4 and hence the sum is greater than or equal to 12. Thus, $|c^3 - a^3| \geqslant 12$. This contradicts the assumption that $c^3 - a^3 = 8$. □

**Theorem 1.15**

If $a$, $b$, and $c$ are odd integers, then $ax^2 + bx + c = 0$ does not have a rational solution.

**PROOF** The proof is by contradiction.

Suppose $a$, $b$, and $c$ are odd integers and $p$ and $q$ are integers such that $p/q$ is a solution of the equation $ax^2 + bx + c = 0$. We can assume that $p$ and $q$ have no common divisors greater than 1. Then

$$a(p^2/q^2) + b(p/q) + c = ap^2/q^2 + bp/q + c = 0$$

so $ap^2 + bpq + cq^2 = 0$.

We will first prove that $p$ and $q$ are both odd. Suppose that $p$ is even. Then $ap^2 + bpq$ is even, so $cq^2$ must be even. But this is a contradiction because $c$ and $q$ are both odd ($q$ is odd because if it were even, then $p$ and $q$ would have a common divisor). Therefore, $p$ is odd. Suppose that $q$ is even. Then $bpq + cq^2$ is even, so $ap^2$ must be even. But this is a contradiction because $a$ and $p$ are both odd. So we have established that $p$ and $q$ are both odd. Therefore, each of $ap^2$, $bpq$, and $cq^2$ is odd. Hence their sum is odd, and this is a contradiction because the sum is 0. □

Proof by contradiction should be used in each of the following exercises. In these exercises, you are free to assume the standard properties of $\leqslant$ for real numbers. Some additional exercises involving proofs by contradiction are given in the next chapter after the notion of set has been introduced.

## EXERCISES 1.5

**75.** Prove, by contradiction, that the sum of two even integers is even.

**76.** Either prove or give a counterexample to the converse of "If $x$ and $y$ are even, then $xy$ is even."

**77.** Either prove or give a counterexample to the conjecture "If $p$, $q$, and $r$ are integers such that $p+q+r$ is odd, then an odd number of $p$, $q$, and $r$ is odd."

**78.** Prove that if $x$ is a positive real number, then $x/(x+1)<(x+1)/(x+2)$.

**79.** Prove that if $x$ and $y$ are positive real numbers and $x \neq y$, then $x+y>4xy/(x+y)$.

**80.** Prove that if $x$ and $y$ are positive real numbers and $x \neq y$, then $x/y+y/x>2$.

**81.** Let $x$, $y$, and $z$ be positive real numbers. Prove that if $x>z$ and $y^2=xz$, then $x>y>z$.

**82.** Let $P_1, P_2, \ldots, P_n$ be primes. Prove that for each $i=1, 2, \ldots, n$, $(P_1 \times P_2 \times \cdots \times P_n)+1$ is not divisible by $P_i$.

**83.** Suppose that $A$, $B$, $P$, $Q$, $R$, and $S$ are integers, where $A=BP+R$, $A=BQ+S$, $0 \leqslant R<|B|$, and $0 \leqslant S<|B|$. Prove that $R=S$ and $P=Q$.

**84.** Let $A$, $B$, and $C$ be integers such that $A^2+B^2=C^2$. Prove that at least one of $A$ and $B$ is even.

**85.** Under the assumption that every natural number greater than 1 has a prime divisor, prove that there are an infinite number of primes. [*Note*: The assumption that every natural number greater than 1 has a prime divisor is proved in Proposition 3.2.]

**86.** Everyone knows that $3^2+4^2=5^2$. Prove that there do not exist three consecutive natural numbers such that the cube of the largest is equal to the sum of the cubes of the other two.

**87.** Prove that the equation $x^5+6x^4+17x^3+3x^2+7x+4=0$ has no positive real solution.

**88.** Is the statement "If $n$ is any prime, then $2^n+1$ is prime" true? If your answer is yes, prove the statement. Otherwise find a counterexample to the statement.

**89.** Is the converse to the statement given in Exercise 88 true? If your answer is yes, prove the statement. Otherwise find a counterexample to the statement.

**90.** For the purpose of this exercise we define a prime to be **average** provided it is the average of two *different* prime numbers (for example, $7=(11+3)/2$ is average). Consider the following propositions:

$P$: Every prime greater than 3 is average.

$Q$: Every even number other than 2 can be written as $x+y$, where $x$ is a prime, $y$ is a prime, and possibly $x=y$ (for example, $4=2+2$, $6=3+3$, $8=5+3$).

$R$: Every even number greater than 6 can be written as the sum of two different prime numbers.

**a)** Prove that $R \rightarrow P \wedge Q$.

**b)** Prove that $P \wedge Q \rightarrow R$.

It is widely believed that propositions $P$, $Q$, and $R$ are true, but they are all unsolved problems. Proposition $Q$ is known as the Goldbach Conjecture [Christian Goldbach (1690–1764)].

**91.** Let $p$ and $q$ be positive integers with $p < q$.

**a)** Prove that $q - p$ divides $p - 1$ if, and only if, $q - p$ divides $q - 1$.

**b)** Let $d$ be a common divisor of $p - 1$ and $q - 1$. Prove that $d$ divides $q - p$.

## 1.6 *More Proofs*

Although it may be desirable to make a list of all the objects in our universe that possess a certain property, sometimes it is impractical or impossible to do so. The list would be too long. For example, if we consider the collection of all natural numbers less than one million, it would not be practical to take the time to list all these objects. Also, if we consider the collection of all positive real numbers that are less than one, it would be impossible to list these objects.

If we want to prove something that involves the quantifier "for all," we need some method to avoid listing all the objects that satisfy the condition. Choosing an object that satisfies the condition and checking to see that the desired result holds for that object is such a method. Let us refer to this method as the **element-chasing method**. The element-chasing method can be illustrated by thinking of it as a "proof machine." In other words, we have a machine that can check any object we choose in order to see that the desired result holds. The proof is a description of *what the machine does* with any object fed into it. Be careful not to confuse the element-chasing method with the *incorrect* method of proof by example.

Relating the element-chasing method to the direct method of proof, we choose one object that satisfies the condition and, using a logical sequence of steps, conclude that the desired result holds. The important thing is that the logical sequence of steps must be capable of being repeated regardless of which object is chosen.

Let's consider an example, Theorem 1.16.

**Theorem 1.16**

If $x$ is any real number such that $x^2 - x - 2 < 0$, then $-1 < x < 2$.

**PROOF** Choose any real number $x$ such that $x^2 - x - 2 < 0$. Note that $x^2 - x - 2 = (x - 2)(x + 1)$. Now the product of two real numbers is negative if and only if one of the numbers is negative and the other one is positive, so we have two cases to consider.

***Case I.*** $x - 2 > 0$ and $x + 1 < 0$. In this case, $x$ must be greater than 2 and less than $-1$. Obviously no real number satisfies both conditions; that is, there is no real number $x$ such that $x > 2$ and $x < -1$.

***Case II.*** $x-2<0$ and $x+1>0$. In this case, $x$ must be less than 2 and greater than $-1$; that is, $-1<x<2$.

We have thus shown that any real number $x$ that satisfies the condition $x^2-x-2<0$ also has the property that $-1<x<2$. Since the proof is independent of the real number $x$ chosen, as long as it satisfies the condition $x^2-x-2<0$, we have proved the theorem for all real numbers $x$. □

We mentioned proof by example, so let's illustrate what we mean. A student may say, "Okay, $x=0$ satisfies the conditions of the theorem; that is, 0 is a real number such that $0^2-0-2<0$. Since $-1<0<2$, the theorem is true." Note that we have *not* proved the theorem because in order to do so, we must show that if $x$ is *any* real number such that $x^2-x-2<0$, then $x$ satisfies the conclusion of the theorem. That is, in order to prove a theorem by the element-chasing method, we cannot specify the element. As we mentioned earlier, the proof must work for an arbitrary object that satisfies the hypothesis of the theorem.

The proof of Theorem 1.16 also illustrates proof by cases. That is, we examined two possibilities. The situation occurs whenever it is necessary to examine several possibilities. We illustrate this with another example.

**Theorem 1.17**

There do not exist natural numbers $m$ and $n$ such that $7/17=1/m+1/n$.

**PROOF** Observe that $1/m$ and $1/n$ are positive numbers and hence neither $m$ nor $n$ can be 2 because $1/2+1/n>7/17$ for each natural number $n$.

Since $1/5+1/5<7/17$ and $1/m+1/n<1/5+1/5$ for all $m$ and $n$ greater than 5, at least one of $m$ and $n$ must be less than 5.

Therefore, one of $m$ and $n$ must be 3 or 4. But $7/17-1/3=4/51$ and $4/51 \neq 1/n$ for any natural number $n$, and $7/17-1/4=11/68$ and $11/68 \neq 1/n$ for any natural number $n$. Therefore, since we have examined all possibilities, there do not exist natural numbers $m$ and $n$ such that $7/17=1/m+1/n$.

□

Until now, we have been concerned with the logic behind mathematical proofs, but we wish to dispel any impression that mathematics is just a study of logic. Here then is a method of proof that has nothing to do with tautologies.

**Definition**

Two integers have the same parity if they are both even or both odd.

The principle of parity is that comparing the parity of two or more integers may provide the key to a proof of a result, even of a result that on the surface has nothing to do with integers.

We illustrate this principle by Example 5 and by some, but not all, of the exercises at the end of this section.

**EXAMPLE 5**

A Hollywood star is building a 12 ft × 20 ft patio in front of her swimming pool. Since one square foot in the lower left corner is needed for the submersible pump and one square foot in the upper right corner is needed for the pipe that feeds the pool's cascading waterfalls, the star has imported 119 handpainted 1 ft × 2 ft porcelain tiles. The workers have laid these tiles haphazardly only to find, after laying 118 tiles, that two separated 1 ft × 1 ft squares have been left uncovered (Figure 1.1). The workers suggest breaking the remaining tile in half, but the movie star finds this suggestion "unthinkable." Is there a pattern of tiling the patio so that all 238 $\text{ft}^2$ of patio can be tiled with 119 unbroken tiles?

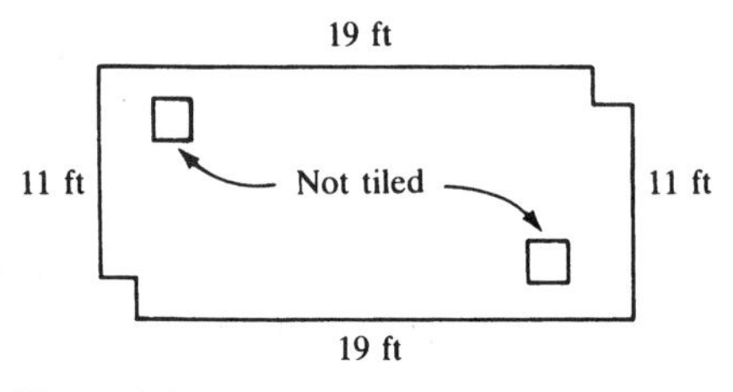


Figure 1.1

The answer is no, and since we are seeking to show that something cannot be done, it is reasonable to consider a proof by contradiction. Imagine overlaying the entire 240 $\text{ft}^2$ with a checkerboard pattern of 240 1 ft × 1 ft squares. As in the rules for both chess and checkers, assume that the lower left-hand square is black. Then it is easy to see that the upper right-hand square is also black. Thus there are two more white squares than black squares to be covered by tiles. We are now able to use the Principle of Parity. Since each tile covers up one white square and one black square, any tiling with 118 tiles will leave two white squares uncovered. No two white squares are adjacent, and so the movie star is out of luck.

Note that we have not mentioned distinguishing between even and odd. In Exercise 98, you will be asked to rework Example 5 in such a way that even and odd integers are used explicitly. ❑

**ADDENDUM**

What distinguishes some athletes is not so much their technical skills as their intuitive understanding of their sport. In baseball, they are the fielders who always get a jump on the ball; in basketball, they are the players who have "court sense." Similarly, the ability to prove theorems is founded as much on developing mathematical intuition as it is on practicing a large array of technical skills. So we pause here to consider how you can begin developing your own mathematical presence of mind. (Our first advice is to seek an expert second opinion and we especially recommend George Pólya's

*How to Solve It* and *Mathematical Discoveries I, II*, Paul Halmos's *How to Write Mathematics*, and Jacques Hadamard's, *An Essay on the Psychology of Invention in the Mathematical Field*.)

There are three states of being unable to prove a theorem.

**1.** You haven't a clue. You see no reason to believe the result, beyond someone else's assurances that it might be so, and you have no idea how to approach the problem.

The conscious mind does its best to dominate the subconscious because it realizes that the subconscious is not very good at making rational decisions such as deciding to get you out of the path of a Mack truck. Although conscious effort is required to prove a theorem, it is often the subconscious that provides the key, so you should look for ways to free your subconscious. Try working hard on the proof and then, as the saying goes, "sleep on it." Drawing a picture will sometimes trick your conscious mind into dropping its guard, and as we shall see in the next chapter, Venn diagrams show that even something so abstract as de Morgan's laws for logic have an associated picture. Turn the problem upside down. Look for a counterexample, and when (presumably) you can't find one, try to figure out why not. Take away one or more hypotheses until you can find an example. Then think how the hypothesis you have deleted can be used in a proof. Can you translate the problem into a new setting? For example, what makes analytic geometry an important part of calculus is that the cartesian coordinate system is essentially a device for translating back and forth between geometry and algebra. Can you relate the problem to some other result you already know? Is there some consequence of the result that you can prove? Sometimes, when you work out part of a result, you will see how the rest follows. And then there is Pólya's advice, "See if you can think of the result as part of a result that comprises it."

**2.** You feel certain that the result is true and you think there is a natural approach, but when you make the argument that seems headed in the right direction you get stuck.

Go back to make sure that you have used all the hypotheses. If so, have you given up too much ground? For example, if one hypothesis is that $x > 5$ and all you use is that $x \neq -1$, then you are probably leaving too much of this hypothesis unused. Obviously you should use all the available information. If after an honest effort you cannot use some part of the hypothesis, you should then consider a contrapositive proof or a proof by contradiction. Such a proof allows you to write down what appears to be a new piece of information and thereby gives you a new starting point.

**3.** You know the result is true because you have thought of an argument that establishes it. Unfortunately, your argument does not persuade a knowledgeable and patient reader such as your teacher or the brightest of your classmates.

For most people this state is not so common as states (1) and (2) and is not really so important a concern. But if you encounter state (3) repeatedly, it is a sign that you are allowing yourself to gloss over gaps in your argument or that you are allowing yourself to write up your results in your own private language as if you were writing a diary rather than a communication. One way of encoding your proof so that it cannot be deciphered, which is particularly irksome, is to use a letter or symbol without disclosing its meaning. You should treat your readers as royalty by introducing each symbol you use into their gracious presence.

Here is how to check for gaps. Look at the last statement first. Is it what you wanted to prove? If your conclusion is that $1 = 1$, you may very well have a sound proof of this equality, but surely you do not have a proof of the proposition you wish to establish. (You will see a subtler example of this gap illustrated by Exercise 31 of Section 3.2.) Once you know that your last conclusion is the result you want, go back to check the first line of your proof. Make sure that it follows from the hypotheses and that it is relevant. Thereafter, it is just a matter of applying a simple principle—if there is going to be a mistake in your proof there has got to be a first mistake. That's the one to avoid. Similarly, if a classmate or teacher cannot follow your argument, you are entitled to know the first time your explanation is deemed unsatisfactory. When looking for the gap in your argument, you should be especially wary of a step that appears obvious. Thus you should view the words *obvious* and *obviously* as warning signs that the step in question is suspect.

We have already given the most important advice about writing up mathematical arguments clearly: read Halmos's *How to Write Mathematics*. There are, however, a few simple tricks worth mentioning here. Watch out for the quantifiers *for each* and *there is*, and watch out for the little words like *the*, *all*, *and*, *or*, and *only*. Don't abuse notation or terminology. For example, respect the distinction between $\in$ and $\subseteq$ and don't use $=$ as if this mathematical symbol were just shorthand—"$A - B \in C \cap B' = B \cup C' \subseteq B \cup A'$" is gibberish.

We have come to the end of our advice, but we pass along invaluable advice given us by Professor William Jenner.

**Jenner's Law: If after working long into the night you think you have discovered a proof, write down the idea and without scrutinizing the argument go to bed.**

If your proof is correct, you will have a better chance of writing up a clear presentation of it in the morning. If the proof is incorrect, you will have maximized the joy and excitement of mathematical discovery, and in the morning the mistake will be obvious. Besides, sometimes your conscious mind will get so distracted by all the excitement that it will allow your subconscious to tell you the next morning both where the mistake is and how to fix it.

## EXERCISES 1.6

**92.** Prove that if $n$ is a natural number, then $n^2 + n$ is even.

**93.** Prove that if $n$ is a natural number greater than 1, then $n! + 1$ is odd.

**94.** Prove that if $a$ and $b$ are rational numbers with $a < b$, then there exists a rational number $r$ such that $a < r < b$.

**95.** Prove that if $x$ is a positive real number, then $x + 1/x \geqslant 2$.

**96.** **a)** Find positive real numbers $x$ and $y$ such that $\sqrt{x+y} \neq \sqrt{x} + \sqrt{y}$.

**b)** Prove that if $x$ and $y$ are positive real numbers, then $\sqrt{x+y} \leqslant \sqrt{x} + \sqrt{y}$.

***97.** **a)** Verify by direct multiplication that for any positive integers $a$ and $k$,

$$(a-1)(a^{2k-4} + a^{2k-5} + a^{2k-6} + \cdots + a^3 + a^2 + a + 1) = a^{2k-3} - 1$$

**b)** Verify by direct multiplication that for any positive integer $k$,

$$2^{4k^2-4k+1} - 1 = (2^{2k-1} - 1)(2^{4k^2-6k+2} + 1) - 2^{2k-1} + 2^{4k^2-6k+2}$$

**c)** Using part (a), prove that for any positive integer $k$, $2^{2k-1} - 1$ divides

$$-2^{2k-1} + 2^{4k^2-6k+2} \; (= -2^{2k-1} + 2^{(2k-1)(2k-2)})$$

**d)** Prove that if $n$ is a perfect square, then $2^n - 1$ is not a prime number.

**98.** Rework Example 5 in such a way that even and odd integers are used explicitly in order to show that the patio cannot be tiled with 1 ft $\times$ 2 ft tiles.

**99.** How many pairs of primes $p$ and $q$ are there such that $q - p = 3$? Prove your answer.

**100.** In Exercise 99, we saw that there is only one pair of primes that differs by 3. Indeed, if $n$ is any odd integer, then the number of pairs of primes that differ by $n$ is either 0 or 1. But suppose that $n$ is even. In particular, how many pairs of primes differ by 2? Such primes, called **twin primes**, have been the object of much research in number theory for half a century. It is, however, still not known how many there are. A related question is concerned with triplets of primes that differ by 2—for example, 3, 5, 7. How many triplets of the form $p$, $p+2$, $p+4$ are there? Prove your answer.

**101.** Prove that if $x$ is any real number greater than 2, then there is a negative real number $y$ such that $x = 2y/(1+y)$.

**102.** Prove that if $p$ is an odd integer, then $x^2 + x - p = 0$ has no integer solution.

**103.** Let $n$ be an odd natural number greater than 1. Suppose that we are given an $n \times n$ table in which each of the first $n$ natural numbers appears exactly once in each row and each column. Suppose further that whenever a number appears in the $i$th row and $j$th column, it also appears in the $j$th row and $i$th column. Prove that for each positive integer $k$, $k \leqslant n$, there is a number $j$ such that $k$ appears in the $j$th row and $j$th column. (In other words, prove that each number appears in the diagonal going from the upper left corner to the lower right corner.)

**104.** A pothole inspector works for the Commonwealth of Pennsylvania. His job is to inspect each major road in his county during each of the winter months to see that no pothole has grown to more than 3 ft in diameter. The inspector is considering applying for work in both of the counties whose road maps are shown on the next page. In each case, the inspector wants to know if there are two towns such that he can begin his inspection in one town, inspect each road exactly once, and end his inspection in the second town. Help the inspector, and if you can, state a general principle that would apply to any road map.

**a)**

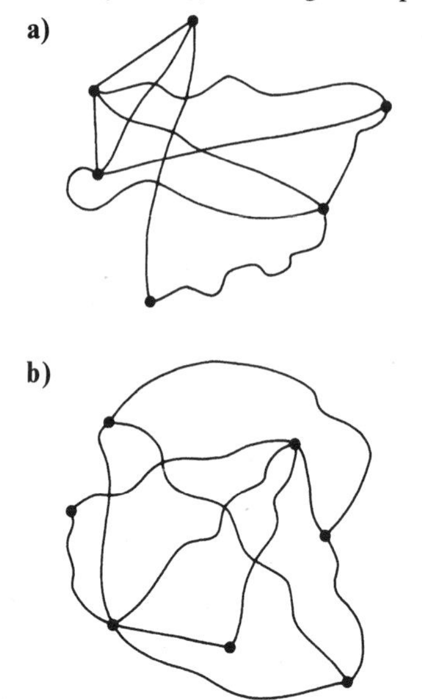

**b)**

## WRITING EXERCISES

1. Look up Roger B. Nelsen's article, "Proof Without Words: The sum of a positive number and its reciprocal is at least two (four proofs)" [*Math. Magazine*, Vol. 67, No. 5, December, 1994, page 374]. Explain the proofs the author has in mind and relate at least one of them to an exercise in Section 1.5.

2. In our text, the authors present the usual two-valued logic in which a statement by definition is true or false but not both. Read pages 8 through 10 of Steven Vicker's *Topology via Logic*, Cambridge University Press, Cambridge, UK (1989). Write an essay in which you try to persuade skeptics (of which there are many) that there are situations in which it is valuable to have a system of logic such as the one Vicker is introducing, in which we are no longer assured that every statement is true or false.

# CHAPTER 2 SETS

The theory of sets is the fundamental system of mathematics on which all other mathematical theory is founded. The first aspect of set theory that one encounters is the friendly face of intuitive or naive set theory; behind it is the austere face of axiomatic set theory. In this chapter, we concentrate on the friendly face; in the last section, we give an extremely brief introduction to axiomatic set theory.

## 2.1 *Introduction*

Intuitively, by a set we mean a collection thought of as a unit, a collection of entities to form a single object. It may happen that for a given set we do not know explicitly which objects belong to the set. For example, it is a routine test of computer power to program a new computer to hunt for members of the set $S$ to which the natural number $n$ belongs, provided that $2^n - 1$ is prime. In all likelihood, no computer will ever find all the members of this set. Nonetheless, $S$ is a perfectly good set. Membership is also an all-or-nothing situation. An object cannot be granted partial membership in a set, as is sometimes done in country clubs where a partial member is allowed to use the tennis courts but not the golf course.

**Definition**

If $A$ is a set and $x$ is an object that belongs to $A$, we say that $x$ is an **element** of $A$ or that $x$ is a **member** of $A$, or that $x$ **belongs to** $A$, and write $x \in A$.

For example, if $A$ is the set of all even natural numbers, then $10 \in A$, but $5 \notin A$ (5 is not an element of $A$).

**Definition**

If $A$ and $B$ are sets, $B$ is said to be a **subset** of $A$, written $B \subseteq A$, provided every member of $B$ is a member of $A$. Sets $A$ and $B$ are said to be **equal**, written $A = B$, provided $A \subseteq B$ and $B \subseteq A$. If $B$ is a subset of $A$ and $A \neq B$, then $B$ is said to be a **proper subset** of $A$.

The proof of the following proposition is left as an exercise.

**Proposition 2.1**

If $A$ is a set that has no members and $B$ is a set that has no members, then $A = B$.

Note that Proposition 2.1 does not say that there is an empty set. It says only that there are not two empty sets. We need to settle whether there is an empty set, but the answer to this question is not subject to proof. It is clear that by a set, we could mean a collection of one or more objects, in which case if someone presented us with an empty set, we would insist that so far as we were concerned this empty collection was not a set at all. Although we are considering sets only from a naive and intuitive point of view, we are already in need of a ruling—that is, we need an axiom.

**Axiom 1** There is an empty set.

Since Axiom 1 guarantees the existence of a set that has no members, the preceding proposition allows us to speak of *the* set that has no members. This set is called the **empty set** and is denoted by the symbol $\varnothing$, which is the last letter of the Danish–Norwegian alphabet. Because this letter $\varnothing$ is for most Americans quite unpronounceable (it sounds somewhat like "er"), it is customary to read "the empty set" in any sentence in which the symbol $\varnothing$ appears. It is evident that some confuse the symbol $\varnothing$ with the Greek letter $\phi$, to which it bears a faint resemblance.

**Definition**

The number of elements in a finite set $A$ is called the **cardinality** of $A$ and is denoted by $|A|$.

Note that $|\varnothing| = 0$, and if $A$ is a nonempty finite set, then $|A|$ is a natural number.

**Proposition 2.2**

Let $A$ be a set. Then $\varnothing \subseteq A$ and $A \subseteq A$.

**PROOF** See Exercise 2. □

**Corollary 2.3**

Every nonempty set has at least two subsets.

**Proposition 2.4**

Let $A$, $B$, and $C$ be sets. If $A \subseteq B$ and $B \subseteq C$, then $A \subseteq C$.

**PROOF** Supopose that $A \subseteq B$ and $B \subseteq C$. If $A = \varnothing$, by Proposition 2.2, $A \subseteq C$. Suppose that $A \neq \varnothing$ and let $x \in A$. Since $A \subseteq B$, $x \in B$. Since $B \subseteq C$, $x \in C$. Therefore, $A \subseteq C$. □

Suppose we wish to describe a set. If the set is finite and small, one obvious way is to list its elements. We're thinking of the set $A = \{2, 3, 5\}$. A second method is to use the following notation, which is called **set-builder notation**. Here, and throughout the text, we use $\mathbb{N}$ to denote the set of all natural numbers. Let

$$A = \{x \in \mathbb{N}: x \text{ is a prime number less than } 7\}$$

The colon is shorthand for "such that," and in words our description reads "$A$ equals the set of all $x$'s in the set $\mathbb{N}$ such that $x$ is a prime number less than seven." Set-builder notation is very useful, but it is not necessarily preferable to the most obvious way of describing a set by using just the English language.

It is easy to confuse $\in$ (is an element of) with $\subseteq$ (is a subset of). Consider the set $\{17\}$. This set is the set whose only member is 17, or as bridge players are wont to call it, "singleton seventeen." Clearly $17 \in \{17\}$, but 17 is not a subset of $\{17\}$. Indeed $\{17\}$ has exactly two subsets. They are $\varnothing$ and the set $\{17\}$ itself. Even though $\varnothing \subseteq \{17\}$, $\varnothing \notin \{17\}$. (We just got through agreeing that $\{17\}$ has only one member and it's 17.) Another distinction between $\in$ and $\subseteq$ is indicated by Proposition 2.4. For any three sets $A$, $B$, and $C$, if $A \subseteq B$ and $B \subseteq C$, then $A \subseteq C$. But if $A = \{3\}$, $B = \{\{3\}, 5\}$, and $C = \{B, 17\}$, then $A \in B$ and $B \in C$, but $A \notin C$. The set $C$ has exactly two members, and it is easy to see that neither of these members is the set $A$.

**Definition**

If $A$ is a set, then the set whose members are the subsets of $A$ is called the **power set** of $A$ and is denoted by $\mathscr{P}(A)$.

For example, if $A = \{2, 3, 4\}$, then $\mathscr{P}(A)$ has exactly eight members: $\varnothing$, $\{2\}$, $\{3\}$, $\{4\}$, $\{2, 3\}$, $\{2, 4\}$, $\{3, 4\}$, and $\{2, 3, 4\}$. The closed interval $[3, 7]$ is a member of the power set of the set of all real numbers. Thus if we let $\mathbb{R}$ denote the set of all real numbers, we can write $[3, 7] \subseteq \mathbb{R}$ or equivalently we can write $[3, 7] \in \mathscr{P}(\mathbb{R})$. We trust you understand why we have not tried to list all the members of $\mathscr{P}(\mathbb{R})$.

## EXERCISES 2.1

**1.** Prove Proposition 2.1.

**2.** Prove Proposition 2.2.

**3.** Prove that if $A \subseteq B$, $B \subseteq C$, and $C \subseteq A$, then $A = C$.

**4.** **a)** Prove that $\{2, 3\} = \{3, 2\}$. **b)** Prove that $\{2, 3, 3\} = \{2, 3\}$.

**5.** Let $A = \{1, 2\}$. List the members of $\mathscr{P}(A)$.

***6.** Let $A = \{1, 2\}$. List all members of $\mathscr{P}[\mathscr{P}(A)]$ that have exactly two members.

**7.** Is there a set with exactly 12 subsets? Explain.

**8.** A certain set $A$ has exactly 15 proper subsets. How many members does $\mathscr{P}(A)$ have?

**9.** A certain set $S$ has exactly 64 subsets. If a new element is added to $S$, how many subsets will the new set have? Explain.

**10.** Determine which of the following statements are true.

**a)** $\varnothing \in \{\varnothing\}$ **b)** $\varnothing \subseteq \{\varnothing\}$ **c)** $\pi/4 \in \{\pi/4\}$

**d)** $\varnothing \in \{\pi/4\}$ **e)** $\pi/4 \in \{\{\pi/4\}\}$ **f)** $\{\pi/4\} \subseteq \{\{\pi/4\}\}$

**g)** $\{\pi/4\} \in \{\{\pi/4\}\}$ **h)** $\varnothing \subseteq \{\{\pi/4\}\}$ **i)** $\{\pi/4\} \subseteq \{\pi/4, \{\pi/2\}\}$

**11.** Determine which of the following statements are true.

**a)** The empty set is a subset of every set.

**b)** If $A$ is a proper subset of $\varnothing$, then $A = \{17\}$.

**c)** If $A \subseteq B$, then $A = B$.

**d)** If $A = B$, then $A \subseteq B$.

**e)** Since $\varnothing$ is a member of $\{\varnothing\}$, $\varnothing = \{\varnothing\}$.

**f)** There is a set that is a member of every set.

**g)** $\{\{3, 5, 7\}\} \in \mathbb{N} \cup \mathscr{P}(\mathbb{N})$

**h)** $\{\{3, 5, 7\}\} \subseteq \mathbb{N} \cup \mathscr{P}(\mathbb{N})$

**12.** Let $X = \{a, b, c, d\}$. List all the members of $\mathscr{P}(X)$. Do you know how many sets you have listed? Why not?

**13.** Prove or give a counterexample. For any two sets $A$ and $B$, at least one of the following statements is true.

**a)** $A \subseteq B$ **b)** $B \subseteq A$ **c)** $A = B$

**14.** Prove or give a counterexample. For any two sets $A$ and $B$, at most one of the following statements is true.

**a)** $A$ is a proper subset of $B$. **b)** $B$ is a proper subset of $A$.

**c)** $A = B$

**15.** For each of the following subsets of $\mathbb{R}$, either list all the members of the set or describe the set in plain English.

**a)** $\{x \in \mathbb{R}: x > 3, x^2 < 5, \text{ and } x \neq 4\}$ **b)** $\{x \in \mathbb{R}: x > 3 \text{ or } -x > 3\}$

**c)** $\{x \in \mathbb{R}: x > 3 \text{ and } -x > 3\}$ **d)** $\{x \in \mathbb{R}: x > 3 \text{ and } -x < 3\}$

**e)** $\{x \in \mathbb{R}: x^2 \neq x\}$

***16.** Find three sets $A$, $B$, and $C$ such that $A \in B$, $B \in C$, and $A \in C$.

**17.** We give two "proofs" and a "counterexample" to the following conjecture: Suppose $A$, $B$, and $C$ are sets. If $A \nsubseteq B$ and $B \nsubseteq C$, then $A \nsubseteq C$.

Which, if any, are correct? Justify your answer.

"Proof 1": Let $x \in A$. Since $A \nsubseteq B$, $x \notin B$. Since $x \notin B$, $x \notin C$. Therefore $x$ is a member of $A$ and $x$ is not a member of $C$, so $A \nsubseteq C$.

"Proof 2": Since $B \nsubseteq C$, there exists $x \in B$ such that $x \notin C$. Since $A \nsubseteq B$, there exists $y \in A$ such that $y \notin B$. Therefore $x \neq y$, and therefore $A \nsubseteq C$.

"Counterexample": Let $A = \{1, 2, 3\}$, $B = \{1, 2, 4, 5\}$, and $C = \{1, 2, 3, 4\}$.

**18.** We give three "proofs" of the following conjecture: If $A$ and $B$ are sets and $A \subseteq B$, then $\mathscr{P}(A) \subseteq \mathscr{P}(B)$.

Which, if any, are correct? Justify your answer.

"Proof 1": Let $x \in \mathscr{P}(A)$. Then $x \in A$. Since $A \subseteq B$, $x \in B$. Therefore $x \in \mathscr{P}(B)$, so $\mathscr{P}(A) \subseteq \mathscr{P}(B)$.

"Proof 2": Let $A = \{1, 2\}$ and $B = \{1, 2, 3\}$. Then $\mathscr{P}(A) = \{\varnothing, \{1\}, \{2\}, A\}$ and $\mathscr{P}(B) = \{\varnothing, \{1\}, \{2\}, \{3\}, \{1, 2\}, \{1, 3\}, \{2, 3\}, B\}$. Therefore $\mathscr{P}(A) \subseteq \mathscr{P}(B)$.

"Proof 3": Let $x \in A$. Since $A \subseteq B$, $x \in B$. Since $x \in A$ and $x \in B$, $\{x\} \in \mathscr{P}(A)$ and $\{x\} \in \mathscr{P}(B)$. Therefore $\mathscr{P}(A) \subseteq \mathscr{P}(B)$.

**19.** Modify an incorrect proof in Exercise 18 to obtain a correct proof of the conjecture in Exercise 18.

**20.** We give two "proofs" of the following conjecture: If $A$ and $B$ are sets such that $\mathscr{P}(A) \subseteq \mathscr{P}(B)$, then $A \subseteq B$.

Which, if either, is correct? Justify your answer.

"Proof 1": Let $X \in \mathscr{P}(A)$. Then $X \subseteq A$. Since $A \subseteq B$, $X \subseteq B$. Therefore $X \in \mathscr{P}(B)$, so $\mathscr{P}(A) \subseteq \mathscr{P}(B)$.

"Proof 2": Let $x \in A$. Then $\{x\} \in \mathscr{P}(A)$. Since $\mathscr{P}(A) \subseteq \mathscr{P}(B)$, $\{x\} \in \mathscr{P}(B)$. Therefore $x \in B$.

## 2.2 Operations on Sets

If for a certain problem all the sets being considered are subsets of a given set $U$, then $U$ is called a **universal set of discourse**. Obviously, different problems may involve different universal sets, and we will see in Section 2.4 that there is no universal set for all sets. Nonetheless, it is common practice to speak of "the" universal set when, of course, there is certainly more than one set that contains all the sets we are considering.

In this section and in the next, we assume that all sets under consideration are subsets of some given universal set $U$. This assumption allows us to draw diagrams that enable us to picture the behavior of sets. For example, if we are given that $A \subseteq B$ and $B \subseteq C$, we can draw Figure 2.1, which records the given information. This figure certainly suggests that Proposition 2.4 ought to be true, but of course it is only a diagram and not a proof. Diagrams such as Figure 2.1 were used explicitly by the Swiss mathematician Leonhard Euler (1707–1783), and it is hard to believe that others had not thought of them previously. They are now called **Venn diagrams** in honor of the English mathematician John Venn (1834–1923). Venn diagrams are instructive, and

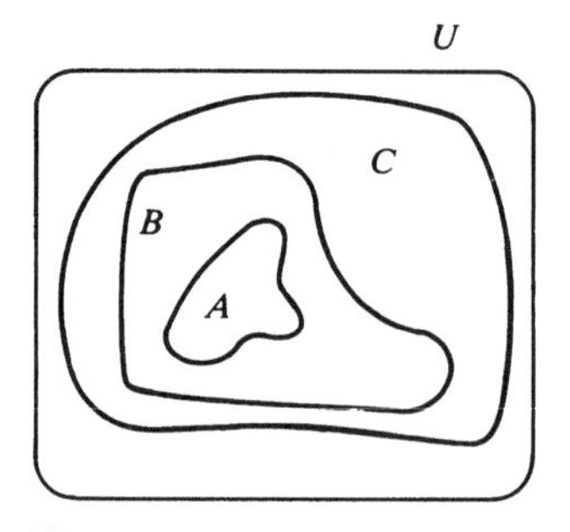


Figure 2.1

you are encouraged to use them to gain insights into problems. But although a Venn diagram may suggest a method of proof, it never constitutes a proof in and of itself.

Given two sets $A$ and $B$, we can form two (presumably) new sets.

**Definition**

If $A$ and $B$ are sets, then the **union** of $A$ and $B$ is the set of all objects that belong to $A$ or to $B$. The union of $A$ and $B$ is denoted by $A \cup B$, which is read "$A$ union $B$." (See Figure 2.2a.) In set-builder notation,

$$A \cup B = \{x \in U : x \in A \text{ or } x \in B\}$$

If $A$ and $B$ are sets, then the **intersection** of $A$ and $B$ is the set of all objects that belong to both $A$ and $B$. The intersection of $A$ and $B$ is denoted by $A \cap B$, which is read "$A$ intersect $B$." (See Figure 2.2b.) In set-builder notation,

$$A \cap B = \{x \in U : x \in A \text{ and } x \in B\}$$

We are, of course, using the word *or* in the definition of $A$ union $B$ in the inclusive sense (see Figure 2.2a).

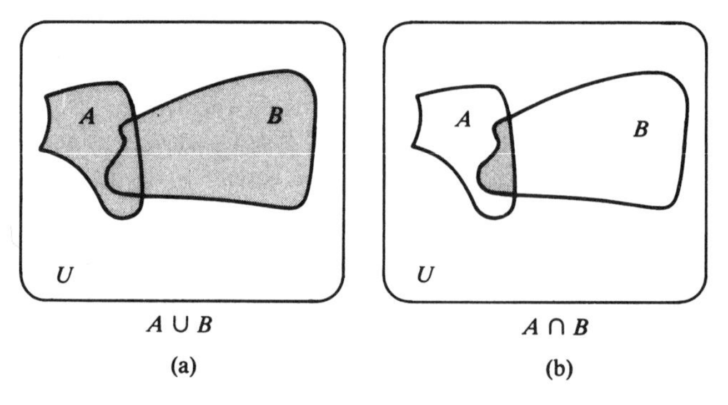


Figure 2.2

**EXAMPLE 1** Let $U = \{1, 2, 3, 4, 5, 6, 7, 8, 9, 10\}$, $A = \{2, 3, 4, 6\}$, and $B = \{3, 4, 7, 9\}$. Then $A \cup B = \{2, 3, 4, 6, 7, 9\}$ and $A \cap B = \{3, 4\}$. ❑

**EXAMPLE 2** Let $\mathbb{R}$ be the set of all real numbers, and let $A = [1, 3)$, $B = (2, 4]$ (that is, $A = \{x \in \mathbb{R}: 1 \leqslant x < 3\}$, and $B = \{x \in \mathbb{R}: 2 < x \leqslant 4\}$). Then $A \cup B = [1, 4]$ and $A \cap B = (2, 3)$. ❑

**EXAMPLE 3** Let $\mathbb{Z}$ be the set of all integers, $A$ the set of all even integers, and $B$ the set of all odd integers. Then $A \cup B = \mathbb{Z}$ and $A \cap B = \varnothing$. ❑

If $A$ and $B$ are sets and $A \cap B = \varnothing$, we say that $A$ and $B$ are **disjoint** sets (illustrated in Figure 2.3). Note that if $A$ is any set whatsoever (even $A = \varnothing$), then $A$ and $\varnothing$ are disjoint. (See Proposition 2.5a.)

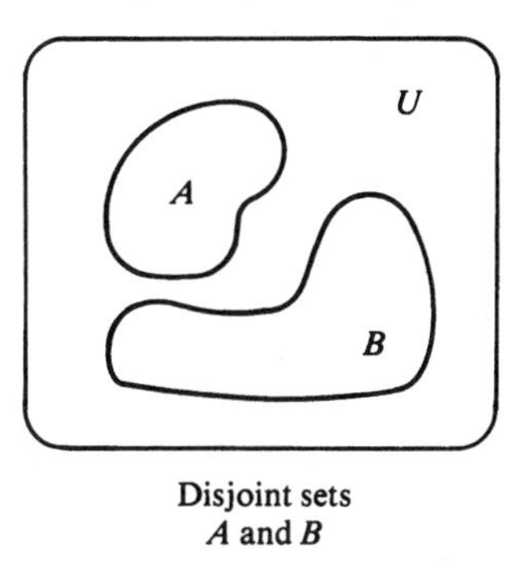


**Disjoint sets**
***A* and *B***

Figure 2.3

Venn diagrams are not the only way to indicate inclusion between sets; Figure 2.4 is self-explanatory.

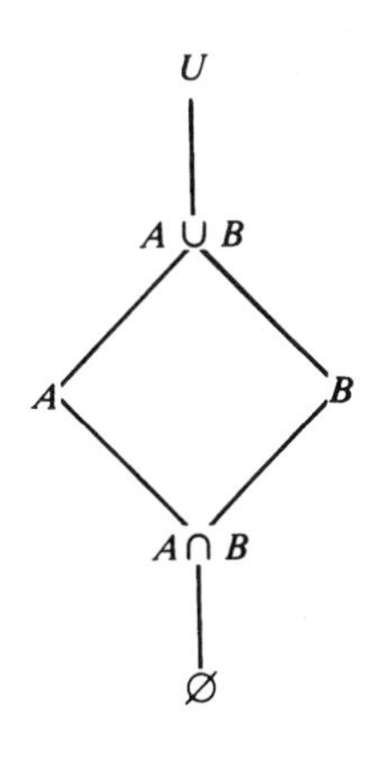


**Diagram of inclusion**

Figure 2.4

We begin this section with a catch-all proposition of intuitively obvious statements involving the union and intersection of sets. We verify two parts of this proposition, and in the exercises that follow we ask you to verify other parts, but the primary purpose of the proposition is to list in one place some set-theoretic results that may be assumed without comment in the proofs of other set-theoretic results.

**Proposition 2.5**

Let $A$, $B$, and $C$ be sets.

**a)** $\varnothing \cap A = \varnothing$ and $\varnothing \cup A = A$

**b)** $A \cap B \subseteq A$

**c)** $A \subseteq A \cup B$

**d)** $A \cup B = B \cup A$ and $A \cap B = B \cap A$

**e)** $A \cup (B \cup C) = (A \cup B) \cup C$ and $A \cap (B \cap C) = (A \cap B) \cap C$

**f)** $A \cup A = A = A \cap A$

**g)** If $A \subseteq B$, then $A \cup C \subseteq B \cup C$ and $A \cap C \subseteq B \cap C$.

**PROOF** We prove parts (f) and (g) and leave the remaining parts as exercises.

**f)** By part (b), $A \cap A \subseteq A$, and, by part (c), $A \subseteq A \cup A$. In light of Exercise 3 in Section 2.1, it suffices to show that $A \cup A \subseteq A \cap A$. Let $x \in A \cup A$. Then $x \in A$ or $x \in A$, so $x \in A$. It folows that $x \in A \cap A$.

**g)** We first prove that $A \cup C \subseteq B \cup C$. Suppose that $A \subseteq B$ and let $x \in A \cup C$. In order to show that $x \in B \cup C$, it suffices to show that if $x \notin C$, then $x \in B$. Suppose, therefore, that $x \notin C$. Then $x \in A$ and, as $A \subseteq B$, $x \in B$ as is required. We now prove that $A \cap C \subseteq B \cap C$. Suppose that $A \subseteq B$ and let $x \in A \cap C$. Then $x \in A$ and $x \in C$. Since $x \in A$ and $A \subseteq B$, $x \in B$. It follows that $x \in B$ and $x \in C$ and so $x \in B \cap C$. □

**Definition**

Let $A$ and $B$ be sets. Then the **complement of $A$ relative to $B$** is the set $\{x \in B : x \notin A\}$.

Since the complement of $A$ relative to $B$ is written $B - A$, it should come as no surprise that everyone says, "$B$ minus $A$," rather than the more cumbersome "complement of $A$ relative to $B$."

**Definition**

If $U$ is the universal set of discourse and $A$ is a subset of $U$, then $U - A$ is written $A'$ and is called the **complement** of $A$.

**EXAMPLE 4** Let $\mathbb{Z}$ be the universal set of discourse, and let $A$ be the set of all odd integers. Then $A'$ is the set of all even integers. ❑

**EXAMPLE 5** Let $U = \{1, 2, 3\}$, $A = \{1, 3\}$, $B = \{2\}$, and $C = \varnothing$. Then $U' = C$, $A' = B$, $B' = A$, and $C' = U$. ❑

**EXAMPLE 6** Let $U$ be the universal set of discourse, and let $A$ and $B$ be subsets of $U$. In Figure 2.5, the first Venn diagram illustrates the complement of $A$ relative to $U$, the second illustrates the complement of $B$ relative to $U$, and the third illustrates the complement of $A$ relative to $B$.

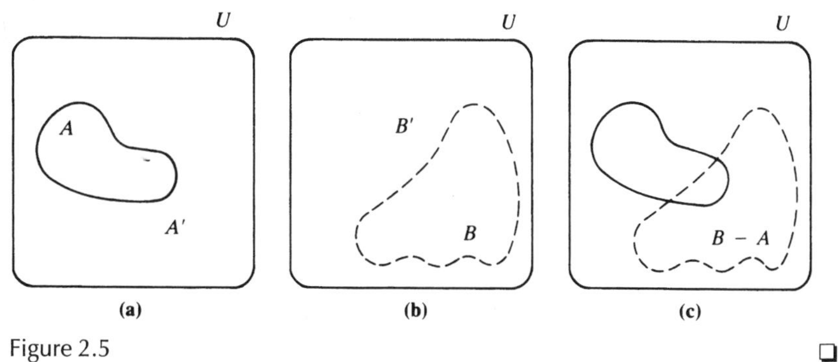


Figure 2.5 ❑

In the following theorem, parts (a) and (c) are called de Morgan's laws. Note the similarity between these equations and the tautologies given in parts (c) and (d) of Proposition 1.1.

**Theorem 2.6**

Let $A$ and $B$ be subsets of some universal set $U$. Then

**a)** $(A \cup B)' = A' \cap B'$

**b)** $(A')' = A$

**c)** $(A \cap B)' = A' \cup B'$

**d)** $A - B = A \cap B'$

**e)** $A \subseteq B$ if and only if $B' \subseteq A'$

**PROOF** We prove part (e) and leave the remaining parts as Exercises 32 and 33.

Suppose that $A \subseteq B$ and let $x \in B'$. Then $x \notin B$ and, because $A \subseteq B$, $x \notin A$. Thus, $x \in A'$ and so $B' \subseteq A'$.

Now suppose that $B' \subseteq A'$. Then $(A')' \subseteq (B')'$. (Why?) It follows immediately from part (b) that $A \subseteq B$. □

Since Theorem 2.6 is both illustrative and fundamental, it deserves special comment. In part (a), we are asked to prove that two sets, the left set and the right set, are equal. There is an obvious approach. We can show that any member of the left set is a member of the right set, whence $(A \cup B)' \subseteq A' \cap B'$, and then we can establish that any member of the right set is a member of the left set, whence $A' \cap B' \subseteq (A \cup B)'$. This approach, which you have already used in Exercise 4, is straightforward, and it is fundamental. Whenever you are asked to show that two sets are equal, the proof just outlined should spring to mind. As Venn diagrams often motivate or guide both directions of such a proof, it is important that you become accustomed to drawing appropriate Venn diagrams. The Venn diagrams associated with Theorem 2.6(a) are shown in Figure 2.6. Because we have already conceded that a Venn diagram does not constitute a proof, in this figure and in Figure 2.7 we use circles to represent sets. It is left as an exercise to draw the Venn diagrams that illustrate the remaining parts of Theorem 2.6.

In part (c) of Theorem 2.6, we are once again asked to prove that two sets are equal, and so we have at our disposal the approach we used in part (a). But parts (c) and (a) look similar, and coincidences in mathematics suggest one of two possibilities: Perhaps there is some general result of which

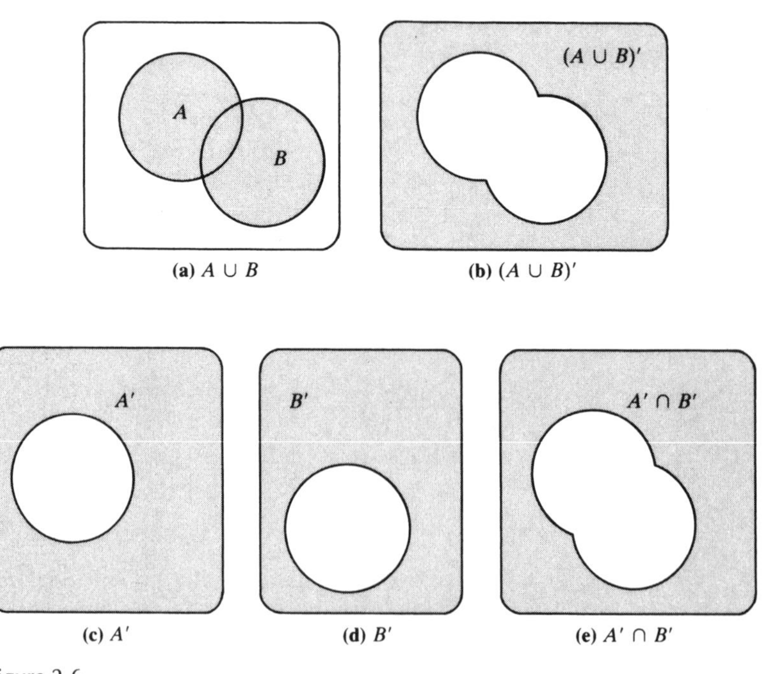


**(a)** $A \cup B$ **(b)** $(A \cup B)'$

**(c)** $A'$ **(d)** $B'$ **(e)** $A' \cap B'$

Figure 2.6

(a) and (c) are special cases, or perhaps (c) can be derived from (a). This idea is pursued in Exercise 38.

Finally, let us observe that, although the formulas for de Morgan's laws are neat, there is no harm in saying them aloud: The complement of the intersection is the union of the complements, and the complement of the union is the intersection of the complements. A pleasant feature of these laws is that there is no way to get them wrong. If you mix up one law, you just get the other. All you need remember is that each involves both a union and an intersection.

A useful property of multiplication of numbers is the distributive law, which says that

$$a \times (b + c) = (a \times b) + (a \times c)$$

It is this law, for example, that allows us to figure a 15% tip by figuring a 10% tip and then adding half again as much. Although the law to which we refer is commonly called the (left) distributive law, what it really says is that multiplication distributes over addition. Does addition distribute over multiplication? In other words, is it true for all numbers $a$, $b$, and $c$ that

$$a + (b \times c) = (a + b) \times (a + c)$$

Of course not! The following theorem says that $\cup$ and $\cap$ are incredibly well behaved in that intersection distributes over union and, what's more, union distributes over intersection. It's just like de Morgan's law—you can't go wrong.

**Theorem 2.7**

Let $A$, $B$, and $C$ be sets. Then the following hold:

$$A \cap (B \cup C) = (A \cap B) \cup (A \cap C) \quad \text{and} \quad A \cup (B \cap C) = (A \cup B) \cap (A \cup C)$$

**PROOF** We prove that $A \cap (B \cup C) = (A \cap B) \cup (A \cap C)$ and leave the remaining equality as Exercise 34.

Let $x \in A \cap (B \cup C)$. Then $x \in A$ and $x \in B \cup C$. Thus, $x \in B$ or $x \in C$. Suppose that $x \notin A \cap B$. Then $x \in A$ and $x \notin B$. It follows that $x \in A \cap C$. Thus, $x \in A \cap B$ or $x \in A \cap C$, so $x \in (A \cap B) \cup (A \cap C)$. We have shown that $A \cap (B \cup C) \subseteq (A \cap B) \cup (A \cap C)$.

Now suppose that $x \in (A \cap B) \cup (A \cap C)$. Then $x \in A \cap B$ or $x \in A \cap C$ and, in either case, $x \in A$. If $x \in A \cap B$, $x \in B \subseteq B \cup C$, and if $x \in A \cap C$, $x \in C \subseteq B \cup C$. In either case, $x \in A$ and $x \in B \cup C$, and we have that $x \in A \cap (B \cup C)$. We have shown that $(A \cap B) \cup (A \cap C) \subseteq A \cap (B \cup C)$. □

We have presented an element-chasing proof of the distributive law $A \cap (B \cup C) = (A \cap B) \cup (A \cap C)$; it is possible to derive this result from the corresponding distributive law, given in Proposition 1.2(h). Let $x$ be an element of the universal set $U$, $P$ be the proposition $x \in A$, $Q$ be the proposition $x \in B$, and $R$ be the proposition $x \in C$. Then $x \in A \cap (B \cup C)$ if and only if $P \wedge (Q \vee R)$ is true and $x \in (A \cap B) \cup (A \cap C)$ if and only if $(P \wedge Q) \vee (P \wedge R)$ is true. It follows from Proposition 1.2(h) that $x \in A \cap (B \cup C)$ if and only if $x \in (A \cap B) \cup (A \cap C)$.

We use Venn diagrams in Figure 2.7 to illustrate what we have just proved—namely, that intersection distributes over union.

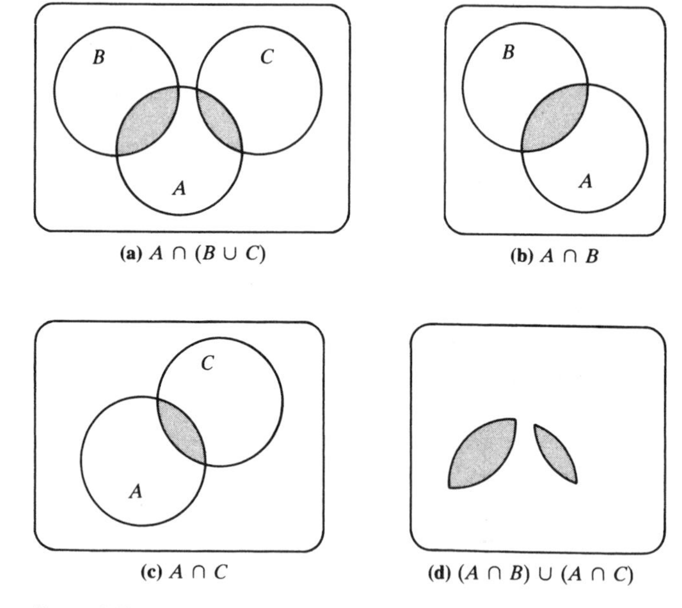


Figure 2.7

**EXAMPLE 7** Let $A$ and $B$ be subsets of a universal set $U$, and let $C$ and $D$ be subsets of $U$ such that $A \cap C = \emptyset$, $B \cup D = U$, and $D \subseteq C$. Prove that $B' \subseteq A'$. ❑

**PROOF** Let $x \in B'$. Since $x \notin B$, $x \in D$ and so $x \in C$. Since $A \cap C = \emptyset$, $x \notin A$ and so $x \in A'$. Thus, $B' \subseteq A'$. □

## EXERCISES 2.2

**21.** Let $U = \{1, 2, 3, 4, 5, 6, 7, 8, 9, 10\}$, $A = \{1, 3, 6\}$, $B = \{3, 4, 5, 6\}$, and $C = \{3, 6, 9\}$. List the members of each of the following sets.

**a)** $A \cup B$ **b)** $(A \cup B)'$ **c)** $A \cap B$ **d)** $(A \cap B)'$
**e)** $(A \cup B) \cap C$ **f)** $(A \cap B) \cup C$ **g)** $(A')'$ **h)** $A \cap (B \cup C)$
**i)** $A' \cup B'$ **j)** $A' \cap B'$

**22.** Let $X = \{1, 2, 3, 4\}$. List all pairs $A$, $B$ of subsets of $X$ such that $A$ and $B$ are disjoint and $A \cup B = X$.

**23.** Prove that $A \cup B = A$ if and only if $B \subseteq A$.

**24.** Prove that $A \cap B = A$ if and only if $A \subseteq B$.

**25.** Let $A$ and $B$ be subsets of a universal set $U$. Prove that $A \cup B = U$ if and only if $A'$ and $B'$ are disjoint.

**26.** Draw Venn diagrams illustrating $(A')'$.

**27.** Draw Venn diagrams illustrating $A \cup (B \cap C)$ and $(A \cup B) \cap (A \cup C)$.

**28.** **a)** Draw the Venn diagrams associated with Theorem 2.6(c).

**b)** Draw the Venn diagrams associated with Theorem 2.6(d).

**29.** Prove each of the following parts of Proposition 2.5.

**a)** Part (a) **b)** Part (b) **c)** Part (c)

**d)** Part (d) **e)** Part (e)

**30.** Prove or find a counterexample to the following statement. For any sets $P$, $Q$, and $R$,

$$(P \cap Q) \cup R = P \cap (Q \cup R)$$

**31.** A student is asked to prove that for any sets $A$, $B$, and $C$,

$$A - (B \cup C) = (A - B) \cap (A - C)$$

The student writes: "Let $x \in A - (B \cup C)$. Then $x \in A$ and $x \notin B$ or $x \notin C$. Therefore, $x \in A - B$ and $x \in A - C$. Thus, $A - (B \cup C) = (A - B) \cap (A - C)$." What, if anything, is wrong with this proof?

**32.** Prove Theorem 2.6(a) and (b).

**33.** Prove Theorem 2.6(c) and (d).

**34.** Prove that for any sets $A$, $B$, and $C$, the following equation holds:

$$A \cup (B \cap C) = (A \cup B) \cap (A \cup C)$$

**35.** Prove that for any sets $A$ and $B$, the following are true.

**a)** $(A \cup B) - (A \cap B) = (A - B) \cup (B - A)$

**b)** $A \cap B$ and $A - B$ are disjoint.

**c)** $A = (A \cap B) \cup (A - B)$

**36.** In former times, it was customary to denote intersection by $\cdot$ and union by $+$. (Because of this notation, a famous result of G. Cantor is still called the Cantor Product Theorem.) To what extent do $\cup$ and $\cap$ mimic the behavior of $+$ and $\cdot$?

**37.** A student is asked to prove that for any two sets $A$ and $B$,

$$(A - B) \cup (B - A) = (A \cup B) - (A \cap B)$$

The student writes: "Let $A \cup B$ be the universal set. Then

$$\begin{aligned}(A \cap B)' &= A' \cup B' \\ &= [(A \cup B) - A] \cup [(A \cup B) - B] \\ &= (B - A) \cup (A - B) \\ &= (A - B) \cup (B - A).\text{''}\end{aligned}$$

What, if anything, is wrong with this proof?

**38.** Use Theorem 2.6(a) and (b) to prove Theorem 2.6(c).

**39.** Let $A$ and $B$ be sets. Prove that if $A \cup B = B$, then $A \cap B = A$. Is the converse true? If so, prove it; if not, find a counterexample.

**40.** Prove that for any sets $A$ and $B$, the following are true.

**a)** $A - (A \cap B') = A \cap B$ **b)** $(A' \cup B)' \cap A = A - B$

**c)** $(A - B)' - (B - A)' = B - A$

**41.** Prove that if $A$ and $B$ are sets such that $A \cup B = A \cap B$, then $A \cap B' = \emptyset$.

**42.** Prove that if $A$ and $B$ are sets such that $(A \cup B)' = A' \cup B'$, then $A = B$.

**43.** Let $A$, $B$, and $C$ be sets. Prove that $A - B \subseteq C$ if and only if $A - C \subseteq B$.

**44.** Prove that for any sets $A$, $B$, and $C$, the following are true.

**a)** $A - (B - C) \subseteq A \cap (B' \cup C)$ **b)** $(A \cap B') \cap C' = (A \cap C') - (B \cap C')$

**c)** $(A - B) - C = (A - C) - (B - C)$ **d)** $A \cap (B - C) = (A \cap B) - (A \cap C)$

**45. a)** Let $A$, $B$, and $C$ be sets. Prove that $(A \cup B) - C \subseteq [A - (B \cup C)] \cup [B - (A \cap C)]$.

**b)** Give an example of sets $A$, $B$, and $C$ for which the reverse containment in part (a) does not hold.

**46.** Prove or find a counterexample to each of the following.

**a)** $(A - B) \cup C = A - (B \cup C)$ **b)** $(A' \cup B) \cap (B' \cup C) \subseteq A' \cup C$

**47.** Let $A$, $B$, and $C$ be sets such that $A \cup B \neq A \cap C$. Prove that $A$ is not a subset of $C$ or $B$ is not a subset of $A$.

**48.** A student wishes to prove that if $A$ and $B$ are sets such that $A - B = B - A$ then $A = B$. The student writes, "Suppose $A - B = B - A$. This means $A - B \subseteq B - A$ and $B - A \subseteq A - B$. Let $x \in A - B$. Then $x \in B - A$. This means $x \in A$ and $x \notin B$, and $x \in B$ and $x \notin A$. Thus, $x \in A$ and $x \in B$, so $A \subseteq B$ and $B \subseteq A$. Therefore, $A = B$." Grade this student's proof.

**49.** Let $A$ and $B$ be sets. Prove that if $\mathscr{P}(A \cup B) = \mathscr{P}(A) \cup \mathscr{P}(B)$, then $A \subseteq B$ or $B \subseteq A$.

**50.** We give a "proof" and a "counterexample" to the following conjecture: "If $A \subseteq B$ and $C \subseteq D$, then $A \cap C \subseteq B \cap D$."

Which, if either, of these is correct?

"Counterexample": Let $A = \{1, 2\}$, $B = \{1, 2, 3, 4\}$, $C = \{\{1\}, \{2\}\}$, and $D = \{1, 2, 3, 4\}$.

"Proof": Let $x \in A \cap C$. Then $x \in A$ and $x \in C$. Since $A \subseteq B$ and $C \subseteq D$, $x \in B$ and $x \in D$. Therefore $x \in B \cap D$, so $A \cap C \subseteq B \cap D$.

**51.** We give a "proof" of the following conjecture: If $A$, $B$, and $C$ are sets such that $A \cap B = A \cap C$ and $A' \cap B = A' \cap C$, then $B = C$.

Is this "proof" correct? Justify your answer.

"Proof": Suppose $x \in B$ and $x \notin C$. Now $x \in A$ or $x \in A'$, so we consider two cases.

*Case 1:* Suppose $x \in A$. Then $x \in A \cap B$, but $x \notin A \cap C$. Therefore $A \cap B \neq A \cap C$.

*Case 2:* Suppose $x \in A'$. Then $x \in A' \cap B$, but $x \notin A' \cap C$. Therefore $A' \cap B \neq A' \cap C$.

**52.** A student wishes to prove that if $A$, $B$, and $C$ are sets and $A \cap B \neq A \cap C$, then $A \nsubseteq C$ or $B \nsubseteq A$. The student writes, "Suppose $A \cup B = A \cap C$. Then $A \subseteq A \cup B = A \cap C \subseteq A$, so $A = A \cup B$. But $B \subseteq A \cup B = A$. Thus if $A \cap B \neq A \cap C$, then $A \nsubseteq C$ or $B \nsubseteq A$." Is this "proof" correct? Justify your answer.

## 2.3 Indexed Families

We have already observed that the members of a set may themselves be sets (as is the case whenever we consider the power set of a set), but in most instances the members of a given set are thought of as basic entities. We use the terms *collection* and *family* in place of the term *set* when we wish to emphasize that the members of a given set are themselves sets. Thus we would probably speak of the set $\{1, 2, 3\}$ and of the family of all straight lines in the plane. There would be nothing wrong with saying "the set of all straight lines in the plane"; the term *family* is a friendly reminder that each straight line is itself a set of points of the plane.

Suppose that for each real number $x$, we let $A_x$ be the closed interval $[-x, x]$. Then it makes perfectly good sense to speak of the family $\{A_x: x$ is a real number$\}$. We call such a family an **indexed family**. (The precise definition of an indexed family involves functions and so we defer the definition until Chapter 5.) We have been considering an indexed family of closed intervals, indexed by the set of all real numbers. It is natural to guess that any infinite family can be indexed by the set of all natural numbers and that any finite set can be indexed by a finite set of natural numbers. It turns out that some infinite families, including the one we have been considering, are too large to be indexed by $\mathbb{N}$, but at least it is true that any family can be indexed by some set. Before we prove this fact, let us consider some examples.

We define the union and intersection of an arbitrary collection of sets, and it is convenient to use the notion of an arbitrary nonempty set as the indexing set. The following examples, however, use the natural numbers, the integers, or the real numbers as the indexing set. Thus, although we discuss arbitrary collections of sets, we encourage you always to think of the indexing set as being one of the familiar sets whose symbols we listed on the inside front cover. For convenience we review that list of symbols.

List of Symbols

- $\mathbb{N}$ The set of all natural numbers
- $\mathbb{Z}$ The set of all integers
- $\mathbb{Q}$ The set of all rational numbers
- $\mathbb{R}$ The set of all real numbers
- $\mathbb{C}$ The set of all complex numbers

**EXAMPLE 8** For each natural number $n$, let $A_n = \{2n - 1, 2n\}$. In order to see what these sets are, let us examine $A_n$ for some specific values of $n$. Now $A_1 = \{1, 2\}$, $A_2 = \{3, 4\}$, and $A_3 = \{5, 6\}$. So we have a family of sets $\{A_n: n \in \mathbb{N}\}$, and each set in the family consists of two consecutive natural numbers. The indexing set is the set of natural numbers. ❑

**EXAMPLE 9** For each integer $n$, let $A_n$ be the closed interval $[n-1, n]$. That is,

$$A_n = \{x \in \mathbb{R}: n-1 \leqslant x \leqslant n\}$$

Once again, we examine $A_n$ for some specific values of $n$. Now $A_1 = [0, 1]$, $A_5 = [4, 5]$, and $A_{-2} = [-3, -2]$. So we have a family of sets

$$\{A_n: n \in \mathbb{Z}\} = \{[n-1, n]: n \in \mathbb{Z}\}$$

and each set in the family consists of all the real numbers in a closed interval of length 1. The indexing set is the set of integers. ❑

**EXAMPLE 10** For each natural number $n$, let $A_n$ be the open interval $(-n, n)$. That is,

$$A_n = \{x \in \mathbb{R}: -n < x < n\}$$

Therefore, $A_1 = (-1, 1)$, $A_4 = (-4, 4)$, and $A_7 = (-7, 7)$. We have a family of sets

$$\{A_n: n \in \mathbb{N}\} = \{(-n, n): n \in \mathbb{N}\}$$

and each set in the family consists of all the real numbers in an open interval. The indexing set is the set of natural numbers. ❑

In the examples that we have been considering, the families have been indexed in such a way that $A_m \neq A_n$ unless $m = n$. This does not always happen.

**EXAMPLE 11** For each natural number $n$, let

$$A_n = \{p \in \mathbb{N}: p \text{ is a prime number that divides } n\}$$

Then $\{A_n: n \in \mathbb{N}\}$ is an indexed family. We note that $A_{12} = A_{24} = \{2, 3\}$ and that if $k$ is a natural number greater than 1, then there are infinitely many $n \in \mathbb{N}$ such that $A_k = A_n$. Finally, we note that $A_1 = \varnothing$. ❑

The union and the intersection of two sets, which were defined in the preceding section, are actually special cases of the following definitions.

**Definition** Let $\mathscr{A}$ be a collection of sets. Then $\cup\mathscr{A}$ is the set to which $x$ belongs provided *there is* a set $A$ belonging to the collection $\mathscr{A}$ such that $x \in A$. If $\mathscr{A} \neq \varnothing$, $\cap\mathscr{A}$ is the set to which $x$ belongs provided $x \in A$ *for each* $A \in \mathscr{A}$.

**Notation**

If $\mathscr{A} = \{A_\alpha: \alpha \in \Lambda\}$, then $\cup\{A_\alpha: \alpha \in \Lambda\}$ and $\cup_{\alpha \in \Lambda} A_\alpha$ are alternative notations for $\cup\mathscr{A}$. Similarly, if $\mathscr{A} \neq \varnothing$ and $\Lambda$ is a nonempty set such that $\mathscr{A} = \{A_\alpha: \alpha \in \Lambda\}$, then $\cap\{A_\alpha: \alpha \in \Lambda\}$ and $\cap_{\alpha \in \Lambda} A_\alpha$ are alternative notations for $\cap\mathscr{A}$.

**Definition**

Let $\mathscr{A}$ be a collection of sets. We say that $\mathscr{A}$ is **pairwise disjoint** provided that if $A$ and $B$ are members of $\mathscr{A}$ such that $A \neq B$, then $A \cap B = \varnothing$.

Certainly if $\mathscr{A}$ is a pairwise disjoint family, then $\cap\mathscr{A} = \varnothing$, but the converse is false (Exercise 72).

**EXAMPLE 12** Let $A_1 = \{12, 13, 14\}$, $A_2 = \{13, 14, 15\}$, and $A_3 = \{12, 13, 14, 15, 16\}$. Let $\mathscr{A} = \{A_1, A_2, A_3\} = \{A_i: i \in \{1, 2, 3\}\}$. Then $\cup\mathscr{A} = A_1 \cup A_2 \cup A_3 = \{12, 13, 14, 15, 16\}$ and $\cap\mathscr{A} = A_1 \cap A_2 \cap A_3 = \{13, 14\}$. Note that $\cup\mathscr{A}$ can be written $\cup\{A_i: i \in \{1, 2, 3\}\}$, but most people write either $A_1 \cup A_2 \cup A_3$ or $\cup\{A_1, A_2, A_3\}$. Also, $\cap$ can be written $\cap\{A_i: i \in \{1, 2, 3\}\}$. ❑

**EXAMPLE 13** Let $A_1$ and $A_2$ be sets. Then

$$A_1 \cup A_2 = \cup\{A_i: i \in \{1, 2\}\} \quad \text{and} \quad A_1 \cap A_2 = \cap\{A_i: i \in \{1, 2\}\}$$ ❑

**EXAMPLE 14** For each natural number $n$, let $A_n = \{x \in \mathbb{R}: -n < x \leqslant 1/n\}$. Then

$$\bigcup_{n \in \mathbb{N}} A_n = \{x \in \mathbb{R}: \text{there is a natural number } n \text{ such that } x \in A_n\} = (-\infty, 1]$$

Moreover,

$$\bigcap_{n \in \mathbb{N}} A_n = \{x \in \mathbb{R}: x \in A_n \text{ for each natural number } n\} = (-1, 0]$$ ❑

**EXAMPLE 15** For each $\alpha \in \mathbb{R}$, let $A_\alpha$ be the union of the circle of radius $\alpha$ and center $(0, 0)$ with its diameter on the $x$-axis. The set $\cup\{A_\alpha: \alpha \in [1, 2]\}$, which is a subset of the plane, is illustrated in Figure 2.8(a). The set $\cap\{A_\alpha: \alpha \in [1, 2]\}$, which is a subset of the $x$-axis, is illustrated in Figure 2.8(b). ❑

The preceding examples illustrate that in working with

$$\bigcup_{\alpha \in \Lambda} A_\alpha \quad \text{and} \quad \bigcap_{\alpha \in \Lambda} A_\alpha$$

there are two crucial facts to keep in mind:

1. $x \in \bigcup_{\alpha \in \Lambda} A_\alpha$ means exactly that *there is* an $\alpha \in \Lambda$ such that $x \in A_\alpha$,
2. $x \in \bigcap_{\alpha \in \Lambda} A_\alpha$ means exactly that *for each* $\alpha \in \Lambda$, $x \in A_\alpha$.

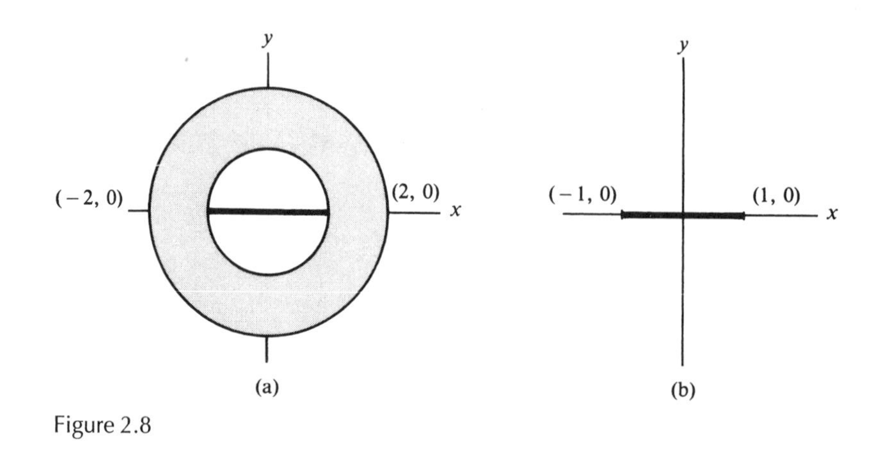


Figure 2.8

With these facts in mind, it is easy to extend formulas relating two sets to formulas involving indexed families. We consider two simple but useful examples. Surely for any two sets $A$ and $B$, we have the facts that $A \subseteq A \cup B$, that $B \subseteq A \cup B$, that $A \cap B \subseteq A$, and that $A \cap B \subseteq B$. Put more succinctly, we are asserting that if we have a family $\mathscr{A} = \{A, B\}$ and a member $X$ of $\mathscr{A}$, then $X \subseteq \cup\mathscr{A}$ and $\cap\mathscr{A} \subseteq X$. Let us prove extensions of these results for any indexed family $\mathscr{A} = \{A_\alpha : \alpha \in \Lambda\}$.

We have stated the theorems in this section for the general case, but it may be helpful if you think of $\Lambda$ as being $\mathbb{N}$ in the first reading of Theorems 2.8 and 2.9.

**Theorem 2.8**

Let $\mathscr{A} = \{A_\alpha : \alpha \in \Lambda\}$ be an indexed family of sets and let $\beta \in \Lambda$. Then

**a)** $A_\beta \subseteq \bigcup_{\alpha \in \Lambda} A_\alpha$, and

**b)** $\bigcap_{\alpha \in \Lambda} A_\alpha \subseteq A_\beta$.

**PROOF**

**a)** Let $x \in A_\beta$. Since $\beta \in \Lambda$, there is an $\alpha \in \Lambda$ such that $x \in A_\alpha$. By the preceding fact (1), $x \in \cup_{\alpha \in \Lambda} A_\alpha$.

**b)** Let $x \in \cap_{\alpha \in \Lambda} A_\alpha$. By the preceding fact (2), for each $\alpha \in \Lambda$, $x \in A_\alpha$. Since $\beta \in \Lambda$, $x \in A_\beta$. □

Among the formulas that can be extended to indexed families are the two distributive laws and de Morgan's laws. We leave most of these extensions as exercises.

**Theorem 2.9**

Let $\{A_\alpha : \alpha \in \Lambda\}$ be an indexed family of sets, and let $B$ be a set. Then the following statements hold:

**a)** $B \cap (\cup\{A_\alpha : \alpha \in \Lambda\}) = \cup\{B \cap A_\alpha : \alpha \in \Lambda\}$

**b)** $B \cup (\cap\{A_\alpha : \alpha \in \Lambda\}) = \cap\{B \cup A_\alpha : \alpha \in \Lambda\}$

**c)** $(\cap\{A_\alpha : \alpha \in \Lambda\})' = \cup\{A'_\alpha : \alpha \in \Lambda\}$

**d)** $(\cup\{A_\alpha : \alpha \in \Lambda\})' = \cap\{A'_\alpha : \alpha \in \Lambda\}$

**PROOF** The proof of part (d) is typical.

Let $x \in (\cup\{A_\alpha : \alpha \in \Lambda\})'$. Then $x \notin \cup\{A_\alpha : \alpha \in \Lambda\}$ and so for each $\alpha \in \Lambda$, $x \notin A_\alpha$. Thus for each $\alpha \in \Lambda$, $x \in A'_\alpha$ and so

$$(\cup\{A_\alpha : \alpha \in \Lambda\})' \subseteq \cap\{A'_\alpha : \alpha \in \Lambda\}$$

The remaining containment is established in a similar manner. □

## EXERCISES 2.3

**53.** For each natural number $n$, let $A_n = \{7, n\}$. Find $\cup_{n \in \mathbb{N}} A_n$ and $\cap_{n \in \mathbb{N}} A_n$.

**54.** For each natural number $n$, let $A_n = \{k \in \mathbb{N} : k \geqslant n\}$. Find $\cup_{n \in \mathbb{N}} A_n$ and $\cap_{n \in \mathbb{N}} A_n$.

**55.** For each natural number $n$, let $A_n = [-1/n, 1/n]$ and $B_n = (-1/n, 1/n)$. Find $\cap_{n \in \mathbb{N}} A_n$ and $\cap_{n \in \mathbb{N}} B_n$.

**56.** For each natural number $n$, let $A_n = [1 - 1/n, 1 + 1/n]$. Find $\cup_{n \in \mathbb{N}} A_n$ and $\cap_{n \in \mathbb{N}} A_n$.

**57.** **a)** For each natural number $n$, let $A_n = (n, n+1)$. Find $\cup_{n \in \mathbb{N}} A_n$ and $\cap_{n \in \mathbb{N}} A_n$.

**b)** For each real number $x$, let $A_x = (x, x+1)$. Find $\cup_{x \in \mathbb{R}} A_x$ and $\cap_{x \in \mathbb{R}} A_x$.

**58.** For each natural number $n$, let $A_n = \{-n\} \cup [1/n, 3n+1)$. Find $\cup_{n \in \mathbb{N}} A_n$ and $\cap_{n \in \mathbb{N}} A_n$.

**59.** For each natural number $n$, let $A_n = (0, 1 + 3/n)$. Find $\cup_{n \in \mathbb{N}} A_n$ and $\cap_{n \in \mathbb{N}} A_n$.

**60.** For each real number $x$, let $A_x = \{3, x+1\}$. Find $\cup_{x \in \mathbb{R}} A_x$ and $\cap_{x \in \mathbb{R}} A_x$.

**61.** For each natural number $n$, let $A_n = (-1/n, (2n-1)/n)$. Find $\cup_{n \in \mathbb{N}} A_n$ and $\cap_{n \in \mathbb{N}} A_n$.

**62.** For each real number $x$, let $A_x = \{3, -2\} \cup \{y \in \mathbb{R} : y > x\}$. Find $\cup_{x \in \mathbb{R}} A_x$ and $\cap_{x \in \mathbb{R}} A_x$.

**63.** Give an example of a collection of sets indexed by a set consisting of 5 members.

**64.** Give an example of a collection of sets indexed by the set of prime numbers.

**65.** Prove Theorem 2.9(c).

**66.** Use Theorem 2.9(c), Theorem 2.6(b), and Exericse 25 to prove Theorem 2.9(d).

**67.** Prove Theorem 2.9(a) and (b).

**68.** Assume that the universe of discourse for Exercises 53 and 54 is the set of all natural numbers. Find $\cup_{n \in \mathbb{N}} A'_n$ and $\cap_{n \in \mathbb{N}} A'_n$ for each exercise.

**69.** Assume that the universe of discourse for Exercises 55, 56, and 57 is the set of all real numbers. Find $\cup_{n\in\mathbb{N}} A_n'$ and $\cap_{n\in\mathbb{N}} A_n'$ for each of these exercises.

**70.** Let $\{A_\alpha: \alpha\in\Lambda\}$ be an indexed family of sets. Prove that $\cup_{\alpha,\beta\in\Lambda,\alpha\neq\beta}(A_\alpha - A_\beta)\subseteq \cup_{\alpha\in\Lambda}A_\alpha - \cap_{\alpha\in\Lambda}A_\alpha$.

**71.** A subset $K$ of $\mathbb{R}$ is called **cofinite** provided $\mathbb{R} - K$ is finite. Let $\mathscr{A}$ be a nonempty collection of cofinite subsets of $\mathbb{R}$. Prove that $\cup\mathscr{A}$ is a cofinite subset of $\mathbb{R}$.

**72.** Give an example of a family $\mathscr{C} = \{C_n: n\in\mathbb{N}\}$ such that $\cap\mathscr{C} = \varnothing$ but, for each $n\in\mathbb{N}$, $\cap\{C_k: 1\leqslant k\leqslant n\}\neq\varnothing$.

## 2.4 *An Axiomatic Approach to Sets*

We have left the impression that any collection of sets is a set and that if we have for each collection $x$ a corresponding statement $P(x)$, then we can form $\{x: P(x) \text{ is true}\}$. Bertrand Russell (1872–1970)* showed that such indiscriminate use of set-builder notation cannot be allowed: Suppose that for each set $x$, we let $P(x)$ be the statement $x = x$ and that we then believe that $S = \{x: P(x) \text{ is true}\}$ is a set. ($S$ would be the "set" of all sets, for certainly no matter what set $x$ we have, the statement $x = x$ is true.) Assuming that $S$ is a set, we observe that $S\in S$. This is a somewhat startling property that the "set" $S$ possesses, so let us consider the "set" $T = \{x: x\notin x\}$. Although most ordinary sets belong to $T$, evidently $S\notin T$. Is $T\in T$? Suppose that $T\in T$. Then $T\notin T$, and if $T\notin T$, then $T\in T$. This is a contradiction. The "set" $T$ cannot be a set at all, and the "set" $S$ is already suspect since it is a member of itself. We are forced to abandon naive set theory and to restrict our notion of set by demanding that sets obey certain axioms.

The axioms most widely accepted are those of Ernst Zermelo (1871–1953) and Abraham Fraenkel (1891–1965) and are known as Zermelo–Fraenkel Set Theory. The rules that we are about to give comprise a jury-rigged axiom system because a real introduction to axiomatic set theory is quite beyond the scope of this introductory section. Although we have based our rules rather loosely on Zermelo–Fraenkel Set Theory, we believe that working with our rules will give you some idea how Zermelo–Fraenkel Set Theory behaves.

Personal Note

### Ernst Zermelo (1871–1956)

It was Zermelo who in 1904 first formulated the Axiom of Choice (Rule 10, page 61). According to Howard Eves [9, 279°], whenever Zermelo was asked about his unusual surname, he would explain that his family name had been *Walzermelodie* ("waltz melody") and that he had lopped off *Wal* and *die*. By

*See Personal Note on Ernst Zermelo.

whatever name, Zermelo's position as one of the founders of modern set theory is assured. Curiously, however, his discovery of one of naive set theory's most basic paradoxes, is not well known. This paradox, now known as Russell's paradox, was established independently by Zermelo and B. Russell and published by Russell in 1903. Zermelo's proof, which is essentially the same as Russell's argument is documented in "Zermelo's discovery of the Russell Paradox" by B. Rung and W. Thomas [*Historia Math*, Vol. 8, 1981, pages 15–22]. The authors of this paper maintain that "Zermelo must have found it [the paradox] by 1900, perhaps as early as 1899, that is, one or two years before Russell."

Here are our rules. **In this chapter, we will consider only the first eight of them, since Rules 9 and 10 involve concepts not yet defined.**

**Rules**

1. The empty set is a set as is every member of a set.
2. If $X$ is a set and, for each $x$ in $X$, $P(x)$ is a proposition, then $\{x \in X: P(x)\}$ is a set.
3. If $X$ is a set and $Y$ is a set, so is $\{X, Y\}$.
4. If $X$ is a set, $\{z: z \in x \text{ for some } x \in X\}$ is a set. This set is denoted by $\cup X$.
5. If $X$ is a nonempty set, $\{z: z \in x \text{ for each } x \in X\}$ is a set. This set is denoted by $\cap X$.
6. If $X$ is a set, $\{z: z \subseteq X\}$ is a set. This set, denoted by $\mathcal{P}(X)$, is called the power set of $X$.
7. The set $\mathbb{N}$ of all natural numbers is a set.
8. No set is a member of itself.
9. If $X$ is a set and $Y$ is a set, so is $X \times Y$.
10. *The Axiom of Choice*: If $X$ is a set and $\varnothing \notin X$, there is a function $f$ with domain $X$ such that $f(x) \in x$ for each $x \in X$.

If the rules you have been given are any good, then we should be able to use them to prove theorems. Let us consider the first six of them and use these six rules to build the first few natural numbers. Of course, we cannot actually build sets, any more than a geometer can "construct" a line perpendicular to a given line. But we can use our rules to discover that certain sets exist, and if it pleases us, we can name the sets that we discover. In reality, our rules are incapable of "building" any infinite set, so we will not be able to "build" all the natural numbers. It turns out, however, that the exercise of "building" the first three natural numbers is already a good exercise in deducing theorems from a given axiom system, and our Rule 7 promises that the collection of all the numbers that we could "build" is a set.

To begin with, note that Rule 1 is the only one of the first six rules that promises that some set actually exists. Hence, our fundamental building block is the empty set. Can be build a second set? Since $\varnothing$ is a set (and $\varnothing$ is a set), by Rule 3, $\{\varnothing, \varnothing\}$ is a set. It follows from the definition of equality of sets that this set has a simpler name. It is the set $\{\varnothing\}$, the set whose only member is the empty set. (See Exercise 78.) Now we have a new set because $\{\varnothing\}$ is nonempty (it has one member, namely, $\varnothing$). Using Rule 3, we can build a third set, $\{\varnothing, \{\varnothing\}\}$. Even at this stage things have gotten a bit mindboggling; it helps to give names to our sets. The set $\{\varnothing\}$ is called the natural number 1, and $\{\varnothing, \{\varnothing\}\}$ is called the natural number 2. Note that "2" has two members and "1" has one member. Also note that "1" is $\{\varnothing\}$ and "2" is $\{\varnothing, 1\}$. It seems reasonable to guess that "3" ought to be $\{\varnothing, 1, 2\}$ (or $\{\varnothing, \{\varnothing\}, \{\varnothing, \{\varnothing\}\}\}$ as it would be written without naming names). Alas, Rule 3 does not allow us to build a set with more than two members and so we are temporarily stymied.

It is Rule 4 that allows us to build the set "3." We have the set $\{\varnothing, 1\}$ and from Rule 3 the set $\{2\}$ and thereafter the set $\mathscr{D} = \{\{\varnothing, 1\}, \{2\}\}$. According to Rule 4,

$$\cup\mathscr{D} = \{z: z \in x \text{ for some } x \in \mathscr{D}\}$$

is again a set. Note that this set is $\{\varnothing, 1, 2\}$ (alias 3). What Rule 4 promises is that once we have a collection $\mathscr{D}$ of sets, we can build a new set $\cup\mathscr{D}$. The requirements for membership in $\cup\mathscr{D}$ are quite straightforward: You get in $\cup\mathscr{D}$ provided that you are a member of at least one of the sets of the collection $\mathscr{D}$.

Incidentally, Rule 5 promises a set $\cap\mathscr{D}$ whose requirements for membership are much more stringent: You are allowed in $\cap\mathscr{D}$ only if you belong to every set in the collection $\mathscr{D}$. Suppose, for example, that $\mathscr{B}$ is the collection with the following three members: $\{1, 2\}$, $\{\varnothing, 2\}$, and $\{1, 2, 3\}$. Then $\cup\mathscr{B} = \{\varnothing, 1, 2, 3\}$ and $\cap\mathscr{B} = \{2\}$.

**EXERCISE**

Our first six rules are sufficient to show that $\mathscr{B} = \{\{1, 2\}, \{\varnothing, 2\}, \{1, 2, 3\}\}$ is really a set. There are two ways to see this. In both methods, we first build the sets 1, 2, and 3.

*Method I.* Using Rule 3, build $\{\varnothing, 2\}$, $\{1, 2\}$, and $\{3\}$. Using Rule 4 with $\{1, 2\}$ and $\{3\}$, build $\{1, 2, 3\}$. Next using Rule 3, build $\{\{1, 2\}, \{\varnothing, 2\}\}$, and $\{\{1, 2, 3\}\}$. Finally, using Rule 4, build $\mathscr{B}$.

*Method II.* Using Rule 3, build $\{\varnothing, 1\}$ and $\{2, 3\}$. With Rule 4, build $X = \{\varnothing, 1, 2, 3\}$. Since $\{1, 2\}$, $\{\varnothing, 2\}$, and $\{1, 2, 3\}$ are all subsets of $X$, Rule 6 promises that each of them belongs to the set $\mathscr{P}(X)$, and, under our rules, you cannot be a member of a set unless you are a set. Once we know that $\{\varnothing, 2\}$, $\{1, 2\}$ and $\{1, 2, 3\}$ are sets, we can proceed as before. Use Rule 3 to build $\{\{1, 2\}, \{\varnothing, 2\}\}$, and $\{\{1, 2, 3\}\}$, and then use Rule 4 to build $\mathscr{B}$.

The point of the preceding exercise is that even as small a collection as $\mathscr{B}$ is not easy to construct. In theory we could use our six rules to build the set of all natural numbers less than 1001, but as the preceding exercise indicates, it would be a time-consuming task. In any case, we could never build all the natural numbers. We need Rule 7 because none of our other rules allows us to build an infinite set, nor to repeat any construction infinitely many times.

To a set theorist, there is no distinction to be made between 3 and $\{\varnothing, 1, 2\}$. Thus a set theorist would say that 3 *is* $\{\varnothing, 1, 2\}$ and that $\{\varnothing\} \in \{1, 5\}$. As you could easily verify by taking a poll among mathematicians from any other branch of mathematics, not everyone is ready to accept this point of view. In this text, we take the set of natural numbers as given and hence do not view 1 as either an element or a subset of 2.

We end this section by justifying notation that we have previously introduced in our discussion of intuitive set theory.

Given sets $X$ and $Y$, we can use Rules 3 and 4 to form $\cup\{X, Y\}$. Hereafter, we denote this set by $X \cup Y$. Note that

$$X \cup Y = \{a: a \in X \text{ or } a \in Y\}$$

Similarly, with Rules 3 and 5, we can form $\cap\{X, Y\}$. Hereafter, we denote this set by $X \cap Y$. Note that

$$X \cap Y = \{a: a \in X \text{ and } a \in Y\}$$

## EXERCISES 2.4

**73.** Without using numbers, list the members of the natural number 4 when this number is considered a set.

**74.** Let $A = \{1, 2, 3, 4\}$ and $B = \{1, 2, 5, 6\}$, and let $\mathscr{C} = \{A, B\}$.

**a)** List the members of $\cup\mathscr{C}$. **b)** List the members of $\cap\mathscr{C}$.

**c)** List the members of $A \cup B$. **d)** List the members of $A \cap B$.

**75. a)** Identify the set $\cup\varnothing$ and identify the set $\cup\{\varnothing\}$.

**b)** List the members of $\cup\{\varnothing, \{\varnothing\}, \{\varnothing, \{\varnothing\}\}\}$.

**c)** By Rule 8, no set is a member of itself. Is there a set $S$ such that $\cup S \in S$? Explain.

**76.** In the statement of Rule 5, it is required that $X$ be a nonempty set. Why do you suppose that Zermelo and Fraenkel would insist that $X$ be nonempty?

**77.** Imagine that we had defined the natural number 1001 (which is the set $\{\varnothing, 1, 2, \ldots, 1000\}$).

**a)** Let $A$ and $B$ be natural numbers belonging to 1001. Explain what $A < B$ means.

**b)** It is customary to denote the minimum of two numbers $m$ and $n$ by $m \wedge n$ and the maximum of $m$ and $n$ by $m \vee n$. Complete the following sentence: For any two numbers $m$ and $n$ belonging to 1001, the minimum of $m$ and $n$ is the set ______ and the maximum of $m$ and $n$ is the set ______.

**c)** Suppose that $\mathscr{B}$ is a nonempty collection of numbers belonging to 1001. What set is the smallest member of this collection? What set is the largest member of this collection?

**78.** Let $x$ and $y$ be sets. Using the definition of equality, prove that $\{x\} = \{x, x\}$ and that $\{x, y\} = \{y, x\}$.

***79.** The Axiom of Foundation states that each set $A$ having at least one set as an element has a member $x$ such that $x$ and $A$ are disjoint sets.

**a)** Prove that Rule 8 is a consequence of the Axiom of Foundation.

**b)** Prove that the Axiom of Foundation implies that there does not exist a family $\{A_n : n \in \mathbb{N}\}$ such that for each $n \in \mathbb{N}$, $A_{n+1} \in A_n$. (In other words, as the term **foundation** implies, every string of memberships must have a leftmost starting set.)

## WRITING EXERCISE

Let $U = 210$ and let $S$ be the set of all positive divisors of $U$. For all $A, B \in S$, define $A \cap B$ to be the largest member of $S$ that divides both $A$ and $B$ (the greatest common divisor of $A$ and $B$) and define $A \cup B$ to be the smallest member of $S$ that both $A$ and $B$ divide (the least common multiple of $A$ and $B$). Also, for each $A \in S$, define $A'$ to be $U/A$ and write $A \subseteq B$ provided $A$ divides $B$. Explain why, with respect to $'$, $\cup$, and $\cap$, $S$ behaves like the power set of a set with 4 members. Is there anything special about the number 210 that is needed to make the system $S$ behave in this way?

CHAPTER

# 3

# MATHEMATICAL INDUCTION

In this chapter, we study the Principle of Mathematical Induction. Like bunting in baseball or the pin in chess, the Principle of Mathematical Induction is a basic technique. There is no area of mathematics that does not occasionally require the use of this principle, and no level of expertise or depth in mathematics at which it can be safely put aside. Just as you might bat .400 and still be required occasionally to bunt, at every level of mathematics you can still run up against theorems that can be proved readily by induction, but which stand undaunted against other attack. Indeed, in mathematical writing, phrases such as "and so on and so forth," "continuing this process, we see," and "aw come on, you see how things are going" mean one of two things:

1. The author has tried many cases and is now willing to make a guess, or
2. The author sees how to construct a proof using mathematical induction but is too lazy to write out the details of the proof by induction that is in the author's mind.

We do not wish to imply that there is anything wrong with trying a few cases to see what is happening. Presumably all important theorems are first guessed at, by analogy to other known theorems, by trial and error, or by some other method of inductive reasoning. But, if you really do have a process that can be continued, and if you really do see how things are going, then you ought to be able to use the Principle of Mathematical Induction to deduce your assertion.

In this chapter, you will be asked to experiment in order to arrive at a reasonable conjecture. Thereafter, you will be asked to prove your assertion, using the Principle of Mathematical Induction.

## 3.1 Proof by Induction

**Definition**

A set $S$ of natural numbers is said to be **inductive** provided that if $n \in S$, then $n+1 \in S$.

**Warning** The empty set is an inductive set, and this fact can easily lead to errors.

**The Principle of Mathematical Induction**

If $S$ is an inductive set and 1 belongs to $S$, then $S$ is the set of all natural numbers.

For the moment, we ask that you accept the Principle of Mathematical Induction as an axiom (that is, without proof). In Section 3.2, we will consider the reasonableness of the Principle of Mathematical Induction in some detail.

We begin the study of mathematical induction with a problem solved by Karl F. Gauss when he was just a child. Because Gauss has the reputation of having published nothing but flawless mathematics, we pause here to point out that even his mathematics was not perfect.

*Personal Note*

### Carl Freidrich Gauss (1777–1855)

Of all the outstanding mathematical achievements of Carl F. Gauss, Gauss himself was proudest of his proof, announced in the press when he was seventeen, that it is possible to construct a regular 17-gon using compass and straightedge alone. This result is the first entry in Gauss's famous mathematical diary, and this result motivated Gauss to request that a regular 17-gon be carved on his tombstone (Gauss's request was not honored because the stonemason believed that the geometric figure would be indistinguishable from a circle). Moreover, it was because of this discovery that Gauss chose to pursue mathematics. What Gauss proved is that a regular $n$-gon is constructible with straightedge and compass alone provided $n$ is of the form $2^k$, $k \geqslant 2$, or $n$ is a product of a nonnegative power of 2 and one or more different primes (such as 17) that are of the form $2^{2^k}+1$. Gauss's first great publication, *Disquitiones arithmeticae* culminates with the assertion that the converse of the above result also holds: "We can prove with all rigor that these equations are unavoidable; ... although the limit of this work does not allow us to give the demonstration, ... nevertheless a warning must be given so that someone should not seek to construct $n$-gons other than those given by our theorem as for example $n=7$, 11, 13, or 19 ... and thereby waste his time." It is ironic, and perhaps comforting to the rest of us, that Gauss never published a proof of this converse and probably never had one.

When Gauss was about ten years old, Master Büttner asked the students in Gauss's class to add the first 100 natural numbers. In a few minutes, Gauss put his slate on his teacher's desk. Although the other students worked on for the rest of the hour, at the conclusion of class, Master Büttner discovered that Karl was the only one with the correct answer. Büttner's math class, we admit, was probably not very exciting, but at least he had the good sense to recognize that he was dealing with a student of exceptional ability.

Can you guess how Gauss arrived at the correct answer so quickly? Perhaps he reasoned that $100 + 1 = 101$, $99 + 2 = 101$, $98 + 3 = 101, \ldots,$ and $51 + 50 = 101$, and since there are 50 such "pairs" in 100, the answer must be $50 \times 101$, or 5050.

Knowing Gauss's result, can you guess a formula for the sum of the first $n$ natural numbers that holds for each natural number $n$? How would you prove or disprove your formula?

In order to make a guess, let's examine the problem more closely. Note that 50 (the number of pairs in 100) is 100/2. So a formula for the sum of the first 100 natural numbers is $(100)(101)/2$. Does this give us a hint about the general formula? Can we replace 100 with $n$? Let's try this and see what happens. We are making a conjecture (a conclusion induced from guesswork) that

$$1 + 2 + 3 + \cdots + n = n(n + 1)/2$$

Now, having arrived at a conjecture, how do we prove or disprove it? We could attempt to make an infinite list of problems, one for each natural number, and then prove each statement separately. Is this feasible? The list would be as follows:

$$\begin{array}{lll}
n = 1: & 1 = 1(1 + 1)/2 & \text{or } 1 = 1 \\
n = 2: & 1 + 2 = 2(2 + 1)/2 & \text{or } 1 + 2 = 3 \\
n = 3: & 1 + 2 + 3 = 3(3 + 1)/2 & \text{or } 1 + 2 + 3 = 6 \\
\vdots & & \\
n: & 1 + 2 + 3 + \cdots + n = n(n + 1)/2 & \\
n + 1: & 1 + 2 + 3 + \cdots + (n + 1) = (n + 1)[(n + 1) + 1]/2 & \\
\vdots & &
\end{array}$$

Induction is a method for proving that each of the statements in the preceding infinite list is true. A proof by induction consists of three steps:

**Step 1.** Define a set $S$ of natural numbers with the property that if $S$ turns out to be $\mathbb{N}$, then the desired result will be known to be true.

**Step 2.** Show that $1 \in S$.

**Step 3.** Show that $S$ is an inductive set.

In the simple problem we are considering, the first step is easy. Let

$$S = \{n \in \mathbb{N}\text{: the sum of the first } n \text{ natural numbers is } n(n + 1)/2\}$$

You cannot always count on this step being easy; in many interesting induction problems, it is the most difficult of the three steps. Many books omit this step entirely and enumerate only two steps to an inductive proof. Indeed, in this first section we consider only straightforward induction problems in which the definition of $S$ is clear. However, by the end of the chapter, we will have given you several problems in which the key to the proof is a clever choice of $S$ (see Examples 7 and 8 in Section 3.3).

The next step is to show that 1 belongs to $S$. This step is easy. [We already saw that $1 = (1 + 1)/2$.]

The last step, which is always more difficult, is to show that $S$ is inductive (that is, to show that for each natural number $n$, $n + 1 \in S$ whenever $n \in S$). We choose a natural number $n$, but instead of showing that $n \in S$, we *assume* that $n$ belongs to $S$ and *show* that $n + 1$ belongs to $S$.

In the example above, we are assuming that

$$1 + 2 + 3 + \cdots + n = n(n + 1)/2$$

and we want to prove that

$$1 + 2 + 3 + \cdots + n + (n + 1) = (n + 1)[(n + 1) + 1]/2$$

How do we do this? There are at least two different algebraic approaches that we can use. However, the use of induction is exactly the same, in that both methods require the use of our assumption that $n \in S$.

***Method I.*** Suppose that $1 + 2 + 3 + \cdots + n = n(n + 1)/2$. Adding $n + 1$ to both sides of this equation, we obtain

$$1 + 2 + 3 + \cdots + n + (n + 1) = n(n + 1)/2 + (n + 1)$$

We observe that

$$\begin{aligned} n(n + 1)/2 + (n + 1) &= [n(n + 1) + 2(n + 1)]/2 \\ &= [(n + 2)(n + 1)]/2 \\ &= (n + 1)[(n + 1) + 1]/2 \end{aligned}$$

Therefore,

$$1 + 2 + 3 + \cdots + n + (n + 1) = (n + 1)[(n + 1) + 1]/2$$

***Method II.*** Suppose that $1 + 2 + 3 + \cdots + n = n(n + 1)/2$. We start with $1 + 2 + 3 + \cdots + n + (n + 1)$ and break it down into $(1 + 2 + 3 + \cdots + n) + (n + 1)$. Then we use the assumption that $1 + 2 + 3 + \cdots + n = n(n + 1)/2$ to write $(1 + 2 + 3 + \cdots + n) + (n + 1) = n(n + 1)/2 + (n + 1)$. We can then use the same algebra as that in Method I to show that

$$n(n + 1)/2 + (n + 1) = (n + 1)[(n + 1) + 1]/2$$

Perhaps Method II will be clarified if we write it as one sequence of equations:

$$\begin{aligned} 1+2+3+\cdots+n+(n+1) &= (1+2+3+\cdots+n)+(n+1) \\ &= n(n+1)/2+(n+1) \\ &= n(n+1)/2+2(n+1)/2 \\ &= [(n+2)(n+1)]/2 \\ &= [(n+1)(n+1+1)]/2 \end{aligned}$$

We have now proved that $1 \in S$ and that $S$ is inductive. By the Principle of Mathematical Induction, $S = \mathbb{N}$, and by step 1 (of the proof by induction), we know that our formula holds for every natural number $n$.

There is a curious feature of mathematical induction that is noteworthy. Proofs by mathematical induction actually establish infinitely many theorems in one fell swoop. Suppose that you need to know the sum of the first 10,000 natural numbers and that you have no interest whatever in the sum of the first $n$ natural numbers for any natural number $n$ other than $n =$ 10,000. It would still be worthwhile to find the formula above and prove by induction that it holds for every natural number $n$, rather than to try to add up the first 10,000 natural numbers — unless of course you happen to hit on Gauss's clever trick. There is an important principle in mathematics, which to our knowledge was first stated by George Pólya (1887–1985).

Pólya's Principle is as follows: If you wish to prove a theorem, it is sometimes easier to prove a harder theorem. Induction provides many dramatic examples in which Pólya's Principle applies. In these examples, the most difficult aspect of a proof by induction is the realization that induction can be made to apply to the problems. In this first section, we consider only straightforward induction problems, but by the end of the chapter we will have given you several problems that encompass Pólya's Principle (see Exercise 70 in Section 3.3).

The third step in a proof by induction, in which one shows that a certain set $S$ is inductive, is called the **inductive step** of a proof by mathematical induction. Always in this step you must drag in, by hook or crook, whatever information is known from the assumption that $n$ belongs to $S$. In our first example, it was fairly obvious how to use the fact that $1+2+3+\cdots+n = n(n+1)/2$. Let us consider an induction problem in which the inductive step is not so obvious.

**EXAMPLE 1** Prove by induction that for any natural number $n$, 6 divides $n^3 - n$.

Let $S = \{n \in \mathbb{N}\colon 6 \text{ divides } n^3 - n\}$. Then $S \subseteq \mathbb{N}$. In order to use the Principle of Mathematical Induction to prove that $S = \mathbb{N}$, we must prove two things:

**a)** $1 \in S$.

**b)** If $n \in S$, then $n+1 \in S$.

First we show that $1 \in S$: If $n = 1$, then $n^3 - n = 1 - 1 = 0$. Since 6 divides 0, $1 \in S$.

Now we prove that if $n \in S$, then $n+1 \in S$. Suppose that $n \in S$. Then 6 divides $n^3 - n$. We want to use this fact to prove that $n+1 \in S$. Note that $n+1 \in S$ if and only if 6 divides $(n+1)^3 - (n+1)$. So we need to prove that if 6 divides $n^3 - n$, then 6 divides $(n+1)^3 - (n+1)$. How do we do this? One method would be to write $(n+1)^3 - (n+1)$ as the sum of $n^3 - n$ and an integer. By the assumption that $n \in S$, we know that 6 divides $n^3 - n$. If we could show that 6 divides an integer, then, by Exercise 62 in Section 1.4, 6 would divide the sum of $n^3 - n$ and an integer. So let's see if we can write $(n+1)^3 - (n+1)$ as the sum of $n^3 - n$ and an integer.

$$\begin{aligned}
(n+1)^3 - (n+1) &= (n+1)(n+1)^2 - (n+1) \\
&= (n+1)[(n+1)^2 - 1] \\
&= (n+1)(n^2 + 2n) \\
&= n^3 + 3n^2 + 2n \\
&= (n^3 - n) + (3n^2 + 3n) \\
&= (n^3 - n) + 3(n^2 + n)
\end{aligned}$$

By the assumption that $n \in S$, we know that 6 divides $n^3 - n$. We need to show that 6 divides $3(n^2 + n)$ or, equivalently, that 2 divides $n^2 + n$. But this is Exercise 92 in Section 1.6. Since 6 divides both $n^3 - n$ and $3(n^2 + n)$, 6 divides the sum. Therefore, 6 divides $(n+1)^3 - (n+1)$, and so $n+1 \in S$. By the Principle of Mathematical Induction, $S = \mathbb{N}$. ❑

Incidentally, in the preceding proof of Example 1, there is an alternative approach to proving that 6 divides $3n^2 + 3n$. Couldn't we prove it by induction? (Try it.)

Let's return to the proof that, for each natural number $n$, $1 + 2 + 3 + \cdots + n = n(n+1)/2$. As indicated earlier, the proof that this is true for $n = 1$ is easy, but the second step is much more difficult. We have observed that many students want to start with the statement

$$1 + 2 + 3 + \cdots + n + (n+1) = (n+1)[(n+1) + 1]/2$$

and using a logical sequence of steps, arrive at the conclusion that

$$1 + 2 + 3 + \cdots + n = n(n+1)/2$$

They then say that they have proved that the statement is true for $n+1$ under the assumption that it is true for $n$. Is this assertion justified? No.

They have proved that the statement is true for $n$ under the assumption that it is true for $n+1$. Compare this to the following proof:

Suppose that $5 = 1$. Subtracting 3 from both sides of the equation, we have $2 = -2$. Squaring both sides of this equation, we have $4 = 4$. We know that the latter is true, so by the same type of reasoning as above, we conclude that $5 = 1$.

You may say, "But that's absurd." Maybe so, but it is precisely the same reasoning you are using if you assert that you have proved that the statement is true for $n+1$ under the assumption that it is true for $n$.

Let's consider another example.

**EXAMPLE 2** Let $a$ be a nonzero real number. Prove by induction that the following two matrices are equal for each natural number $n$:

$$\begin{bmatrix} a & 1 \\ 0 & a \end{bmatrix}^n = \begin{bmatrix} a^n & na^{n-1} \\ 0 & a^n \end{bmatrix}$$

Let

$$S = \left\{ n \in \mathbb{N} : \begin{bmatrix} a & 1 \\ 0 & a \end{bmatrix}^n = \begin{bmatrix} a^n & na^{n-1} \\ 0 & a^n \end{bmatrix} \right\}$$

In order to use the Principle of Mathematical Induction to prove that $S = \mathbb{N}$, we need to prove two things:

**a)** $1 \in S$.

**b)** If $n \in S$, then $n+1 \in S$.

In order to show that $1 \in S$, we are faced with the problem of showing that

$$\begin{bmatrix} a & 1 \\ 0 & a \end{bmatrix}^1 \quad \text{is equal to} \quad \begin{bmatrix} a^1 & 1a^0 \\ 0 & a^1 \end{bmatrix}$$

But this is obvious because each is equal to

$$\begin{bmatrix} a & 1 \\ 0 & a \end{bmatrix}$$

Now we prove that if $n \in S$, then $n+1 \in S$. Suppose that $n \in S$. Then

$$\begin{bmatrix} a & 1 \\ 0 & a \end{bmatrix}^n = \begin{bmatrix} a^n & na^{n-1} \\ 0 & a^n \end{bmatrix}$$

We want to prove that $n+1 \in S$, so we want to prove that

$$\begin{bmatrix} a & 1 \\ 0 & a \end{bmatrix}^{n+1} = \begin{bmatrix} a^{n+1} & (n+1)a^n \\ 0 & a^{n+1} \end{bmatrix}$$

We start with

$$\begin{bmatrix} a & 1 \\ 0 & a \end{bmatrix}^n = \begin{bmatrix} a^n & na^{n-1} \\ 0 & a^n \end{bmatrix}$$

and multiply each side of this equation by

$$\begin{bmatrix} a & 1 \\ 0 & a \end{bmatrix}$$

in order to get the desired matrix on the left side of the equation. This yields

$$\begin{bmatrix} a & 1 \\ 0 & a \end{bmatrix}^n \begin{bmatrix} a & 1 \\ 0 & a \end{bmatrix} = \begin{bmatrix} a^n & na^{n-1} \\ 0 & a^n \end{bmatrix} \begin{bmatrix} a & 1 \\ 0 & a \end{bmatrix}$$

The left side of this equation is obviously equal to

$$\begin{bmatrix} a & 1 \\ 0 & a \end{bmatrix}^{n+1}$$

We perform ordinary matrix multiplication on the right side of the equation and obtain the matrix

$$\begin{bmatrix} a^n \times a & a^n + [na^{n-1} \times a] \\ 0 & a^n \times a \end{bmatrix}$$

This matrix is equal to

$$\begin{bmatrix} a^{n+1} & a^n + na^n \\ 0 & a^{n+1} \end{bmatrix} \quad \text{or} \quad \begin{bmatrix} a^{n+1} & (n+1)a^n \\ 0 & a^{n+1} \end{bmatrix}$$

Therefore, we conclude that

$$\begin{bmatrix} a & 1 \\ 0 & a \end{bmatrix}^{n+1} = \begin{bmatrix} a^{n+1} & (n+1)a^n \\ 0 & a^{n+1} \end{bmatrix}$$

So $n + 1 \in S$.

By the Principle of Mathematical Induction, $S = \mathbb{N}$. ❑

The last example has been contrived by us to show that a proposition that sounds mindboggling sometimes yields easily to a proof by induction.

**EXAMPLE 3**

Let's call a natural number a *fiver* if all its digits are 5's. Prove that a fiver is divisible by the first 5 odd primes provided the number of its digits is a multiple of 6.

Let

$$S = \{n \in \mathbb{N}: \text{the fiver with exactly } 6n \text{ digits is divisible by 3, 5, 7, 11, and 13}\}$$

$$1 \in S: 555{,}555 = (3)(5)(7)(11)(13)(37)$$

Suppose that $n \in S$ and let $j$ denote the fiver with exactly $6n$ digits. Since $n \in S$, $j$ is divisible by 3, 5, 7, 11, and 13.

To show that $n+1 \in S$, consider the fiver with exactly $6(n+1)$ digits. This number equals $1{,}000{,}000j + 555{,}555$, and since both $j$ and 555,555 are divisible by 3, 5, 7, 11, and 13, it follows from Exercise 56 in Section 1.4 that $n+1 \in S$. By the Principle of Mathematical Induction, $S = \mathbb{N}$, and so the result holds. ❑

---

## EXERCISES 3.1

**1.** Prove by induction that for each natural number $n$, each of the following is true.

**a)** $2+4+6+\cdots+2n = n(n+1)$

**b)** $3+6+9+\cdots+3n = 3n(n+1)/2$

**c)** $4+8+12+\cdots+4n = 2n(n+1)$

**d)** $5+10+15+\cdots+5n = 5n(n+1)/2$

**e)** $2+5+8+\cdots+(3n-1) = n(3n+1)/2$

**f)** $5+7+9+\cdots+(2n+3) = n(n+4)$

**g)** $1(1+1)+2(2+1)+3(3+1)+\cdots+n(n+1) = n(n+1)(n+2)/3$

**h)** $1^2+2^2+3^2+\cdots+n^2 = n(n+1)(2n+1)/6$

**i)** $5^n - 2^n$ is divisible by 3.

**j)** $7^n - 2^n$ is divisible by 5.

**k)** $x^n - y^n$ is divisible by $x-y$, where $x$ and $y$ are any two distinct integers.

**l)** $3^n - 1$ is divisible by 2.

**m)** $4^n - 1$ is divisible by 3.

**n)** $n^3 - n$ is divisible by 3.

**o)** $2^{2n-1} + 3^{2n-1}$ is divisible by 5.

**p)** $n^4 - 6n^3 + 23n^2 - 18n$ is divisible by 24.

**q)** $\left(1+\frac{1}{2}\right)^n \geqslant 1 + \frac{n}{2}$

**2.** Prove by induction that for each natural number $n$,

$$(\cos x + i \sin x)^n = \cos(nx) + i \sin(nx).$$

Assume that $i^2 = -1$ and also assume the following two trigonometric identities:

$$\cos(x+y) = \cos x \cos y - \sin x \sin y \quad \text{and} \quad \sin(x+y) = \sin x \cos y + \cos x \sin y$$

**3.** Prove by induction that, for each natural number $n$, if $X$ is a set consisting of $n$ elements, then $\mathscr{P}(X)$ is a set consisting of $2^n$ elements.

**4.** Two students agree that in order to do Exercise 3, it is necessary to prove that if $S = \{n \in \mathbb{N}$: any $n$-element set has $2^n$ subsets$\}$ then $S$ is the set of all natural numbers. The first student argues that $1 \in S$ as follows: $\{1\}$ has two subsets, namely $\{1\}$ and $\varnothing$, and since $2 = 2^1$, $1 \in S$. The second student argues that the first student is using what he is trying to prove and therefore his argument that 1 belongs to $S$ is completely wrong. Explain what irks the second student. Is the point well taken?

5. Prove that the product of any three consecutive natural numbers is divisible by 6.
6. Find and verify, by induction, a formula for the sum of the first $n$ odd natural numbers.
7. Take an equilateral triangle, divide each side into $n$ equal segments, and connect the division points with all possible segments that are parallel to the original sides. Find a formula for relating the number $n$ to the number of small triangles and prove, by induction, that your formula holds. (For example, if $n = 3$, there are nine small triangles, as shown in the accompanying figure.)

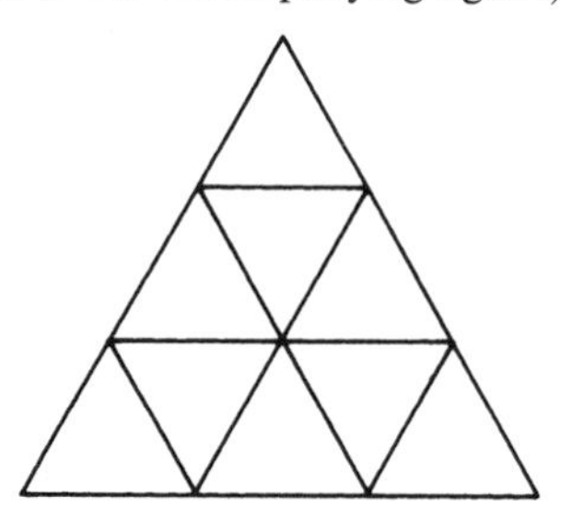

8. What is wrong with the following proof that for each natural number $n > 1$,

$$3(1) + 3(2) + 3(3) + \cdots + 3(n) = (3n^2 + 3n + 2)/2$$

**PROOF** Let

$$S = \{n \in \mathbb{N}: n = 1 \text{ or } 3(1) + 3(2) + 3(3) + \cdots + 3(n) = (3n^2 + 3n + 2)/2\}$$

Clearly $1 \in S$. (We put it there.) Suppose that $n \in S$. Then

$$3(1) + 3(2) + 3(3) + \cdots + 3(n) = (3n^2 + 3n + 2)/2$$

So

$$\begin{aligned} &3(1) + 3(2) + 3(3) + \cdots + 3(n) + 3(n+1) \\ &\quad = [3(1) + 3(2) + 3(3) + \cdots + 3(n)] + 3(n+1) \\ &\quad = (3n^2 + 3n + 2)/2 + 3(n+1) \\ &\quad = (3n^2 + 3n + 2 + 6n + 6)/2 \\ &\quad = [(3n^2 + 6n + 3) + (3n + 3) + 2]/2 \\ &\quad = [3(n^2 + 2n + 1) + 3(n+1) + 2]/2 \\ &\quad = [3(n+1)^2 + 3(n+1) + 2]/2 \end{aligned}$$

Therefore, $n + 1 \in S$. By the Principle of Mathematical Induction, $S = \mathbb{N}$. □

9. What is wrong with the following proof that all members of the human race are of the same sex?

**PROOF** Let

$S = \{n \in \mathbb{N}$: in any set of $n$ people, all the members of the set are the same sex$\}$.

If we have a set consisting of one person, then clearly all the members of the set are of the same sex, so $1 \in S$. Suppose that $n \in S$. Then in any set of $n$ people, all the members of the set are of the same sex. In order to show that $n+1 \in S$, we need to show that in any set of $n+1$ people, all the members of the set are of the same sex. Let $A$ be a set consisting of $n+1$ people. Let's call these people $A_1, A_2, A_3, \ldots, A_n, A_{n+1}$. If we send one person out of the room, say $A_1$, we have a set of $n$ people left in the room, so by the assumption that $n \in S$, they are all of the same sex. Now let's bring $A_1$ back into the room and send $A_2$ out. Again, there is a set of $n$ people left in the room, so by the assumption that $n \in S$, they are all of the same sex. Now observe that everyone in the original set of $n+1$ people is of the same sex as $A_3$, so they are all of the same sex. Therefore, $n+1 \in S$. By the Principle of Mathematical Induction, $S = \mathbb{N}$. □

**10.** What, if anything, is wrong with the following proof that for each natural number $n$, $4 + 10 + 16 + \cdots + (6n-2) = n(3n+1)$?

**PROOF** Let $S = \{n \in \mathbb{N}: 4 + 10 + 16 + \cdots + (6n-2) = n(3n+1)\}$. Since $1(3 \cdot 1 + 1) = 4$, $1 \in S$. Suppose that $n \in S$. We have

$$4 + 10 + 16 + \cdots + (6n-2) + [6(n+1)-2] = (n+1)[3(n+1)+1]$$

$$4 + 10 + 16 + \cdots + (6n-2) + (6n+4) = (n+1)(3n+4)$$

$$4 + 10 + 16 + \cdots + (6n-2) + (6n+4) = n(3n+1) + (6n+4)$$

Subtracting $6n+4$ from each side of the equation, we obtain

$$4 + 10 + 16 + \cdots + (6n-2) = n(3n+1)$$

Since $n \in S$, the latter equation is true. Therefore $n+1 \in S$, so by the Principle of Mathematical Induction, $S = \mathbb{N}$. □

**11.** What, if anything, is wrong with the following proof that for each natural number $n$, $7^n - 2^n$ is divisible by 5?

**PROOF** Let $S = \{n \in \mathbb{N}: 7^n - 2^n$ is divisible by $5\}$. Since $7^1 - 2^1 = 5$, $1 \in S$. Suppose that $n \in S$. There is an integer $p$ such that $7^{n+1} - 2^{n+1} = 5p$. Therefore $7 \cdot 7^n - 2 \cdot 2^n - 7 \cdot 2^n + 7 \cdot 2^n = 5p$, and so $7(7^n - 2^n) + 2^n(7-2) = 5p$. Thus $7(7^n - 2^n) = 5(p - 2^n)$, and hence $7(7^n - 2^n)$ is divisible by 5. Since 7 is not divisible by 5, 5 divides $7^n - 2^n$. Therefore $n+1 \in S$, and so by the Principle of Mathematical Induction, $S = \mathbb{N}$. □

**12.** What, if anything, is wrong with the following proof that for each natural number $n$, in any set of $n$ cars, all cars are of the same make.

**PROOF** Let $S = \{N \in \mathbb{N}$: in any set of $n$ cars, all cars are of the same make$\}$. Obviously $1 \in S$. Suppose that $n+1 \in S$. Then any set of $n+1$ cars are of the same make. To see that $n \in S$, take any set of $n$ cars. Add a car to this set. Then the new set has $n+1$ cars, so the cars are of the same make. It follows that all the cars in the set with $n$ cars are of the same make. Therefore $n \in S$, and since $S$ is an inductive set that contains 1, $S = \mathbb{N}$. □

**13.** Either prove or disprove that, for each natural number $n$, $n^2 + n + 41$ is a prime.

**14.** Prove by induction that the sum of any three consecutive positive cubes is a multiple of 9. (For example, $3^3 + 4^3 + 5^3 = 216 = 9(24)$.)

**15.** Given that $x$ is a real number and $x > -1$, prove by induction that for each natural number $n$, $(1+x)^n \geqslant 1 + nx$. Explain where you used the assumption that $x > -1$.

**16.** Prove that for each even natural number $n$,

$$\left(1-\frac{1}{2}\right)\left(1+\frac{1}{3}\right)\left(1-\frac{1}{4}\right)\cdots\left(1-\frac{(-1)^n}{n}\right)=\frac{1}{2}$$

## 3.2 Other Principles of Induction

The Principle of Mathematical Induction is well suited to deal with propositions that hold for every natural number, but there are interesting propositions about the natural numbers that hold only if we are willing to overlook a few maverick cases. Let us begin with such a proposition, which at first appears unsuited to mathematical induction.

**EXAMPLE 4**

Prove that for every natural number $n \geqslant 5$, $n^2 < 2^n$.

First observe that if $n = 3$, then $n^2 = 9$ and $2^n = 8$, so $n^2 \nless 2^n$. It is clear that we cannot use the Principle of Mathematical Induction in the same manner as we have been doing, so let us define $S$ in a slightly different manner.

Let $S = \{n \in \mathbb{N} : n^2 < 2^n\} \cup \{1, 2, 3, 4\}$. Then, if we can show that $S = \mathbb{N}$, we will have proved that $n^2 < 2^n$ for all $n \geqslant 5$. Clearly, $1 \in S$. Suppose that $n \in S$. If $n = 1$, 2, or 3, then clearly $n + 1 \in S$. If $n = 4$, then $n + 1 = 5$, $5^2 = 25$, and $2^5 = 32$, so $n + 1 \in S$. Suppose that $n > 4$. Then

$$(n+1)^2 = n^2 + 2n + 1 \quad \text{and} \quad 2^{n+1} = 2^n \times 2$$

so we want to show that $n^2 + 2n + 1 < 2 \times 2^n$. Since $n \in S$ and $n > 4$, $n^2 < 2^n$. Therefore, $2 \times n^2 < 2 \times 2^n$, and it is sufficient to show that $n^2 + 2n + 1 < 2 \times n^2$.

Observe that $n(n-2) > 1$ because each of $n$ and $n - 2$ is greater than 1. This means that $n^2 - 2n > 1$ and so $n^2 > 2n + 1$. Adding $n^2$ to both sides of this equation, we obtain $2n^2 > n^2 + 2n + 1$. This completes the proof of Example 4. ❑

Now was it really necessary to define $S$ in the manner that we did? Observe that we proved that $5 \in \{n \in \mathbb{N} : n^2 < 2^n\}$ and if $n \geqslant 5$ and $n \in \{n \in \mathbb{N} : n^2 < 2^n\}$, then

$$n + 1 \in \{n \in \mathbb{N} : n^2 < 2^n\}$$

This gives us some idea of how an extended version of the Principle of Mathematical Induction should be stated in order to cover the preceding example.

## The Extended Principle of Mathematical Induction

> Let $k \in \mathbb{N}$ and let $S$ be a subset of $\mathbb{N}$ such that
>
> **a)** $k \in S$, and
>
> **b)** If $n \geqslant k$ and $n \in S$, then $n+1 \in S$.
>
> Then $\{n \in \mathbb{N}: n \geqslant k\} \subseteq S$.

**EXAMPLE 5** Use the Extended Principle of Mathematical Induction to prove that $2n < 2^n - 1$ for all $n \geqslant 3$. Let $S = \{n \in \mathbb{N}: n \geqslant 3 \text{ and } 2n < 2^n - 1\}$. Since $6 < 2^3 - 1$, $3 \in S$. Suppose that $n \in S$. Then

$$2(n+1) = 2n + 2 < 2^n - 1 + 2 = 2^n + 1$$

By Exercise 36, $2^n + 1 < 2^{n+1} - 1$, so $n+1 \in S$. Therefore, by the Extended Principle of Mathematical Induction, $\{n \in \mathbb{N}: n \geqslant 3\} \subseteq S$. ❑

It is clear that the Extended Principle of Mathematical Induction implies the Principle of Mathematical Induction. (We simply take $k$ to be 1.)

The proof that $n^2 < 2^n$ for all $n > 4$ indicates how to prove that the Principle of Mathematical Induction implies the Extended Principle of Mathematical Induction.

**Theorem 3.1**

> The Principle of Mathematical Induction implies the Extended Principle of Mathematical Induction.

**PROOF** Let $k \in \mathbb{N}$ and let $S$ be a subset of $\mathbb{N}$ such that

**a)** $k \in S$, and

**b)** If $n \geqslant k$ and $n \in S$, then $n+1 \in S$.

Define a set $S'$ by

$$S' = \{n \in \mathbb{N}: n < k\} \cup \{n \in \mathbb{N}: n \in S\}$$

In order to prove that $\{n \in \mathbb{N}: n \geqslant k\} \subseteq S$, it is sufficient to show that $S' = \mathbb{N}$. Note that $S'$ has the property that $1 \in S'$; and if $n \in S'$, then $n+1 \in S'$. Therefore, by the Principle of Mathematical Induction, $S' = \mathbb{N}$. □

We have need of a second principle of induction, which is often called the Strong Principle of Induction. The terminology "strong" is a bit misleading, since (as we will see) this second principle is actually a consequence of the original principle of induction.

## The Second Principle of Mathematical Induction

Let $S$ be a set of natural numbers with the following properties:

**a)** $1 \in S$.

**b)** For each $n \in \mathbb{N}$, if $\{1, 2, 3, \ldots, n\} \subseteq S$, then $n+1 \in S$.

Then $S = \mathbb{N}$.

To see the need for the Second Principle of Mathematical Induction, we consider the proposition that every natural number greater than 1 is either a prime number or the product of prime numbers. We begin naively by attempting a proof using the ordinary Extended Principle of Mathematical Induction. Let $S = \{n \in \mathbb{N}: n > 1$ and $n$ is prime or $n$ is the product of primes$\}$. Clearly, $2 \in S$. Suppose that $n \in S$. If $n+1 \notin S$, then $n+1$ is the product of two smaller numbers $a$ and $b$. Since $n+1 = ab$, in order to establish that $n+1 \in S$, it is enough to know that both $a$ and $b$ belong to $S$. Unfortunately we know only that $n \in S$ and there is no reason to believe that either $a$ or $b$ is $n$. Thus there is no reason for us to believe that $a$ and $b$ belong to $S$. We are stuck. It is the Second Principle of Mathematical Induction that rescues us.

**Proposition 3.2**

Every natural number greater than 1 is either a prime number or the product of prime numbers.

**PROOF** Let $S = \{n \in \mathbb{N}: n = 1$, $n$ is a prime number, or $n$ is the product of prime numbers$\}$. Clearly, $1, 2 \in S$. Suppose that $\{1, 2, \ldots, n\} \subseteq S$ and that $n+1 \notin S$. Then $n+1$ is not a prime number, so there are natural numbers $a \neq 1$ and $b \neq 1$ such that $n+1 = ab$. It follows that both $a$ and $b$ are less than $n+1$ and so belong to $\{1, 2, \ldots, n\} \subseteq S$. But then $n+1$, being the product of prime numbers, also belongs to $S$. Consequently, by the Second Principle of Mathematical Induction, $S = \mathbb{N}$. □

It seems a bit much to expect you to accept the Second Principle of Mathematical Induction as obvious or natural; indeed, its usefulness in situations where the first principle does not readily apply should make the second principle suspect. Happily, the last induction-type principle that we ask you to accept is, for most people, quite acceptable.

## The Least-Natural-Number-Principle

Every nonempty set of natural numbers has a least member.

In some books, the Least-Natural-Number Principle is called the Well-Ordering Principle. However, we prefer not to use this name because traditionally the name Well-Ordering Principle has been used to mean that "*Every* set can be well ordered" (see Chapter 4), whereas the Least-Natural-Number Principle asserts only that $\mathbb{N}$ with the natural order is a well-ordered set.

The restriction in the Least-Natural-Number Principle to nonempty sets is, of course, necessary because the empty set has no members at all. Note too that there is neither a "least-integer principle" nor a "least-positive-real-number principle," because the set of all integers less than $-5$ is a nonempty set of integers that has no least member and the set of all real numbers greater than $+5$ is a nonempty set of positive real numbers that has no least member.

We now give two proofs that illustrate the power of the Least-Natural-Number Principle. The first result has already been proved by induction, and the new proof is given only as a simple illustration of the use of the Least-Natural-Number Principle. The second result, whose proof is considerably more involved, will be useful to us in later work.

**EXAMPLE 6**

For any natural number $n$, 6 divides $n^3 - n$.

The proof is by contradiction.

Suppose that $\{n \in \mathbb{N}: 6 \text{ does not divide } n^3 - n\}$ is nonempty. By the Least-Natural-Number Principle, this set has a least member $n$. Certainly 6 divides $1^3 - 1$, so $n \neq 1$, and thus there is a natural number $k$ such that $k = n - 1$. Then 6 divides $k^3 - k$ (why?), but $n = k + 1$ and

$$\begin{aligned} n^3 - n &= (k+1)^3 - (k+1) \\ &= k^3 + 3k^2 + 3k + 1 - k - 1 \\ &= (k^3 - k) + 3(k)(k+1) \end{aligned}$$

We see that $k(k+1)$ is even. It follows that 6 divides $3(k)(k+1)$. Since 6 divides both $k^3 - k$ and $3(k)(k+1)$, 6 divides the sum, which is $n^3 - n$ (see Exercise 62 in Section 1.4). We have reached a contradiction. ❑

We hope you will compare Example 6 to Example 1. Otherwise, there will have been no point in having proved the same result twice.

**Theorem 3.3**

**The Division Algorithm for Integers**

Let $a$ be an integer and let $b$ be a natural number. Then there are integers $q$ and $r$ such that $a = bq + r$ and $0 \leqslant r < b$. Furthermore, the numbers $q$ and $r$ exist uniquely in the sense that if $q'$ and $r'$ are integers such that $a = bq' + r'$ and $0 \leqslant r' < b$, then $q = q'$ and $r = r'$.

**PROOF** We first show that $q$ and $r$ exist. Let

$$S = \{s \in \mathbb{Z}: \text{there is an integer } t \text{ such that } s = a - bt\}$$

and let $S' = S \cap \mathbb{N}$. In order to use the Least-Natural-Number Principle to get the smallest member of $S'$, we need to show that $S' \neq \varnothing$. We consider three cases:

**a)** Suppose that $a > 0$. Then $t = 0$ yields $a \in S'$.

**b)** Suppose that $a = 0$. Then $t = -1$ yields $b \in S'$.

**c)** Suppose that $a < 0$. Then $t = a$ yields $a - ba \in S$. (But we need something in $S'$.) Recall that $b \geqslant 1$. If $b > 1$, then $a - ba = a(1 - b)$ is the product of two negative numbers, so it is positive and hence a member of $S'$. If $b = 1$, then $t = a - 1$ yields $1 \in S'$.

Let $s'$ be the smallest member of $S'$. Note that $s'$ is the smallest natural number that differs from $a$ by a multiple of $b$. If $a$ itself is a multiple of $b$, then by definition there is an integer $q$ such that $a = qb$, so we can take $r = 0$ to get $a = qb + r$ with $0 \leqslant r < b$. If $a$ is not a multiple of $b$, then $s' = a - bt'$ for some integer $t'$, so taking $r = s'$ and $q = t'$, we have $a = bq + r$. Moreover, $0 < r$. It only remains to show that $r < b$. Suppose that $r \geqslant b$. Since $a$ is assumed not to be a multiple of $b$, $r \neq b$. Thus $r > b$, and

$$a - b(q + 1) = a - bq - b = r - b > 0$$

Therefore, $a - b(q + 1) \in S'$, which contradicts the choice of $r$ as the least element of $S'$.

Uniqueness was proved in Exercise 83 in Section 1.5. However, for completeness, we include a proof. Assume that $a = bq + r$, $0 \leqslant r < b$ and $a = bq' + r'$, $0 \leqslant r' < b$. We must show that $q = q'$ and $r = r'$. We have $bq + r = bq' + r'$, so $bq - bq' = r' - r$, and hence $b(q - q') = r' - r$. Thus, $b$ divides $r' - r$. But $0 \leqslant r < b$ and $0 \leqslant r' < b$, so $-b < r' - r < b$. But the only multiple of $b$ between $-b$ and $b$ is the multiple 0. Therefore, $r' - r = 0$ and $r = r'$. Consequently, $b(q - q') = 0$ and since $b \neq 0$, $q = q'$. □

We now show that the Principle of Mathematical Induction is equivalent to both the Least-Natural-Number Principle and the Second Principle of Mathematical Induction.

**Theorem 3.4**

> The Principle of Mathematical Induction and the Least-Natural-Number Principle are equivalent statements.

**PROOF** Suppose that the Principle of Mathematical Induction holds. Let $T \subseteq \mathbb{N}$, and suppose that $T$ has no smallest member. Let $S = \{n \in \mathbb{N}: \{1, 2, \ldots, n\} \subseteq T'\}$. Since $T$ has no smallest member, $1 \in S$. Suppose that $n \in S$. Then $\{1, 2, \ldots, n\} \subseteq T'$, so if $n + 1 \notin S$, then $n + 1$ is the smallest member of $T$. Therefore, $n+1 \in S$ and by the Principle of Mathematical Induction, $S = \mathbb{N}$. Therefore, $T = \varnothing$.

Now assume the Least-Natural-Number Principle. Let $S$ be an inductive set containing 1. In order to show that $S = \mathbb{N}$, we show that $\mathbb{N} - S = \varnothing$. Suppose that $\mathbb{N} - S \neq \varnothing$. By the Least-Natural-Number Principle, $\mathbb{N} - S$ has a least member $n$. Since $n-1 \notin \mathbb{N} - S$, either $n-1 \in S$ or $n-1 \notin \mathbb{N}$. Suppose that $n-1 \in S$. Since $S$ is inductive, $n = (n-1)+1 \in S$, a contradiction. Suppose that $n-1 \notin \mathbb{N}$. Then $n = 1$, but $1 \in S$ and $n \notin S$, a contradiction. Thus the assumption that $\mathbb{N} - S \neq \varnothing$ has led to a contradiction. Since $\mathbb{N} = S$, the Principle of Mathematical Induction holds. □

**Theorem 3.5**

The Principle of Mathematical Induction and the Second Principle of Mathematical Induction are equivalent statements.

**PROOF** For obvious reasons, it is convenient to refer to the Principle of Mathematical Induction as the First Principle of Mathematical Induction. It should be clear that the Second Principle of Mathematical Induction implies the first principle. But experience has shown that it is not. Suppose, therefore, that we believe the second principle and suppose that $S$ is an inductive set containing 1. Let $n \in \mathbb{N}$. In order to apply the second principle, we assume that $\{1, 2, 3, \ldots, n\} \subseteq S$ (and we must show that $n+1 \in S$). What luck! We know that $n \in S$, and because $S$ is inductive, it follows that $n+1 \in S$. Thus by the Second Principle of Mathematical Induction, $S = \mathbb{N}$, so the First Principle of Mathematical Induction holds.

Now assume that the First Principle of Mathematical Induction holds. Then, by Theorem 3.4, the Least-Natural-Number Principle holds. Let $S$ be a set of natural numbers containing 1 with the property that for each $n \in \mathbb{N}$, if $\{1, 2, 3, \ldots, n\} \subseteq S$, then $n+1 \in S$. The proof is completed by showing that $S = \mathbb{N}$ (see Exercise 34). □

We state, without proof, the obvious analogue of the Extended Principle of Mathematical Induction.

### The Extended Second Principle of Mathematical Induction

Let $k \in \mathbb{N}$ and let $S$ be a subset of $\mathbb{N}$ such that

**a)** $k \in S$, and

**b)** If $n \geqslant k$ and $\{k, k+1, \ldots, n\} \subseteq S$, then $n+1 \in S$.

Then $\{n \in \mathbb{N} : n \geqslant k\} \subseteq S$.

In the problems to follow, there is no need to use the Second Principle of Mathematical Induction. We are confident, however, that our assertion that the Second Principle of Mathematical Induction is useful will be proved correct by those who work the problems at the end of Section 3.3.

## EXERCISES 3.2

**17.** Prove that for each natural number $n \geqslant 2$,

$$[(2^2-1)/2^2][(3^2-1)/3^2][(4^2-1)/4^2]\cdots[(n^2-1)/n^2]=(n+1)/2n$$

**18.** Prove that for each natural number $n \geqslant 2$,

$$[2^2/(1\cdot 3)][3^2/(2\cdot 4)][4^2/(3\cdot 5)]\cdots[n^2/(n-1)(n+1)]=2n/(n+1)$$

**19.** Prove that for each odd natural number $n \geqslant 3$,

$$(1+1/2)(1-1/3)(1+1/4)\cdots(1+(-1)^n/n)=1$$

**20.** Prove that for each natural number $n \geqslant 6$, $(n+1)^2 \leqslant 2^n$.

**21.** Prove that for each natural number $n \geqslant 2$, $n^3+1>n^2+n$.

**22.** Prove that for each natural number $n \geqslant 2$, $2^{n+1} \leqslant 3^n$.

**23.** Prove that for each natural number $n \geqslant 3$, $(1+1/n)^n<n$.

**24.** Prove that every natural number $n \geqslant 5$ can be written as the sum of natural numbers, each of which is a 2 or a 3.

**25.** Prove that every natural number $n \geqslant 6$ can be written as the sum of natural numbers, each of which is a 2 or a 5.

**26.** Prove that for each natural number $n \geqslant 5$, $4^n>n^4$.

**27.** Prove by induction that for all natural numbers $n \geqslant 10$, $2^{n-10}(1000)<2^n-2^{n-6}$.

**28.** Use the Least-Natural-Number Principle to establish Proposition 3.2.

**29.** **a)** For $n \geqslant 2$, find a formula for $(1-1/4)(1-1/9)(1-1/16)\cdots(1-1/n^2)$.

**b)** Use the Extended Principle of Mathematical Induction to prove that your formula holds.

**30.** **a)** For which natural numbers $n$ is it true that $n^3<2^n$?

**b)** Prove your assertion.

**31.** What, if anything, is wrong with the following argument that the Least-Natural-Number Principle is a consequence of the Principle of Mathematical Induction? *Argument*: Let

$S=\{n\in\mathbb{N}$: any set of natural numbers with exactly $n$ elements has a least member$\}$

Obviously, $1\in S$. Suppose that $n\in S$. Let $A$ be a set of natural numbers with exactly $n+1$ members. *Let* $q\in A$ and let $p$ be the least member of $A-\{q\}$. Then the minimum of $p$ and $q$ is the least member of $A$. Therefore, $n+1\in S$ and by the Principle of Mathematical Induction, $S=\mathbb{N}$, as required.

**32.** In Chapter 9, we give a formal definition of a finite set. In this problem, assume that a set is finite provided it has exactly $n-1$ members for some $n \in \mathbb{N}$. Prove that every nonempty finite set of real numbers has a *greatest* member and a *smallest* member. [*Hint*: The Fletcher–Patty Principle, which is not as well known as Pólya's Principle, states that an incorrect proof can often be modified to provide a correct proof of a different theorem from the one you set out to establish. Consider Exercise 31.]

**33.** **a)** Find a natural number $k$ such that $k \neq 1$ and $k^4 < 2^k$.

**b)** If $S = \{n \in \mathbb{N}: n \neq 1 \text{ and } n^4 < 2^n\}$, find the smallest member of $S$.

**c)** If $m$ is the smallest member of $S$, prove that $S = \{n \in \mathbb{N}: n \geqslant m\}$.

**34.** Complete the proof of Theorem 3.5.

**35.** *Cutting up a pie.* Place $n$ distinct dots on a circle and connect each pair of dots with the chord determined by the pair of points. (See the accompanying figures.)

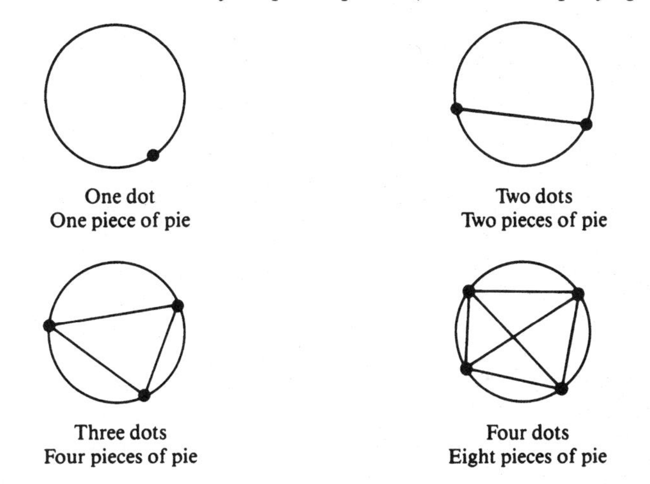


**a)** Guess (without drawing anything) how many pieces of pie you would get with five dots. How confident are you that your guess is correct? Would you bet Fletcher or Patty your bottom dollar? a small Coke? Check to see if your guess is correct.

**b)** Guess (without drawing anything) how many pieces of pie you would get with six dots. How confident are you that your guess is correct? Would you bet Fletcher or Patty a large Coke? Check to see if your guess is correct.

**36.** Prove that $2^n + 2 < 2 \cdot 2^n$ for all natural numbers $n \geqslant 2$.

**37.** **a)** Prove that for each natural number $n$, any set with $n$ members has $n(n-1)/2$ 2-element subsets.

**b)** Use the result of (a) to show that for each natural number $n$, any set with $n$ members has $n(n-1)(n-2)/(1 \cdot 2 \cdot 3)$ 3-element subsets.

***38.** Let $S = \{n \in \mathbb{N}$: for any $n$-element set of natural numbers and any natural number $r \leqslant n$, the number of $r$-element subsets of the given set is $n(n-1)\cdots(n-(r-1))/(1 \cdot 2 \cdot \cdots \cdot r)\}$. Prove that $S$ is inductive.

## Induction and Recursion

For each nonnegative integer $n$, let $A_n$ denote the number of subsets of a set with $n$ elements. We have already seen that $A_0 = 1$ and that for each $n \in \mathbb{N}$, $A_{n+1} = 2A_n$. This is an example of a recurrence relation. In other words, in this case, we do not have an explicit formula for $A_{n+1}$, but the recurrence relation defines an algorithm that shows us how to compute $A_{n+1}$.

Another example of a recurrence relation is the famous sequence investigated by Leonardo of Pisa (1170–1250), who is better known as Fibonacci. The Fibonacci numbers are defined as follows: $f_1 = 1$, $f_2 = 1$, and $f_{n+2} = f_{n+1} + f_n$ for each $n \in \mathbb{N}$. In Exercise 75, you are asked to verify an explicit formula for the $n$th term of the sequence. However, even without verifying this formula, if we wanted a particular term—say, the sixth term—we could calculate it using the recurrence relations, as demonstrated by the following:

$$f_3 = 1 + 1 = 2 \qquad f_4 = 2 + 1 = 3$$

$$f_5 = 3 + 2 = 5 \qquad f_6 = 5 + 3 = 8$$

This is an interesting sequence. Sunflower and daisy florets, pinecones, pineapples, and even patterns of paving stones and mating habits of bees and rabbits exhibit properties given by the Fibonacci sequence. The seeds of sunflowers and daisies have a pattern consisting of two sets of spirals, one turning clockwise and the other counterclockwise. In a given floret, the number of clockwise spirals and the number of counterclockwise spirals are usually consecutive Fibonacci numbers. For example, in sunflowers, there are usually 34 clockwise spirals and 55 counterclockwise spirals. In daisies, there are usually 21 clockwise spirals and 34 counterclockwise spirals.

Fibonacci posed and solved the following problem: It is known that rabbits breed rapidly. Suppose a pair of adult rabbits produces a pair of young rabbits each month and that newborn rabbits become adults in two months and produce another pair of rabbits. Starting with an adult pair of rabbits, how many pairs of rabbits will there be after a certain month?

The number of young pairs at any given time is the number just born plus the number born one month ago, which is the number of current adult pairs plus the number of adult pairs one month ago. The number of adult pairs at any given time is the number of adult pairs one month ago plus the number of pairs born two months ago, which is the sum of the number of adult pairs of the two preceding months. It is only after we derive these results from the model that it becomes clear that the recurrence relation for the Fibonacci numbers is indeed the one that provides a solution. The solution for the first 12 months is given in the following table.

| Months | Adult pairs | Young pairs | Total pairs |
|---|---|---|---|
| 1 | 1 | 1 | 2 |
| 2 | 1 | 2 | 3 |
| 3 | 2 | 3 | 5 |
| 4 | 3 | 5 | 8 |
| 5 | 5 | 8 | 13 |
| 6 | 8 | 13 | 21 |
| 7 | 13 | 21 | 34 |
| 8 | 21 | 34 | 55 |
| 9 | 34 | 55 | 89 |
| 10 | 55 | 89 | 144 |
| 11 | 89 | 144 | 233 |
| 12 | 144 | 233 | 377 |

The Fibonacci sequence is also of importance in computer science, because many problems involving the maximum number of calculations in an algorithm can be expressed in terms of the Fibonacci sequence.

Given the recurrence relations, is the $n$th Fibonacci number defined for each natural number $n$? Let $S = \{n \in \mathbb{N} : f_n \text{ is defined}\}$. Clearly, $1 \in S$, since $f_1 = 1$. Suppose that $k \in \mathbb{N}$ and $\{1, 2, \ldots, k\} \subseteq S$. Then $f_{k+1} = f_k + f_{k-1}$, so $k + 1 \in S$. Therefore, by the Second Principle of Mathematical Induction, $S = \mathbb{N}$.

Perhaps you are familiar with summation notation, which is often used to express sums compactly. This is another example of a recurrence relation. For each natural number $n$, let $a_n$ be a real number. The sum of the first $n + 1$ real numbers in this sequence is defined recursively as follows:

$$\sum_{i=1}^{1} a_i = a_1 \quad \text{and} \quad \sum_{i=1}^{n+1} a_i = \left(\sum_{i=1}^{n} a_i\right) + a_{n+1}$$

Using these formulas, we see that

$$\sum_{i=1}^{2} a_i = \sum_{i=1}^{1} a_i + a_2 = a_1 + a_2$$

$$\sum_{i=1}^{3} a_i = \sum_{i=1}^{2} a_i + a_3 = (a_1 + a_2) + a_3$$

Let

$$S = \left\{n \in \mathbb{N} : \sum_{i=1}^{n} a_i \text{ is a real number}\right\}$$

Then $1 \in S$ because

$$\sum_{i=1}^{1} a_i = a_1$$

Suppose that $k \in S$. Then

$$\sum_{i=1}^{k} a_i \text{ is a real number}$$

Since

$$\sum_{i=1}^{k+1} a_i = \left(\sum_{i=1}^{k} a_i\right) + a_{k+1}$$

$k + 1 \in S$. Therefore, by the Principle of Mathematical Induction, $S = \mathbb{N}$.

In the two examples that we have just considered, we used induction to indicate that our function was defined for each natural number. Actually there is a subtle gap in our inductive argument that the functions are defined. This problem is resolved by a theorem known as the **Principle of Inductive Definition**, which ensures that the functions we have been considering are defined unambiguously. We do not include this theorem in the text, but will send its statement and proof, to be found in references [16] and [33], on request. Hereafter in this text, if recurrence relations define a first term and give a formula for the $n$th term in terms of the first $n - 1$ terms, then we assume that the $n$th term is defined for each natural number $n$. So we just assume that every term is defined and proceed. Recursion is really a method of defining a sequence inductively.

From time to time, mathematicians have supplemented their meager incomes by successfully marketing mathematical toys. One recent example is *Rubik's Cube*. The Irish have William Rowan Hamilton's *Around the World Game*, and Americans have Sam Loyd's (Brooklyn, N.Y.) 15-*puzzle*. We conclude this section with the discussion of a puzzle that is over a century old and whose solution involves recursion (induction).

The puzzle was invented by the French mathematician Edouard Lucas (1842–1891), who called it the *Tower of Hanoi*. It consists of a board with three upright pegs and seven rings with different outside diameters (see Figure 3.1). The seven rings are all on one peg with the largest on the bottom, as shown, to form a pyramid. The object is to transfer all the rings, one at a time, to the third peg to form an identical pyramid. During the transfers, we are not permitted to place a larger ring on a smaller ring, and this is

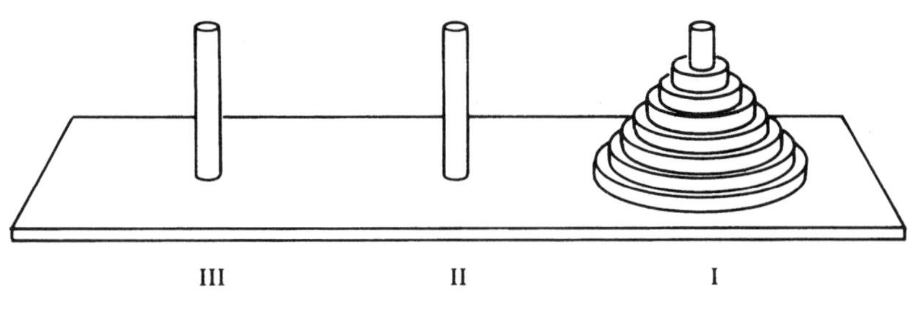


Figure 3.1

why we need the second peg. See if you can do this problem. If so, how many moves did you make in order to complete the transfer?

If you can't solve this problem, try to solve a simpler one—that is, use only five rings rather than seven. If you can solve this problem, how many moves did you make in order to complete the transfer?

If you can't solve either of these two problems, try still a simpler problem—use only four rings. In any case, continue removing one ring at a time until you find a problem you can solve.

Suppose you were not able to solve the problem until you had removed all the rings but one. Then the answer is simply one move—you pick up the one ring and place it on the third peg.

Now we have an induction problem. How do we go from one ring to two rings? Suppose we have two rings on the first peg, and, according to the rules of the game, we want to move them to the third peg. What is our first move? There are only two possibilities. We can move the smaller ring to the second peg or to the third peg. Which do we choose? If we place the smaller ring on the third peg, we cannot place the larger ring on the third peg without first removing the smaller ring. So if we are going to be efficient, our first move should be to place the smaller ring on the second peg—first move. Then we place the larger ring on the third peg—second move, and finally the smaller ring on the third peg—third move. So it takes three moves to transfer the rings to the third peg according to the rules of the game.

Now, how do we go from two rings to three rings? Here we can see the induction process more clearly. We can move the two smaller rings to the second peg in three moves. Why the second peg? Now we move the largest ring to the third peg—one move. Finally we know that we can move the two smaller rings to the third peg in three moves. So we have moved the three rings from the first peg to the third peg in seven moves.

Now you should try going from three rings to four rings (see Exercise 39a). If you are successful with four rings, try five (see Exercise 39b). Then try six (see Exercise 39c) and seven (see Exercise 39d).

Obviously there is nothing magical about the number seven. If $n$ is any natural number, we can have a Tower of Hanoi puzzle with $n$ rings. You should guess a formula for the smallest number of moves required for $n$ rings and then use induction to prove it. (See Exercise 40.)

Incidentally, as Lucas told the story, his game was just a model to illustrate the "real" Tower of Hanoi, which had 64 disks. The real tower was attended by a mystic order of monks who moved 1 disk a minute according to the long-established ritual that no larger disk could be placed on a smaller disk. According to legend, so Lucas's story went, as soon as all 64 disks were transferred, the earth would collapse in a cloud of dust. Even if Lucas had been telling the truth, there would be no cause for alarm: It would take the monks an incredibly long time to move all 64 disks.

We now give some examples of proof by induction that illustrate clever choices of $S$, the common usage of recursion formulas, Pólya's Principle, and the Second Principle of Mathematical Induction.

**EXAMPLE 7** Let $x_1 = 1$, $x_2 = 1$, and for $n \geqslant 2$, let $x_{n+1} = x_n + 2x_{n-1}$. Prove that $x_n$ is divisible by 3 if and only if $n$ is divisible by 3.

This is one of the examples we promised in Section 3.1, where the key to the proof is a clever choice of $S$. Before proceeding, you might try some logical choices of $S$ and see if proofs can be constructed using these choices.

Let $S = \{n \in \mathbb{N}: x_{n+6} - x_n \text{ is divisible by } 3\}$. Since $x_7 - x_1 = 43 - 1$ and $x_8 - x_2 = 85 - 1$, then $1, 2 \in S$. Let $n \geqslant 2$ and assume that $\{1, 2, \ldots, n\} \subseteq S$. Then

$$\begin{aligned} x_{n+1+6} - x_{n+1} = x_{n+7} - x_{n+1} \\ &= x_{n+6} + 2x_{n+5} - x_n - 2x_{n-1} \\ &= x_{n+6} - x_n + 2(x_{k+6} - x_k) \end{aligned}$$

where $k = n - 1$. Since both $k$ and $n$ belong to $S$, both $2(x_{k+6} - x_k)$ and $x_{n+6} - x_n$ are divisible by 3, and so $n + 1 \in S$. By the Second Principle of Mathematical Induction, $S = \mathbb{N}$, and it follows from this equality that $x_n$ is divisible by 3 if and only if $n$ is divisible by 3 (see Exercise 50). ❑

**EXAMPLE 8** Prove that the sum of the odd integers up to and including the odd integer $n$ is $(n+1)^2/4$.

This is another example in which the key to the proof is a clever choice of $S$, and it is also an example that uses the Second Principle of Mathematical Induction. Once again, we urge you to try some logical choices of $S$ before proceeding.

Let

$S = \{n \in \mathbb{N}: n$ is even, or $n$ is odd and the sum of the odd integers up to and including the odd integer $n$ is $(n+1)^2/4\}$

Since $(1+1)^2/4 = 1$, $1 \in S$. Suppose that $n \in \mathbb{N}$ and $\{1, 2, \ldots, n\} \subseteq S$. If $n + 1$ is even, then $n + 1 \in S$. Suppose that $n + 1$ is odd. Then $n - 1$ is odd, and because $n - 1 \in S$,

$$1 + 3 + 5 + \cdots + (n-1) = n^2/4$$

Adding $n + 1$ to both sides of the equation, we obtain

$$\begin{aligned} 1 + 3 + 5 + \cdots + (n-1) + (n+1) &= n^2/4 + (n+1) \\ &= [n^2 + 4(n+1)]/4 \\ &= (n^2 + 4n + 4)/4 \\ &= (n+2)^2/4 \end{aligned}$$

Therefore, $n + 1 \in S$, and hence by the Second Principle of Mathematical Induction, $S = \mathbb{N}$. ❑

**EXAMPLE 9** Define a sequence as follows: $F_1 = 1$, $F_2 = 1$, and $F_{n+2} = F_{n+1} + F_n + F_{n+1}F_n$ for each $n \in \mathbb{N}$. The sequence begins 1, 1, 3, 7, 31,.... Prove that for each natural number $n$, $F_n = 2^{f(n)} - 1$, where $f(n)$ is the $n$th Fibonacci number.

Let $S = \{n \in \mathbb{N}: F_n = 2^{f(n)} - 1\}$. Since

$$2^{f(1)} - 1 = 2^{f(2)} - 1 = 1$$

$1, 2 \in S$. Suppose that $n \geqslant 2$ and $\{1, 2, \ldots, n\} \subseteq S$. Now

$$\begin{aligned} F_{n+1} &= F_n + F_{n-1} + F_n F_{n-1} \\ &= 2^{f(n)} - 1 + 2^{f(n-1)} - 1 + (2^{f(n)} - 1)(2^{f(n-1)} - 1) \\ &= 2^{f(n)} + 2^{f(n-1)} + 2^{f(n)+f(n-1)} - 2^{f(n)} - 2^{f(n-1)} - 1 \\ &= 2^{f(n+1)} - 1 \end{aligned}$$

Hence $n + 1 \in S$ and therefore, by the Second Principle of Mathematical Induction, $\mathbb{N} = S$. ❑

**EXAMPLE 10** Prove that

$$1/\sqrt{1} + 1/\sqrt{2} + 1/\sqrt{3} + \cdots + 1/\sqrt{1996} > \sqrt{1996}$$

Pólya's Principle applies. We prove that

$$1/\sqrt{1} + 1/\sqrt{2} + 1/\sqrt{3} + \cdots + 1/\sqrt{n} > \sqrt{n} \quad \text{for each } n \geqslant 2$$

Let

$$S = \{n \in \mathbb{N}: n \geqslant 2 \text{ and } 1/\sqrt{1} + 1/\sqrt{2} + 1/\sqrt{3} + \cdots + 1/\sqrt{n} > \sqrt{n}\}$$

Since $1/\sqrt{1} + 1/\sqrt{2} > 1.7 > \sqrt{2}$, then $2 \in S$. Suppose that $n \in S$. Then

$$1/\sqrt{1} + 1/\sqrt{2} + 1/\sqrt{3} + \cdots + 1/\sqrt{n} > \sqrt{n}$$

Therefore,

$$\begin{aligned} 1/\sqrt{1} + 1/\sqrt{2} + 1/\sqrt{3} + \cdots + 1/\sqrt{n} + 1/\sqrt{n+1} &> \sqrt{n} + 1/\sqrt{n+1} \\ &= (\sqrt{n}\sqrt{n+1} + 1)/\sqrt{n+1} \\ &> (n+1)/\sqrt{n+1} \\ &= \sqrt{n+1} \end{aligned}$$

Thus, $n + 1 \in S$, and so by the Extended Principle of Mathematical Induction, $\{n \in \mathbb{N}: n \geqslant 2\} \subseteq S$. ❑

There is a proof that $1/\sqrt{1} + 1/\sqrt{2} + 1/\sqrt{3} + \cdots + 1/\sqrt{n} > \sqrt{n}$ for each $n \geqslant 2$ that does not require induction. In Exercise 80, you are asked to give such a proof.

## EXERCISES 3.3

39. a) Write a description of how to move four rings from one peg to another peg using the rules of the Tower of Hanoi puzzle.
    b) Repeat (a) with five rings rather than four.
    c) Repeat (a) with six rings rather than four.
    d) Repeat (a) with seven rings rather than four.
40. With respect to the Tower of Hanoi puzzle, guess a formula for the smallest number of moves required to move $n$ rings from one peg to another peg according to the rules, and then use induction to prove that your formula is correct.
41. Let $x_1 = 1$, and for $n > 1$, let $x_n = \sqrt{3x_{n-1} + 1}$. Prove that $x_n < 4$ for all $n \in \mathbb{N}$.
42. Prove that for each $n \in \mathbb{N}$,
$$1 \cdot 1! + 2 \cdot 2! + 3 \cdot 3! + \cdots + n \cdot n! = (n+1)! - 1$$
43. Prove that for each even natural number $n$,
$$(1 - 1/2)(1 + 1/3)(1 - 1/4) \cdots (1 - (-1)^n/n) = 1/2$$
44. Prove that for each natural number $n$,
$$3 + 6 + 12 + \cdots + 3(2^{n-1}) = 3(2^n - 1)$$
45. Prove by induction that for all natural numbers $n \geqslant 4$, $n! > 2^n$.
46. Prove by induction that for all natural numbers $n > 8$, $n! > 4^n$.
47. Prove by induction that for all natural numbers $n \geqslant 6$, $2^{n+2} \leqslant n!$.
48. Prove by induction that for each natural number $n$, $n^n \geqslant n!$.
49. Prove by induction that for each natural number $n$, $9^n - 8n - 1$ is divisible by 64.
50. Let $n$ be a natural number and let $x_n$ be defined as in Example 7. Use the result established in Example 7 to show that $x_n$ is divisible by 3 if and only if $n$ is divisible by 3.
51. Prove by induction that for each natural number $n$,
$$\sum_{k=1}^{n} 2^{k-1} = 2^n - 1$$
52. Prove by induction that for each natural number $n$,
$$\sum_{k=1}^{n} \left(\frac{1}{2}\right)^k = 1 - \left(\frac{1}{2}\right)^n$$
53. Let $x \in \mathbb{R}$ such that $x \neq 1$. Prove by induction that for each natural number $n$,
$$\sum_{k=1}^{n} x^{k-1} = \frac{x^n - 1}{x - 1}$$
54. Let $a_1 = 1$, and for each natural number $n > 1$, let $a_n = 3a_{n-1} - 1$. Prove that for each natural number $n$, $a_n = \frac{1}{2}(3^{n-1} + 1)$.
55. Let $a_1 = 1$ and $a_2 = 3$, and for each natural number $n > 2$, let $a_n = 3a_{n-1} - 2a_{n-2}$. Prove that for each natural number $n$, $a_n = 2^n - 1$.
56. Let $a_1 = a_2 = 1$, and for each natural number $n \geqslant 2$, let
$$a_{n+1} = \frac{1}{2}\left(a_n + \frac{2}{a_{n-1}}\right)$$
Prove that for each natural number $n$, $1 \leqslant a_n \leqslant 2$.

**57.** Let $a_1 = a_2 = a_3 = 1$, and for each natural number $n > 3$, let

$$a_n = a_{n-1} + a_{n-2} + a_{n-3}$$

Prove that for each natural number $n > 1$, $a_n \leqslant 2^{n-2}$.

**58.** Let $a_1 = a_2 = 1$, and for each natural number $n > 2$, let $a_n = 2a_{n-1} + 3a_{n-2}$.

**a)** Prove that for each natural number $n > 2$, $a_n > 3^{n-2}$.

**b)** Prove that for each natural number $n > 2$, $a_n < 2 \cdot 3^{n-2}$.

**59.** Let $a_1 = 3$, and for each natural number $n > 1$, let $a_n = -a_{n-1}$.

**a)** Find a formula for the $n$th term.

**b)** Prove that your formula holds.

**60.** Let $a_1 = 2$ and $a_2 = 1$, and for each natural number $n > 2$, let $a_n = 2a_{n-1} - a_{n-2}$.

**a)** Find a formula for the $n$th term.

**b)** Prove that your formula holds.

**61.** For each natural number $n$, let $f(n)$ denote the number of subsets of $\{1, 2, 3, \ldots, n\}$ that do not contain two consecutive numbers.

**a)** Find a pattern for $f(n)$. (Don't forget to count the empty set.)

**b)** Use the Second Principle of Mathematical Induction to prove that your pattern is correct.

**62.** Let $t_1 = 1$, $t_2 = 2$, and $t_3 = 3$. Then for each $k \in \mathbb{N}$, let $t_{k+3} = t_{k+2} + t_{k+1} + t_k$. Prove by induction that for each $n \in \mathbb{N}$, $t_n < 2^n$.

**63.** Let $f_1, f_2, f_3, \ldots$ be the Fibonacci numbers. Prove by induction that for each natural number $n > 1$,

$$\sum_{i=1}^{n-1} f_i = f_{n+1} - 1$$

**64.** Let $f_1, f_2, f_3, \ldots$ be the Fibonacci numbers. Prove by induction that for each natural number $n$:

**a)** $f_1 + f_3 + f_5 + \cdots + f_{2n-1} = f_{2n}$

**b)** $f_2 + f_4 + f_6 + \cdots + f_{2n} = f_{2n+1} - 1$

**65.** For each natural number $i$, let $f_i$ be the $i$th Fibonacci number.

**a)** For each natural number $n$, prove that $f_{3n}$ is an even integer.

**b)** For each $m \in \mathbb{N}$ such that $m$ is not a multiple of 3, prove that $f_m$ is an odd integer.

**66.** Let $f_1, f_2, f_3, \ldots$ be the Fibonacci numbers. Prove by induction that for each natural number $n$, $f_{5n}$ is a multiple of 5.

**67.** Let $f_1, f_2, f_3, \ldots$ be the Fibonacci numbers. Prove by induction that for each natural number $n$,

$$\sum_{i=1}^{n} f_i^2 = f_n f_{n+1}$$

**68.** Let $f: \mathbb{N} \to \mathbb{N}$ be a function such that for all $m, n \in \mathbb{N}$, $f(m + n) = f(m) + f(n)$. Prove that $f(170{,}000) = 10{,}000 \times f(17)$. (No proof with thousands of missing steps is deemed acceptable.)

**69.** Prove that for each natural number $n$,

$$\sum_{i=1}^{n} i^3 = \left(\sum_{i=1}^{n} i\right)^2$$

**70.** This problem is not intended as an induction problem; it is intended to illustrate Pólya's Principle that it is sometimes easier to work a harder problem. Prove that every perfect cube is the difference of two perfect squares. For example $3^3 = 6^2 - 3^2$.

**71.** The "name-one-thousand" game is a two-player game. The first player names 1, 2, or 3. Thereafter, each player in turn adds 1, 2, or 3 to the previous total. The first player to name 1000 wins. Prove by induction that the second player has a winning strategy.

**72.** In the Open Tennis Tournament, every player plays every other player exactly once and either wins or loses. Define a Dean to be a player who, for every other player A, either beats A or beats a player B who beats A.

**a)** Prove that there can be more than one Dean.

**b)** Prove that there is at least one Dean.

**73.** Prove by induction that if $n$ people ($n \in \mathbb{N}$) stand in line at a ticket counter, and if the line begins with a woman and ends with a man, then somewhere in the line there is a man standing directly behind a woman.

**74.** The P & W Manufacturing Company makes bubblegum dispensers, and their machines dispense bubblegum balls in pairs (that is, for a nickel each customer gets two bubblegum balls). The machines are of different sizes and capacities, so the number of bubblegum balls in a given machine is not known. However, it is known that each machine contains an odd number of green bubblegum balls and an odd number of red bubblegum balls. Prove by induction that before a dispenser is empty, at least one pair will be dispensed that consists of one green bubblegum ball and one red bubblegum ball.

**75.** For each natural number $i$, let $f_i$ be the $i$th Fibonacci number and let

$$a_i = \{[(1+\sqrt{5})/2]^i - [(1-\sqrt{5})/2]^i\}/\sqrt{5}.$$

Prove by induction that for each $i \in \mathbb{N}$, $f_i = a_i$.

**76.** For each natural number $n$, prove that $n^5/5 + n^4/2 + n^3/3 - n/30$ is an integer.

**77.** Show that every natural number greater than 2 can be written as a sum of distinct Fibonacci numbers.

**78.** For each natural number $n$, let $f_n$ be the $n$th Fibonacci number. Prove that $f_{n+1}^2 + f_n^2 = f_{2n+1}$ for each $n \in \mathbb{N}$.

**79.** A coin is tossed $n(n > 1)$ times. Prove that the probability that two heads will turn up in succession somewhere in the sequence of throws is $1 - f_{n+2}/2^n$, where $f_i$ is the $i$th Fibonacci number.

**80.** Without using induction, prove that $1/\sqrt{1} + 1/\sqrt{2} + 1/\sqrt{3} + \cdots + 1/\sqrt{n} > \sqrt{n}$ for each $n \geqslant 2$.

**81.** Find some properties of Fibonacci numbers that are not given in this text and prove these properties.

## 3.4 Consequences of the Division Algorithm

Let $b$ be an integer and let $n$ be a natural number. Recall that $n$ *divides* $b$ provided there is an integer $x$ such that $b = nx$.

We note that if $b = 0$, then every natural number divides $b$. If $b \neq 0$ and $n$ is a natural number that divides $b$, then $n \leqslant |b|$. In particular, if $b$ is a nonzero integer, then there are only finitely many natural numbers that divide $b$.

**Definition**

Let $a$ and $b$ be integers, not both 0. Then the largest natural number that divides both $a$ and $b$ is called the **greatest common divisor** of $a$ and $b$ and is denoted by gcd $(a, b)$.

Since the set of natural numbers that divide both $a$ and $b$ is finite, by Exercise 32 in Section 3.2, gcd $(a, b)$ exists.

The purpose of this section is to find an algorithm for determining gcd $(a, b)$ for integers $a$ and $b$ that are not both 0 and to prove that the number gcd $(a, b)$ is the only natural number $n$ with the following properties:

**1.** $n$ divides both $a$ and $b$.

**2.** If $d$ is an integer that divides both $a$ and $b$, then $d$ divides $n$.

**Lemma 3.6**

Let $a$ and $b$ be integers, not both 0. Suppose that $d_1$ and $d_2$ are natural numbers that divide both $a$ and $b$, and suppose that any integer that divides both $a$ and $b$ divides both $d_1$ and $d_2$. Then $d_1 = d_2$.

**PROOF** By hypothesis, $d_1$ divides both $a$ and $b$, and so $d_1$ divides $d_2$. Similarly, $d_2$ divides $d_1$. Since both $d_1$ and $d_2$ are natural numbers, $d_1 = d_2$. □

**Lemma 3.7**

Let $b$ be a nonzero integer. Then gcd $(0, b) = |b|$ and gcd $(0, b)$ is the only natural number $n$ with the following properties:

**a)** $n$ divides both 0 and $b$.

**b)** If $d$ is an integer that divides both 0 and $b$, then $d$ divides $n$.

**PROOF** Since every natural number divides 0, the largest natural number that divides both 0 and $b$ is $|b|$. Thus, gcd $(0, b) = |b|$ and so condition (a) holds. Now suppose that $d$ is a natural number that divides both 0 and $b$. Then there is an integer $x$ such that $dx = b$ and so $d|x| = |d||x| = |dx| = |b|$.

Since $d$ divides $|b|$, condition (b) is satisfied. We have shown that gcd $(0, b)$ is a natural number that satisfies conditions (a) and (b), and since the previous lemma asserts that there are not two natural numbers that satisfy conditions (a) and (b), we are finished. □

**Lemma 3.8**

> Let $a$ be an integer and let $b$ be a nonzero integer.
> Then $\gcd(a, b) = \gcd(a, |b|)$.

**PROOF** See Exercise 85. □

The proof of the following classic result of number theory (Theorem 3.9) contains a technique for finding the greatest common divisor of two integers, not both of which are 0. Although the proof makes repeated use of the Least-Natural-Number Principle, it is important that you not get bogged down in the details. The proof should be read through with the conviction that each of the details can be verified. Only after you have grasped the main idea is it worthwhile to consider one or two of the indicated exercises.

**Theorem 3.9**

> Let $a$ and $b$ be integers, not both 0. Then $\gcd(a, b)$ is the only natural number $d$ such that
>
> **a)** $d$ divides both $a$ and $b$, and
>
> **b)** If $c$ is an integer such that $c$ divides both $a$ and $b$, then $c$ divides $d$.

**PROOF** By Lemma 3.6, there do not exist two natural numbers that satisfy (a) and (b) as just given, and so it suffices to show that $\gcd(a, b)$ is a number that satisfies (a) and (b). Furthermore, by Lemma 3.7, we can assume that $b \neq 0$.

We first consider the case where $b$ is positive. By the Division Algorithm, there are integers $q_1$ and $r_1$ such that $a = bq_1 + r_1$ and $0 \leqslant r_1 < b$. If $r_1 = 0$, then $b$ divides $a$, and $b$ will satisfy the conditions for $d$ in parts (a) and (b).

Suppose that $r_1 \neq 0$. By the Division Algorithm, there are integers $q_2$ and $r_2$ such that $b = r_1 q_2 + r_2$ and $0 \leqslant r_2 < r_1$. If $r_2 = 0$, then $r_1$ divides $b$, and

$$a = bq_1 + r_1 = (r_1 q_2)q_1 + r_1 = r_1(q_2 q_1 + 1)$$

Therefore, $r_1$ divides $a$, so $r_1$ satisfies the conditions for $d$ in part (a). Suppose that $c$ is an integer such that $c$ divides $a$ and $c$ divides $b$. Since $r_1 = a - bq_1$, $c$ divides $r_1$. So $r_1$ satisfies the conditions for $d$ in part (b).

Suppose that $r_2 \neq 0$. By the Division Algorithm, there are integers $q_3$ and $r_3$ such that $r_1 = r_2 q_3 + r_3$ and $0 \leqslant r_3 < r_2$. If $r_3 = 0$, we can show that $r_2$ satisfies the conditions for $d$ in parts (a) and (b). (You are asked to prove this in Exercise 87.)

If $r_3 \neq 0$, repeated applications of the Division Algorithm produce a sequence of pairs of integers $q_4, r_4, q_5, r_5, q_6, r_6, \ldots$, such that

$$r_2 = r_3 q_4 + r_4 \quad \text{and} \quad 0 \leqslant r_4 < r_3$$

$$r_3 = r_4 q_5 + r_5 \quad \text{and} \quad 0 \leqslant r_5 < r_4$$

$$r_4 = r_5 q_6 + r_6 \quad \text{and} \quad 0 \leqslant r_6 < r_5$$

$$\vdots$$

Note that we have a decreasing sequence of nonnegative integers $r_1 > r_2 > r_3 > \cdots$, so there exists a natural number $p+1$ such that $r_{p+1} = 0$. (See Exercise 88.)

Once we have some natural number $p+1$ such that $r_{p+1} = 0$, our algorithm comes to a shuddering halt. Rather than prove this fact, we set $S = \{k \in \mathbb{N} : r_k = 0\}$. By the Least-Natural-Number Principle, $S$ has a smallest member. Let $p+1$ be the smallest member of $S$. We first show that $r_p$ satisfies the conditions for $d$ in parts (a) and (b).

Note that $r_p$ divides $r_{p-1}$ because $r_{p-1} = r_p q_{p+1}$. But then $r_p$ divides $r_{p-2}$, because $r_{p-2} = r_{p-1} q_p + r_p$ and $r_p$ divides both $r_{p-1}$ and $r_p$. Continuing in this manner, we see that $r_p$ divides $r_{p-1}, r_{p-2}, r_{p-3}, \ldots$. After a *finite* number of steps, we see that $r_p$ divides $r_1$, that $r_p$ divides $b$, and finally that $r_p$ divides $a$. So $r_p$ satisfies the conditions for $d$ in part (a). (See Exercise 89.)

Suppose that $c$ divides $a$ and $c$ divides $b$. Then $c$ divides $r_1$ because $r_1 = a - bq_1$. Since $c$ divides $b$ and $c$ divides $r_1$, $c$ divides $r_2$ because $r_2 = b - r_1 q_2$. Continuing in this manner, we obtain $c$ divides $r_1$, $c$ divides $r_2$, $c$ divides $r_3, \ldots$. After a *finite* number of steps, we conclude that $c$ divides $r_p$. Therefore, $r_p$ satisfies the conditions for $d$ in part (b). (See Exercise 90.)

We wish to show that $\gcd(a, b)$ satisfies conditions (a) and (b) and, under the assumption that $b > 0$, we have just shown that $r_p$ is the only natural number that satisfies (a) and (b). It remains to show that $\gcd(a, b) = r_p$. Since $r_p$ satisfies property (a), $r_p \leqslant \gcd(a, b)$. Since $r_p$ satisfies property (b), $\gcd(a, b)$ divides $r_p$, and since $r_p > 0$, it follows that $\gcd(a, b) \leqslant r_p$. Therefore, $r_p = \gcd(a, b)$. This concludes the proof for the case that $b$ is positive.

Now suppose that $b < 0$. As above, we can prove that $\gcd(a, |b|)$ is the only natural number that divides both $a$ and $|b|$ and has the property that if an integer $c$ divides $a$ and $|b|$, then $c$ divides $\gcd(a, |b|)$. By Lemma 3.8, $\gcd(a, |b|) = \gcd(a, b)$. Moreover, an integer $c$ divides $b$ if and only if it divides $|b|$. It follows that $\gcd(a, b)$ satisfies the required conditions (a) and (b). (See Exercise 91.) □

We illustrate the usefulness of Theorem 3.9 by considering the following problem: Find the greatest common divisor of 2047 and 1633, and find integers $m$ and $n$ such that $\gcd(2047, 1633) = 2047m + 1633n$. Following the proof of Theorem 3.9, we calculate

$$2047 = 1(1633) + 414 \qquad 0 \leqslant 414 < 1633$$

$$1633 = 3(414) + 391 \qquad 0 \leqslant 391 < 414$$

$$414 = 1(391) + 23 \qquad 0 \leqslant 23 < 391$$

$$391 = 17(23) + 0 \qquad 0 \leqslant 0 < 23$$

The theorem assures us that gcd (2047, 1633) is the last nonzero remainder, namely, 23. Moreover,

$$\begin{aligned} 23 &= 414 - 1(391) \\ &= 414 - [1633 - 3(414)] \\ &= 4(414) - 1633 \\ &= 4(2047 - 1633) - 1633 \\ &= 4(2047) - 5(1633) \end{aligned}$$

Theorem 3.10 in Section 3.5 shows that if $a$ and $b$ are integers, not both 0, then there are always integers $m$ and $n$ such that $\gcd(a, b) = ma + nb$. As we have just illustrated, Theorem 3.9 has hidden in it a method for finding such integers.

The algorithm of Theorem 3.9 is called the **Euclidean Algorithm** and can be found in Euclid's *Elements.* Presumably the mathematics contained in Euclid's *Elements* is the work of many generations of mathematicians. In any case, *Elements* remains a classic (as well as classical) compendium of mathematics. A student of Isaac Newton once reported that in the five years he worked with Newton, he heard Newton laugh only once. That was when another student asked Newton if there was any longer any point in reading Euclid's *Elements.*

## EXERCISES 3.4

**82.** **a)** Use the method given by the proof of Theorem 3.9 to find gcd (219, 69).

**b)** Find integers $m$ and $n$ such that $\gcd(219, 69) = 219m + 69n$.

**83.** A certain rare goblet is supposed to weigh 43 ounces. Explain how to check the weight of this goblet given a balance scale and 1000 each of 7-oz and 11-oz weights.

**84.** For each of the following pairs of natural numbers $a, b$, use the Euclidean algorithm to find $\gcd(a, b)$ and to find integers $m$ and $n$ such that $ma + nb = \gcd(a, b)$. Show your work.

**a)** $a = 901$, $b = 952$.

**b)** $a = 4199$, $b = 1748$.

**c)** $a = 377$, $b = 233$.

**85.** Prove Lemma 3.8.

**86.** Let $S$ be the set of all natural numbers that divide 720. For $a, b \in S$, let $a \cap b = \gcd(a, b)$ and let $a' = 720/a$.

**a)** Find $(a \cap b)'$ when $a = 72$ and $b = 12$; when $a = 30$ and $b = 9$; when $a = 1$ and $b = 720$.

**b)** On the basis of the experimentation of part (a), what seems to be a reasonable interpretation of $a \cup b$ for $a, b \in S$? (Assume that de Morgan's laws hold.)

**87.** In the proof of Theorem 3.9, show that if $r_3 = 0$, then $r_2$ satisfies the conditions for $d$ in parts (a) and (b).

**88.** Let $r_1, r_2, r_3, \ldots$ be a sequence of nonnegative integers such that $r_1 > r_2 > r_3 > \cdots$. Prove that there is a natural number $k$ such that $r_k = 0$.

**89.** Mathematicians are famous for saying, "Continue in this manner." What they usually mean is that they could prove it by induction but do not wish to take the time to do so. In this exercise, you are asked to prove by induction that indeed $r_p$ does divide each of $a$ and $b$. (Each of $r_p$, $a$, and $b$ is as given in the proof of Theorem 3.9.)

**90.** Once again we refer to the proof of Theorem 3.9—in particular, to the place where reference is given to this exercise. Prove by induction that $c$ does indeed divide $r_p$.

**91.** Complete the proof of Theorem 3.9 for the case where $b < 0$.

**92.** Let $b$ be a nonzero integer and let $a$, $q$, and $r$ be integers such that $a = bq + r$. Prove that $\gcd(a, b) = \gcd(b, r)$.

**93.** Suppose $a$, $b$, and $c$ are integers with $a$ and $b$ not both 0 and that $d = \gcd(a, b)$. Prove that if $d$ does not divide $c$, then the equation $ax + by = c$ has no integer solutions for $x$ and $y$.

**94.** Let $a$ and $b$ be nonzero integers. Prove that there is a natural number $m$ such that (1) $a$ divides $m$ and $b$ divides $m$, and (2) if $c$ is an integer such that $a$ divides $c$ and $b$ divides $c$, then $m$ divides $c$. [*Note*: $m$ is called the **least common multiple** of $a$ and $b$, and is denoted by $\operatorname{lcm}(a, b)$.]

**95.** Prove that the terminology "the" least common multiple of $a$ and $b$ is justified. That is, prove that if $m'$ is a natural number satisfying conditions (1) and (2) of Exercise 94, then $m = m'$, where $m$ is the natural number of Exercise 94.

**96.** If $a, b \in \mathbb{N}$, then prove that $\gcd(a, b) \times \operatorname{lcm}(a, b) = ab$.

**97.** Let $n$ be a natural number, and let $S$ be the set of all natural numbers that divide $n$. For $a, b \in S$, let $\gcd(a, b) = a \cap b$ and $\operatorname{lcm}(a, b) = a \cup b$. For each $x \in S$, let $x'$ denote $n/x$. Do de Morgan's laws hold for this system?

## 3.5 Number Theory

Although many people associate Euclid's *Elements* with geometry, this text also contains basic results of number theory. One purpose of this section is to prove two results about prime numbers, which first appeared in Euclid's *Elements*—namely, the result that there are infinitely many prime numbers and the result known as the Fundamental Theorem of Arithmetic, which states that any number greater than 1 that is not prime can be factored into prime numbers and that, except for the order in which the primes appear, this factorization is unique. Another purpose of this section is to prove a result that tells us when a diophantine equation has a solution. The latter result is used in Section 4.5.

**Theorem 3.10**

Let $a$ and $b$ be integers, not both 0, and let $S = \{ma + nb: m \text{ and } n \text{ are integers}\}$. Then $\gcd(a, b)$ is the smallest natural number belonging to $S$.

**PROOF** Let $T$ be the set of positive numbers that belongs to $S$. Since $a^2 + b^2 \in T$, $T \neq \emptyset$, and by the Least-Natural-Number Principle, $T$ has a smallest member $d$. Since $d \in T$, there exist integers $x$ and $y$ such that $d = xa + yb$. By the Division Algorithm, there exist integers $q$ and $r$ such that $a = qd + r$, where $0 \leqslant r < d$. So $r = a - qd = a - q(xa + yb) = a(1 - qx) + b(-qy)$. Therefore, if $r > 0$, then $r \in T$. But since $r < d$ and $d$ is the smallest member of $T$, $r$ must be 0. Hence $a = qd$, and $d$ is a divisor of $a$. It follows in a similar manner than $d$ divides $b$ (see Exercise 114). Suppose that $c$ is an integer that divides both $a$ and $b$. Then there exist integers $u$ and $v$ such that $a = uc$ and $b = vc$. Thus, $d = xa + yb = x(uc) + y(vc)$, so $c$ divides $d$. By Theorem 3.9, $d$ is the greatest common divisor of $a$ and $b$. We complete the proof by observing that $d$ is the smallest natural number belonging to $S$. □

**Definition**

Two nonzero integers $a$ and $b$ are **relatively prime** provided $\gcd(a, b) = 1$.

**Corollary 3.11**

If $a$, $b$, and $c$ are integers such that $a$ and $b$ are relatively prime and $a \mid bc$, then $a \mid c$.

**PROOF** See Exercise 98. □

**Theorem 3.12**

If $p$ is a prime and $a$ is an integer such that $p$ does not divide $a$, then $a$ and $p$ are relatively prime.

**PROOF** See Exercise 99. □

**Corollary 3.13**

Let $a$ and $b$ be integers, and let $p$ be a prime number. If $p$ divides $ab$, then $p$ divides $a$ or $p$ divides $b$.

**PROOF** See Exercise 100. □

**Theorem 3.14**

There are infinitely many primes.

**PROOF** The proof is by contradiction. Suppose that there are only finitely many primes, say, $p_1, p_2, \ldots, p_n$. Let $N = p_1 p_2 \cdots p_n + 1$ and note that $N \neq 1$. By Proposition 3.2, $N$ has a prime divisor $q$. Since $q$ divides both $N$ and $N-1$, by Exercise 62 in Section 1.4, $q$ divides 1. This is a contradiction because no prime number divides 1. □

We are ready to prove the Fundamental Theorem of Arithmetic. Proposition 3.2 is half of the theorem.

**Theorem 3.15**

**The Fundamental Theorem of Arithmetic**

Let $a$ be a natural number that is larger than 1. Then

**a)** $a$ is a prime or $a$ is the product of finitely many primes, and

**b)** If $a = p_1 p_2 \cdots p_r = q_1 q_2 \cdots q_s$, where $p_1, p_2, \ldots, p_r, q_1, q_2, \ldots, q_s$ are primes, then $r = s$ and each $p_i$ $(1 \leqslant i \leqslant r)$ can be paired with a $q_j$ $(1 \leqslant j \leqslant s)$ so that the paired primes are equal.

**PROOF** As indicated, we have proved part (a). The proof of (b) is by contradiction. Let $S = \{n \in \mathbb{N} : n > 1$ and there are primes $p_1, p_2, \ldots, p_r, q_1, q_2, \ldots, q_s$ such that $n = p_1 p_2 \cdots p_r = q_1 q_2 \cdots q_s$ but the $p_i$'s and $q_i$'s cannot be paired off so that paired primes are equal$\}$, and suppose that $S \neq \varnothing$. By the Least-Natural-Number Principle, $S$ has a least member $a$. Since $a \in S$, there are primes $p_1, p_2, \ldots, p_r, q_1, q_2, \ldots, q_s$ such that $a = p_1 p_2 \cdots p_r = q_1 q_2 \cdots q_s$, but the $p_i$'s and $q_j$'s cannot be paired off so that paired primes are equal. By Exercise 101, there exists $k$ $(1 \leqslant k \leqslant s)$ such that $p_1$ divides $q_k$. But $p_1$ and $q_k$ are primes, so $p_1 = q_k$. Therefore, $p_2 p_3 \cdots p_r = q_1 q_2 \cdots q_{k-1} q_{k+1} \cdots q_s$. But this natural number is smaller than $a$, and hence it does not belong to $S$. Thus each $p_i$ $(2 \leqslant i \leqslant r)$ can be paired with a $q_j$ $(1 \leqslant j < k$ or $k < j \leqslant s)$ so that paired primes are equal. Therefore, $a \notin S$ and we have a contradiction. □

**Definition**

An equation whose solutions are required to be integers is called a **diophantine equation**.

Diophantine equations are named in honor of the Greek mathematician Diophantus (about 300 B.C.).

**Definition**

If $a$, $b$, and $c$ are integers, the diophantine equation $ax + by = c$ is called a **linear diophantine equation in two variables.**

The following theorem tells us when a linear diophantine equation has solutions, and it explicitly describes the solutions when they exist.

**Theorem 3.16**

Let $a, b, c \in \mathbb{Z}$ and let $d = \gcd(a, b)$.

**a)** If $d$ does not divide $c$, then $ax + by = c$ has no integral solution.

**b)** If $d|c$, then $ax + by = c$ has an infinite number of integral solutions. Moreover, if $x = x_0$ and $y = y_0$ is a solution, then all solutions are given by $x = x_0 + (b/d)n$, $y = y_0 - (a/d)n$, where $n$ is an integer.

**PROOF**

**a)** The proof is by contrapositive. Suppose that $d$ does not divide $c$, and that $x$ and $y$ are integers such that $ax + by = c$. Since $d|a$ and $d|b$, by Exercise 62 in Section 1.4, $d|c$. This is a contradiction.

**b)** Suppose that $d|c$. By Theorem 3.10, there are integers $s$ and $t$ such that $d = as + bt$. Since $d|c$, there is an integer $e$ such that $de = c$. Therefore, $c = de = (as + bt)e = a(se) + b(te)$. Thus, one solution of the equation $ax + by = c$ is $x_0 = se$ and $y_0 = te$.

Let $n$ be an integer, and let $x = x_0 + (b/d)n$ and $y = y_0 - (a/d)n$. Then

$$ax + by = ax_0 + a(b/d)n + by_0 - b(a/d)n = ax_0 + by_0 = c$$

Therefore, the equation $ax + by = c$ has an infinite number of integral solutions.

We now show that every solution of the equation $ax + by = c$ must be of the form that we have described. Suppose that $x$ and $y$ are integers such that $ax + by = c$. Since $ax_0 + by_0 = c$, we have $(ax + by) - (ax_0 + by_0) = 0$. Therefore, $a(x - x_0) + b(y - y_0) = 0$, and so $a(x - x_0) = b(y_0 - y)$. Hence, $(a/d)(x - x_0) = (b/d)(y_0 - y)$. Therefore,

$$\frac{a}{d} \,\Big|\, \left(\frac{b}{d}\right)(y_0 - y)$$

Since $d = \gcd(a, b)$, by Exercise 103, $\gcd(a/d, b/d) = 1$. Thus, by

Exercise 98, $(a/d)|(y_0 - y)$. Therefore, there is an integer $n$ such that $y_0 - y = (a/d)n$, and so $y = y_0 - (a/d)n$. Substituting this value of $y$ into the equation $a(x - x_0) = b(y_0 - y)$, we find that $a(x - x_0) = b(a/d)n$. This implies that $x = x_0 + (b/d)n$. □

Personal Note

## Sophie Germain (1776–1831)

It might be supposed that Gauss's *Disquitiones arithmeticae* earned him immediate recognition as a great mathematician, but the world was more taken by his calculation of the orbit of the newly discovered asteroid, Ceres, and with one notable exception Gauss's important work on number theory was not appreciated until decades after its publication. The exception was the French mathematician, Sophie Germain, who studied Gauss's work and then wrote to him using the pseudonym, Monsieur Leblanc. Sophie Germain learned mathematics by collecting notes of courses taught at L'École Polytechnique, because at that time L'École Polytechnique refused admission to women. Germain first used her pseudonym when she sent Professor Lagrange a paper related to a course he had taught. Although Germain's use of a masculine pseudonym was prompted by the prejudices of her times, both Lagrange and Gauss, when learning her true sex, gave her encouragement and support. Her letter to Gauss began a long-lived correspondence between them, but Gauss never met Sophie Germain and she died before the University of Göttingen could confer an honorary degree that Gauss had recommended for her.

The reception of Gauss's *Disquitiones arithmeticae* parallels the reaction of the mathematical community to Germain's number theory. As with Gauss, the world was first impressed by Germain's applied mathematics, for she became famous only after winning the French Academy's *grand prix* for her *Memoir on the Vibrations of Elastic Plates.* But her correspondence with Gauss centered around number theory, and today Sophie Germain is known for her contributions to that field. The most obvious contribution is Sophie Germain's Theorem, but more importantly there is an ongoing reassessment of Germain's other contributions to the study of Fermat's Last Theorem.

## EXERCISES 3.5

**98.** Prove Corollary 3.11.

**99.** Prove Theorem 3.12.

**100.** Prove Corollary 3.13.

**101.** Let $p$ be a prime, let $n \in \mathbb{N}$, and let $a_1, a_2, a_3, \ldots, a_n$ be integers such that $p$ divides $a_1 a_2 \cdots a_n$. Prove by induction that there exists $i$ $(1 \leqslant i \leqslant n)$ such that $p$ divides $a_i$.

**102.** Let $n$ be a natural number, let $a$ be an integer such that $\gcd(a, n) = 1$, and let $b$ be an integer. Prove that there is an integer $x$ such that $ax - b$ is divisible by $n$.

**103.** Let $a$ and $b$ be integers not both 0, and let $d$ be a natural number such that $d$ divides $a$ and $d$ divides $b$. Prove that $\gcd(a, b) = d$ if and only if $\gcd(a/d, b/d) = 1$.

**104.** A fraction $a/b$ is said to be in **lowest terms** provided $\gcd(a, b) = 1$. Two fractions $a/b$ and $c/d$ are said to be **equivalent** provided $ad = bc$. Prove that every fraction is equivalent to a fraction in lowest terms.

**105.** Find a fraction equivalent to 1739/4042 that is written in lowest terms. (No fair using a calculator or a computer!)

**106.** Let $a$, $b$, and $c$ be integers such that $\gcd(a, c) = \gcd(b, c) = 1$. Prove that $\gcd(ab, c) = 1$.

**107.** Let $n \in \mathbb{N}$, for each $i = 1, 2, \ldots, n$, let $a_i$ be an integer, and let $b \in \mathbb{Z}$ such that for each $i = 1, 2, \ldots, n$, $\gcd(a_i, b) = 1$. Prove that $\gcd(a_1 a_2 \cdots a_n, b) = 1$.

**108.** Let $a$, $b$, and $c$ be integers such that $a$ and $b$ are relatively prime and $c$ divides $a + b$. Prove that $\gcd(a, c) = \gcd(b, c) = 1$.

**109.** Show that $\gcd(5n + 2, 12n + 5) = 1$ for each integer $n$.

**110.** Let $p, q \in \mathbb{Z}$ such that 3 divides $p^2 + q^2$. Prove that 3 divides $p$ and 3 divides $q$.

**111.** Let $p, q, r \in \mathbb{Z}$ such that 5 divides $p^2 + q^2 + r^2$. Prove that 5 divides at least one of $p$, $q$, or $r$.

**112.** Show that there are infinitely many primes of the form $4n + 3$.

**113.** Show that there are infinitely many primes of the form $6n + 5$.

**114.** In the proof of Theorem 3.10, show that $d$ divides $b$.

**115.** The natural number 1 is not prime. State a result we have proved that would not hold if 1 were a prime number.

**116.** Prove that the diophantine equation $6x + 15y = 83$ does not have a solution.

**117.** Solve the diophantine equation $20x + 50y = 510$.

**118.** For each of the following diophantine equations, either show that no solutions exist or find all solutions.

**a)** $17x + 13y = 100$
**b)** $21x + 14y = 147$
**c)** $60x + 18y = 97$
**d)** $738x + 621y = 45$

## WRITING EXERCISES

1. Let $T = \{n \in \mathbb{N} : n \text{ is odd and } 8 \text{ does not divide } n^2 - 1\}$. Since $1^2 - 1 = 0$, $3^2 - 1 = 8$, $5^2 - 1 = 24$, and $7^2 - 1 = 48$, it is obvious that 1, 3, 5 and 7 do not belong to $S$. Use the Least-Natural-Number Principle to prove that $T = \emptyset$. Use the Second Principle of Induction to prove that $T = \emptyset$. Compare these two proofs. Which would you prefer to present to the man on the street? [*Warning:* This exercise is intended to compare the Second Principle of Induction with the Least-Natural-Number Principle, so it's not fair using that for each natural number $k$, $(2k - 1)^2 - 1 = 4k(k - 1)$].

2. Look up James O. Chilaka's paper, "Proofs without words: Sum of products of consecutive integers [*Math. Magazine*, Vol. 67, No. 5, December 1994, page 365]. Explain the proof Chilaka presents and relate his article to an exercise in Section 3.1.

CHAPTER

# 4 Relations and Orders

In mathematics, we are concerned with the way in which two or more mathematical objects are related. For example, $\triangle ABC \cong \triangle DEF$, $5 > 4$, 3 divides 9, and even $4 \neq 9$ are all statements that relate two mathematical objects. The relationships among mathematical objects play an increasingly important part in advanced mathematics even to the point that the relationships among objects become as important as the objects themselves. In this chapter, we study relations, rather than the objects they relate, and we develop methods of building new relations from given ones.

## 4.1 *Relations*

Suppose that we are given a set $S$, told that some or all the members of $S$ are related to each other in some way, and asked to determine a new set that indicates how the members of $S$ are related. We might consider the relation that holds between members of $S$ to be determined by the set of all subsets of $S$ of the form $\{a, b\}$, where $a$ is related to $b$. If $S = \{2, 4, 6, 8, 10\}$ and we are told that two members of $S$ are related if they have the same number of divisors, then

$$\{\{2, 2\}, \{4, 4\}, \{6, 6\}, \{6, 8\}, \{6, 10\}, \{8, 8\}, \{8, 10\}, \{10, 10\}\}$$

would indicate which members of $S$ are related. Of course, many relations are more one-sided than the example we have just given. For example, consider the set of all people and the relation "is the mother of."

We need the notion of an ordered pair $(a, b)$ so that we can say that $a$ is related to $b$ without being forced to say that $b$ is related to $a$. Intuitively, of course, the ordered pair $(a, b)$ is just a set in which $a$ comes first and $b$ comes second. No harm will come from relying on this intuitive description. Indeed, almost everyone thinks of an ordered pair in this manner. However, we give a precise definition, which leads to an interesting exercise (Exercise 2). The definition, which is due to Kazimierz Kuratowski (1896–1980), is just one of several ways of making the notion of ordered pair precise.

**Definition**

Let $S$ be a set, and let $a$ and $b$ be members of $S$. The **ordered pair** $(a, b)$ is the set $\{\{a\}, \{a, b\}\}$. The element $a$ is called the **first term** of $(a, b)$, and the element $b$ is called the **second term** of $(a, b)$.

Kuratowski's definition is justified by the following theorem, whose proof is left as Exercise 2.

**Theorem 4.1**

Let $(a, b)$ and $(c, d)$ be ordered pairs. Then $(a, b) = (c, d)$ if and only if $a = c$ and $b = d$.

It should be noted that a consequence of this theorem is that $(b, a) \neq (a, b)$ unless $a = b$.

We return to our original problem. We are given a set $S$, told that some or all the members of $S$ are related, and asked to determine a new set that indicates how the members of $S$ are related. The solution is simple. Our new set is the set of ordered pairs to which the ordered pair $(a, b)$ belongs, provided $a \in S$, $b \in S$, and $a$ is related to $b$.

The concept of *relation* is important in mathematics, and the following definitions should be studied with care.

**Definition**

A **relation** is a collection of ordered pairs. If $S$ is a relation, then the collection of first terms of all the ordered pairs of $S$ is called the **domain** of $S$ and the collection of all the second terms of the ordered pairs of $S$ is called the **range** of $S$. If $x$ belongs to the domain of $S$, then $S[x]$ is defined to be $\{y: (x, y) \in S\}$. When the relation $S$ is the only relation under discussion, the symbol $S[x]$ is often abbreviated to $[x]$. Note that $y \in S[x]$ means that $(x, y) \in S$. The notation $x\ S\ y$ is used to mean the same thing; that is, $x\ S\ y$ means $y \in S[x]$. If $A$ and $B$ are sets, then $A \times B$ is the relation $\{(a, b): a \in A \text{ and } b \in B\}$. The relation $A \times B$ is called the **Cartesian product** of $A$ and $B$, in honor of René Descartes (1596–1650).

**EXAMPLE 1**

Let

$$S = \{(-1, 2), (3, 1/2), (3, \pi), (-2, 2), (1, 1), (-1, 3), (-1, 1)\}$$

Then $S$ is a relation. The domain of $S$ is $\{-1, 3, -2, 1\}$, and the range of $S$

is $\{2, 1/2, \pi, 1, 3\}$. Also, $S[-1] = \{2, 3, 1\}$, $S[3] = \{1/2, \pi\}$, $S[-2] = \{2\}$, and $S[1] = \{1\}$. ❑

**EXAMPLE 2** If $A = \{3, 1, -2\}$ and $B = \{1, 4\}$, then

$$A \times B = \{(3, 1), (3, 4), (1, 1), (1, 4), (-2, 1), (-2, 4)\}$$ ❑

**EXAMPLE 3** Let

$$A = \{x \in \mathbb{R}: 1 \leqslant x \leqslant 2\}$$

$$B = \{0, 1\}$$

$$C = \{x \in \mathbb{R}: 1 < x < 2\}$$

$$D = \{x \in \mathbb{R}: 2 \leqslant x \leqslant 3\}$$

The sets $A \times B$, $B \times A$, $A \times C$, $A \times \mathbb{R}$, and $\mathbb{R} \times D$ are shown sketched in Figure 4.1.

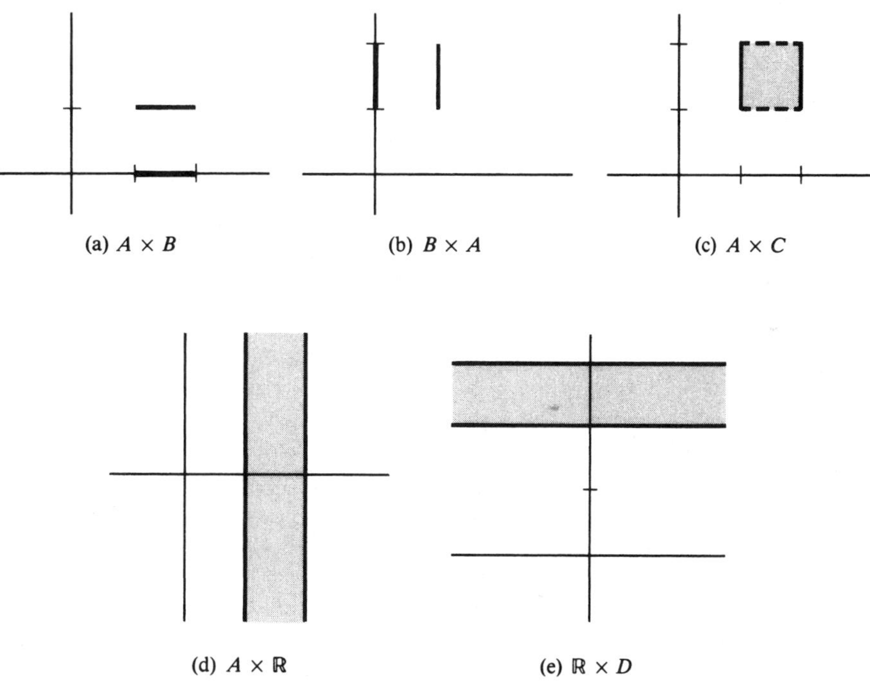


Figure 4.1 ❑

Graphs of relations suffer from two shortcomings. The first is that if the domain or the range of a relation is unbounded, then we must leave off sketching, as in the graphs of $A \times \mathbb{R}$ (Figure 4.1d) and $\mathbb{R} \times D$ (Figure 4.1e). The second shortcoming was pointed out by the King in *Through the Looking-glass*, who complained: "I must really get a thinner pencil. I can't manage this one a bit. It writes all manner of things I don't intend." Under these circumstances, it is hard to say what it is we are doing when we sketch the graph of a relation, but there is no doubt that graphing is an aid to intuition. Therefore, although we offer no definition of graphing relations, we encourage you to graph relations whenever possible. For the following two relations we have sketched, in Figure 4.2, the graph of the one for which the task seems fair—namely, $R_1$:

$$R_1 = \{(x, y) \in \mathbb{R} \times \mathbb{R}: x^2 + y^2 < 1 \text{ and } 0 < y < x\}$$

$$R_2 = \{(x, y) \in \mathbb{R} \times \mathbb{R}: x \text{ is rational and } y \text{ is irrational}\}$$

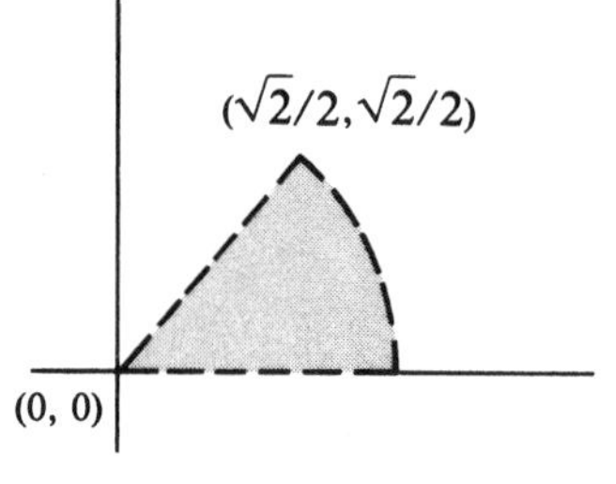


Figure 4.2

**Definition**

If $R \subseteq A \times B$ and $A$ is the domain of $R$, then $R$ is said to be a **relation from $A$ to $B$**. A subset of $A \times A$ is called a **relation on $A$**. Clearly, if $R$ is a relation from $A$ to $B$, then for each $x \in A$, $R[x] \subseteq B$.

Let us consider the relation with which you are probably most familiar—namely, the *plane*. It is $\mathbb{R} \times \mathbb{R}$, where $\mathbb{R}$ denotes the set of all real numbers. If we let $>$ denote the subset of the plane consisting of all points (that is, ordered pairs of real numbers) lying below the line $y = x$, then $>$ is a relation and $(5, 4) \in >$. With this notation, the following statements all say the same thing: (a) $5 > 4$, (b) $(5, 4) \in >$, and (c) $4 \in >[5]$. The notation of (a) is clear. The notation of (b) is considerably less sensible, and the notation of (c) is so obscure as to be absurd. As we will see in the following sections, there are situations in which notation (c) is a natural way to indicate that one object is related to another with respect to a given relation. The following table provides a summary of the three ways used by the mathematical community to indicate that an object $a$ is related to an object $b$ by a relation $R$.

**Given $R \subseteq A \times B$**

$$\left.\begin{array}{l}(a, b) \in R \\ b \in R[a] \\ a\,R\,b\end{array}\right\} \quad \begin{array}{l}b \text{ is an element in the range of } R \\ \text{that is related to the element} \\ a \text{ in the domain of } R\end{array}$$

If $A$ and $B$ are sets, then a member of $A \times B$ is an ordered pair. Recall that if we want to show that $A \subseteq B$, we must show that if $x \in A$, then $x \in B$. Suppose that $C$ and $D$ are also sets and we want to show that $A \times B \subseteq C \times D$. Then we must show that if $(x, y) \in A \times B$, then $(x, y) \in C \times D$.

What does it mean to say that $(x, y) \notin A \times B$? Since $(x, y) \in A \times B$ means $x \in A$ *and* $y \in B$, $(x, y) \notin A \times B$ means $x \notin A$ *or* $y \notin B$.

We illustrate the use of these ideas in the following proposition.

**Proposition 4.2**

For any two sets $A$ and $B$, $(A - B) \times B = (A \times B) - (B \times B)$.

**PROOF** Let $x \in (A - B) \times B$. Then there exist $p \in A - B$ and $q \in B$ such that $x = (p, q)$. Since $p \in A$ and $q \in B$, $(p, q) \in A \times B$. Moreover, since $p \notin B$, $(p, q) \notin B \times B$. Therefore, $x = (p, q) \in (A \times B) - (B \times B)$ and so $(A - B) \times B \subseteq (A \times B) - (B \times B)$.

Now let $x \in (A \times B) - (B \times B)$. There exist $p \in A$ and $q \in B$ such that $x = (p, q)$. Since $x \notin B \times B$ and $q \in B$, then $p \notin B$. Thus, $p \in A - B$ and $x = (p, q) \in (A - B) \times B$. It follows that $(A \times B) - (B \times B) \subseteq (A - B) \times B$. Combining the containments

$$(A - B) \times B \subseteq (A \times B) - (B \times B) \quad \text{and} \quad (A \times B) - (B \times B) \subseteq (A - B) \times B$$

we obtain the stated equality. □

## EXERCISES 4.1

1. Prove that the ordered pair $(17, 17) = \{\{17\}\}$.
2. Prove that $(a, b) = (c, d)$ if and only if $a = c$ and $b = d$.
3. Let $A = \{1\}$, $B = \{2\}$, and $C = \{3\}$. Show that $A \times (B \times C) \neq (A \times B) \times C$.
4. Sketch the graph of each of the following relations. For each relation, state its domain and its range.
   **a)** $S = \{(x, y) \in \mathbb{R} \times \mathbb{R} : x^2 + y^2 = 16\}$
   **b)** $S = \{(x, y) \in \mathbb{R} \times \mathbb{R} : y^2 = 2x\}$
   **c)** $S = \{(x, y) \in \mathbb{R} \times \mathbb{R} : x^2 = 2y^2\}$
   **d)** $S = \{(x, y) \in \mathbb{R} \times \mathbb{R} : x^2 = y^2\}$
   **e)** $S = \{(x, y) \in \mathbb{R} \times \mathbb{R} : |x| \leqslant 1 \text{ and } |y| > 3\}$
   **f)** $S = \{(x, y) \in \mathbb{R} \times \mathbb{R} : |x| = 1 \text{ and } 3 < y \leqslant 5\}$
5. For each of the relations of Exercise 4, evaluate $S[1]$.
6. Let $S = \{1, 2\}$. List all the relations on $S$.
7. Let $A = \{x \in \mathbb{R} : 1 \leqslant x \leqslant 2\}$, $B = \{x \in \mathbb{R} : 3 < x < 4\}$, and $C = \{x \in \mathbb{R} : 2 \leqslant x \leqslant 4\}$.
   **a)** Sketch the graph of $C \times (A \cup B)$.
   **b)** Sketch the graph of $(C \times A) \cup (C \times B)$.
   **c)** Sketch the graph of $B \times (A \cap C)$.

**d)** Sketch the graph of $(B \times A) \cap (B \times C)$.

**e)** Sketch the graph of $A \times C$.

**f)** Sketch the graph of $C \times A$.

**8.** Give an example of two relations $S$ and $T$ such that $\mathbb{R}$ = the domain of $S$ = the domain of $T$ = the range of $S$ = the range of $T$, but $S \cap T = \varnothing$.

**9.** Give a proof of, or a counterexample to, each of the following statements.

**a)** For any three sets $A$, $B$, and $C$,

$$A \times (B \cup C) = (A \times B) \cup (A \times C)$$

**b)** For any three sets $A$, $B$, and $C$,

$$A \times (B \cap C) = (A \times B) \cap (A \times C)$$

**10.** Given that $A$, $B$, and $C$ are sets, prove that

$$(A \cup B) - (A \cap B \cap C) = [A - (B \cap C)] \cup [B - (A \cap C)]$$

**11. a)** Let $S$ and $T$ be relations. Prove that if $S \subseteq T$ and $x$ is a member of the domain of $S$, then $S[x] \subseteq T[x]$.

**b)** Let $A$ and $B$ be nonempty sets. Prove that

$$(A \times B)[x] = B \text{ if } x \in A \quad \text{and} \quad (A \times B)[x] = \varnothing \text{ if } x \notin A$$

**12.** Let $\mathscr{A} = \{A_i : i \in \Lambda\}$ be a collection of relations and suppose that $x$ belongs to the domain of $A_i$ for each $i \in \Lambda$. Prove that

**a)** $(\cup\{A_i : i \in \Lambda\})[x] = \cup\{A_i[x] : i \in \Lambda\}$;

**b)** $(\cap\{A_i : i \in \Lambda\})[x] = \cap\{A_i[x] : i \in \Lambda\}$.

**13.** Let $\mathbb{N}$ denote the set of all natural numbers. Let $R = \{(a, b) \in \mathbb{N} \times \mathbb{N} : a \text{ divides } b\}$. List five members of $R[7]$, and list five members of $R[14]$. For which $n \in \mathbb{N}$ is it true that $R[n] = \mathbb{N}$?

**14.** Let $X$ be the closed interval $[0, 1]$, and let $A$ be a nonempty subset of $X$. Prove that

$$X \times X - (A \times A) = [(X - A) \times X] \cup [X \times (X - A)]$$

**15.** Let $A$ and $B$ be nonempty sets. Prove that $A \times \varnothing = \varnothing \times A$ and that $A \times \varnothing = B \times \varnothing$.

**16.** Let $A$ and $B$ be nonempty sets. Prove that $A \times B = B \times A$ if and only if $A = B$.

## 4.2 *Cartesian Graphs and Directed Graphs*

Just as Venn diagrams are used to picture sets, Cartesian graphs and directed graphs can be used to depict relations. To begin, let us consider a very simple finite relation,

$$\{(1, 1), (2, 3), (1, 3), (4, 3), (2, 2), (1, 2), (4, 1)\}$$

We can certainly circle members of this relation when it is considered as a subset of the plane.

|  |  |  |  |  |  |
|---|---|---|---|---|---|
|  | · | · | · | · | (4, 4) |
|  | ⊙ | ⊙ | · | ⊙ |  |
|  | ⊙ | ⊙ | · | · |  |
| (1, 1) | ⊙ | · | · | ⊙ |  |

There is nothing different here from the usual graphing of relations in the plane except that ordinarily we graph relations that have infinitely many ordered pairs. Nonetheless, we are not really restricted to graphing relations whose domain and range are subsets of $\mathbb{R}$. Given a set $\{a, b, c, d\}$ with four members, we can still graph the relation

$$S = \{(a, a), (b, c), (a, c), (d, c), (b, b), (a, b), (d, a)\}$$

Indeed, since we can list the members of the domain of $S$ in any order, there are 24 different ways of drawing a Cartesian graph associated with $S$. We have indicated two of these graphs in the following figures. The term *Cartesian graph* is used to honor Descartes, who along with Pierre Fermat (1601–1665) invented analytic geometry. All the graphs of inequalities that we recall sketching in analytic geometry were Cartesian graphs.

|  |  |  |  |  |
|---|---|---|---|---|
| $d$ | · | · | · | · |
| $c$ | ⊙ | ⊙ | · | ⊙ |
| $b$ | ⊙ | ⊙ | · | · |
| $a$ | ⊙ | · | · | ⊙ |
|  | $a$ | $b$ | $c$ | $d$ |

|  |  |  |  |  |
|---|---|---|---|---|
| $a$ | ⊙ | · | · | ⊙ |
| $b$ | · | · | ⊙ | ⊙ |
| $c$ | ⊙ | · | ⊙ | ⊙ |
| $d$ | · | · | · | · |
|  | $d$ | $c$ | $b$ | $a$ |

Often we are interested in obtaining some information about relations that may have infinite domains. For example, suppose that before setting out to prove Proposition 4.2, which asserts that for any two sets $A$ and $B$,

$$(A - B) \times B = (A \times B) - (B \times B)$$

we want to sketch a typical Cartesian graph to see whether the alleged proposition seems plausible. We could consider the special case that $A$ is the closed interval $[1, 4]$ and $B$ is the closed interval $[3, 5]$ (Figure 4.3). In this special case, we would have the shaded areas representing $(A - B) \times B$ and $(A \times B) - (B \times B)$, respectively. These graphs prove nothing, but as in Venn diagrams they suggest that the proposition in question is reasonable and even give some hint how the element-chasing proof will proceed.

There is a second, and in some ways more natural, way to graph a relation. Suppose that we are given a relation $R$ on a set $\{a, b, c, d\}$ with four elements. We first choose four points to represent $a$, $b$, $c$, and $d$, and for

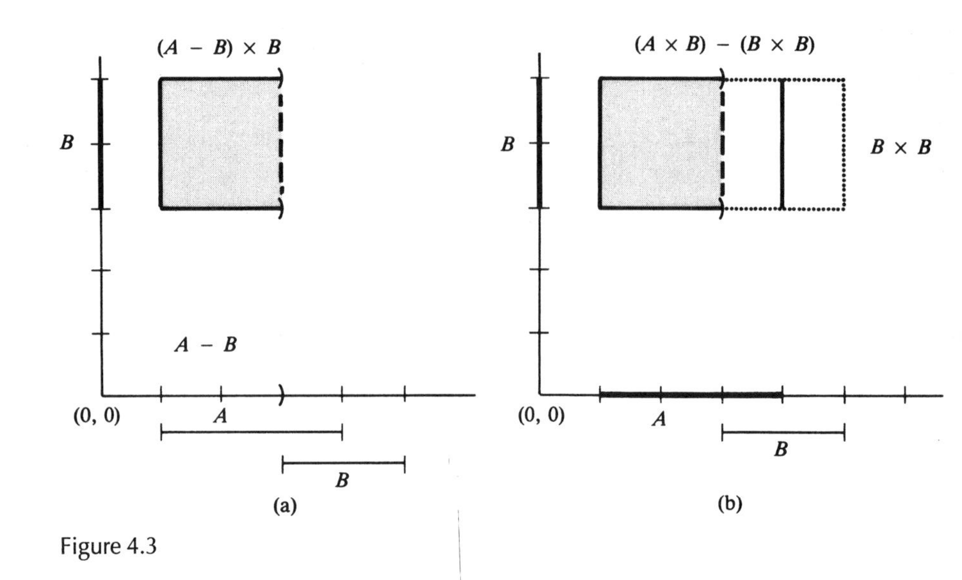


Figure 4.3

each $(x, y) \in R$ we draw an arrow starting at $x$ and pointing to $y$. If $(x, x)$ belongs to $R$, we simply draw a loop at the point $x$. The representation of a relation in this manner is called a **directed graph** (or **digraph** for short). A directed graph of the relation on $\{a, b, c, d\}$ that we considered previously is shown in Figure 4.4.

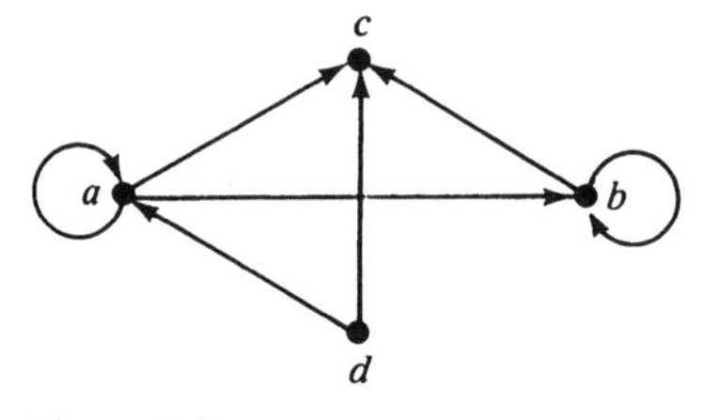


Figure 4.4

In a directed graph, the points are called **vertices** and the arrows are called **directed edges**. Clearly if $R$ is a relation on a nonempty set $S$, then there is an associated directed graph that represents the relation $R$. In the associated directed graph, $S$ is the *set of vertices* of the graph, and if $x \in S$ we draw a directed edge from $x$ to each member of $R[x]$. Of course, if $S$ is infinite or even a large finite set, we may not actually be able to draw the directed graph associated with $R$, but we could still define it to be $(S, E)$, where $S$ is the set of vertices and $E$ is the set of directed edges (to which the directed edge $\overrightarrow{xy}$ belongs exactly when $(x, y) \in R$).

We have been emphasizing that directed graphs come from relations, but surely we might be led to draw a directed graph without ever having a relation in mind. For example, in Exercise 72 in Section 3.3, we are asked to find a tournament in which there is more than one Dean. It seems natural to consider a directed graph in which the vertices represent the players in the tournament and each directed edge from one vertex to another represents a victory in the match between the represented players. Thus, the directed graph in Figure 4.5 indicates among other things that player $b$ beat player $d$ but lost to player $c$. Recall that a Dean is defined to be a player who, for each other player $A$, either beats $A$ or beats a player who beats $A$. Does the directed graph in Figure 4.5 indicate a tournament in which there is more than one Dean?

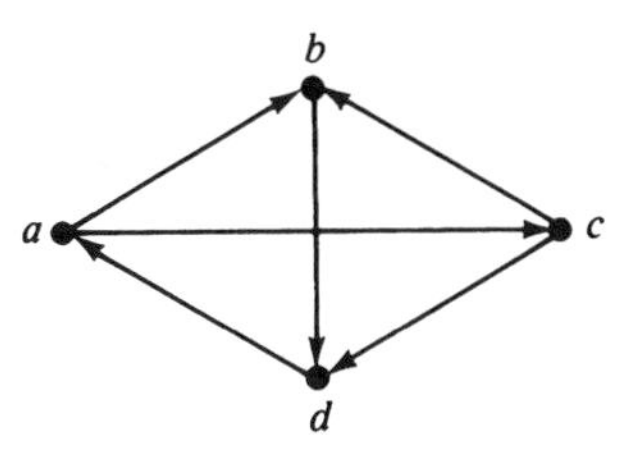


Figure 4.5

Associated with the directed graph in Figure 4.5 is a relation $R$ on the set $S = \{a, b, c, d\}$. The relation $R$ is

$$\{(a, b), (a, c), (c, b), (c, d), (d, a), (b, d)\}$$

In general, whenever we are given a directed graph, we can associate a relation $R$ on the set of vertices of the graph by agreeing that $(x, y) \in R$ exactly when the directed edge $\overrightarrow{xy}$ belongs to the graph. We have the following situation. Every time we are given a relation, we can associate a directed graph with the given relation, and every time we are given a directed graph, we can associate a relation with the given graph. Moreover, if we start with a relation, draw the associated graph, and then find the relation associated with the graph we have just drawn, we get back our original relation. In the same way, if we start with a graph, obtain its associated relation, and then draw the associated graph, we will wind up drawing the graph we started with. The upshot is that the study of relations and the study of directed graphs are coextensive. Any result obtained in the one area can be translated to a corresponding result in the other area. Roughly speaking, the setting of relations lends itself to algebraic intuition whereas the setting of directed graphs lends itself to geometric intuition.

We illustrate the difference in point of view with a simple property of relations.

**Definition**

Let $R$ be a relation. Then $R^{-1}$ (called ***R*-inverse**) is the set of ordered pairs to which $(x, y)$ belongs if and only if $(y, x) \in R$. A relation $R$ is **symmetric** provided that $R = R^{-1}$.

**Proposition 4.3**

> Let $R$ and $S$ be relations. Then each of the following statements is true.
>
> **a)** $R = (R^{-1})^{-1}$
>
> **b)** $R \subseteq S$ if and only if $R^{-1} \subseteq S^{-1}$.
>
> **c)** $(R \cup S)^{-1} = R^{-1} \cup S^{-1}$
>
> **d)** $(R \cap S)^{-1} = R^{-1} \cap S^{-1}$
>
> **e)** $R \cup R^{-1}$ is symmetric.
>
> **f)** $R \cap R^{-1}$ is symmetric.

**PROOF** We prove (c) and (f) and leave the remaining parts as exercises.

**c)** Let $(x, y) \in (R \cup S)^{-1}$. Then $(y, x) \in R \cup S$. If $(y, x) \in R$, then $(x, y) \in R^{-1}$, and if $(y, x) \in S$, then $(x, y) \in S^{-1}$. In either case, $(x, y) \in R^{-1} \cup S^{-1}$. Thus, $(R \cup S)^{-1} \subseteq R^{-1} \cup S^{-1}$. These steps can be reversed to show that $(R^{-1} \cup S^{-1}) \subseteq (R \cup S)^{-1}$. Alternatively, we observe by part (b) that $R^{-1} \subseteq (R \cup S)^{-1}$ and that $S^{-1} \subseteq (R \cup S)^{-1}$. Thus, $R^{-1} \cup S^{-1} \subseteq (R \cup S)^{-1}$.

**f)** We have by part (d) that $(R \cap R^{-1})^{-1} = R^{-1} \cap (R^{-1})^{-1}$ and by part (a) that $R^{-1} \cap (R^{-1})^{-1} = R^{-1} \cap R = R \cap R^{-1}$. Since $(R \cap R^{-1})^{-1} = R \cap R^{-1}$, $R \cap R^{-1}$ is symmetric. □

Now let us consider Proposition 4.3 in the setting of directed graphs. Suppose that the associated graph of a relation $R$ is given in Figure 4.6(a). We might imagine the associated graph as the street pattern of a town with a severe traffic problem. There is a traffic circle at $b$ and a two-way street between $e$ and $d$. Also, $R^{-1}$ (Figure 4.6b) is the same street system in which the direction of traffic has been reversed. In $R \cup R^{-1}$ (Figure 4.6c), every existing street is made into a two-way street. (In some sense, the traffic circle already is a two-way street.) In $R \cap R^{-1}$ (Figure 4.6d), the town has taken the drastic step of tearing out all one-way streets except the traffic circle at $b$.

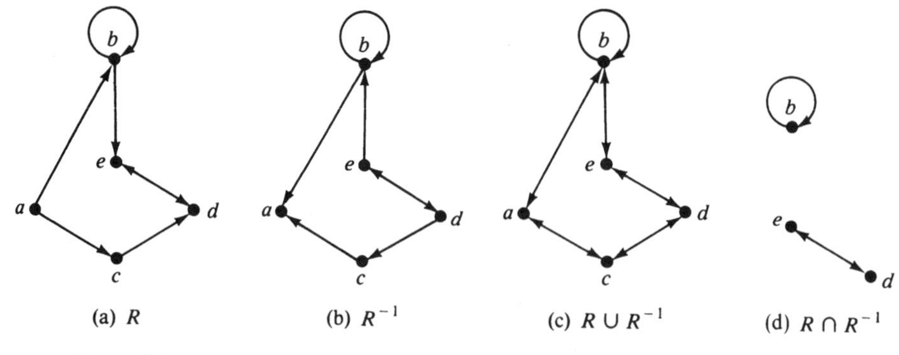


Figure 4.6

In the setting of directed graphs, Proposition 4.3 seems painfully obvious. Part (e) of the proposition, whose proof was somewhat involved, seems to say merely that in a traffic pattern in which only two-way streets are used, all streets are two-way streets. Of course, our examples do not constitute a proof of Proposition 4.3.

We now reconsider the relation $R$, this time drawing its Cartesian graph.

$R$

| | | | | | | ↙ |
|---|---|---|---|---|---|---|
| $e$ | · | ⊙ | · | ⊙ | · | |
| $d$ | · | · | ⊙ | · | ⊙ | |
| $c$ | ⊙ | · | · | · | · | |
| $b$ | ⊙ | ⊙ | · | · | · | |
| $a$ | · | · | · | · | · | |
| ↗ | $a$ | $b$ | $c$ | $d$ | $e$ | |

$R^{-1}$

| | | | | | | ↙ |
|---|---|---|---|---|---|---|
| | · | · | · | ⊙ | · | |
| | · | · | · | · | ⊙ | |
| | · | · | · | ⊙ | · | |
| | · | ⊙ | · | · | ⊙ | |
| | · | ⊙ | ⊙ | · | · | |
| ↗ | | | | | | |

$R \cup R^{-1}$

| | | | | | | ↙ |
|---|---|---|---|---|---|---|
| | · | ⊙ | · | ⊙ | · | |
| | · | · | ⊙ | · | ⊙ | |
| | ⊙ | · | · | ⊙ | · | |
| | ⊙ | ⊙ | · | · | ⊙ | |
| | · | ⊙ | ⊙ | · | · | |
| ↗ | | | | | | |

$R \cap R^{-1}$

| | | | | | | ↙ |
|---|---|---|---|---|---|---|
| | · | · | · | ⊙ | · | |
| | · | · | · | · | ⊙ | |
| | · | · | · | · | · | |
| | · | ⊙ | · | · | · | |
| | · | · | · | · | · | |
| ↗ | | | | | | |

Given the Cartesian graph of a relation, we have a simple test for symmetry. The relation is symmetric if and only if the graph is symmetric about the diagonal indicated by the arrows. Intuitively, we obtain the Cartesian graph

Figure 4.7

of $R^{-1}$ by spinning the graph of $R$ about this line. Our phrase "spinning the graph" is admittedly flippant, because it is an instruction that is impossible to carry out. But the intuition is correct, and there is a bit of mathematical chicanery based on this intuition. Suppose we are given some complicated relation $R$ in the plane (Figure 4.7) and asked to draw $R^{-1}$. The trick is to draw $R$ in the lower right-hand corner of a page and then fold over the lower right-hand corner parallel to the line $y = x$. The graph that you see through the back of the page is $R^{-1}$.

## EXERCISES 4.2

**17.** Which of the following relations are symmetric?

**a)** $A \cup B$, where $A = \{(x, y) \in \mathbb{R} \times \mathbb{R} : x = 0\}$ and $B = \{(x, y) \in \mathbb{R} \times \mathbb{R} : y = 0\}$

**b)**

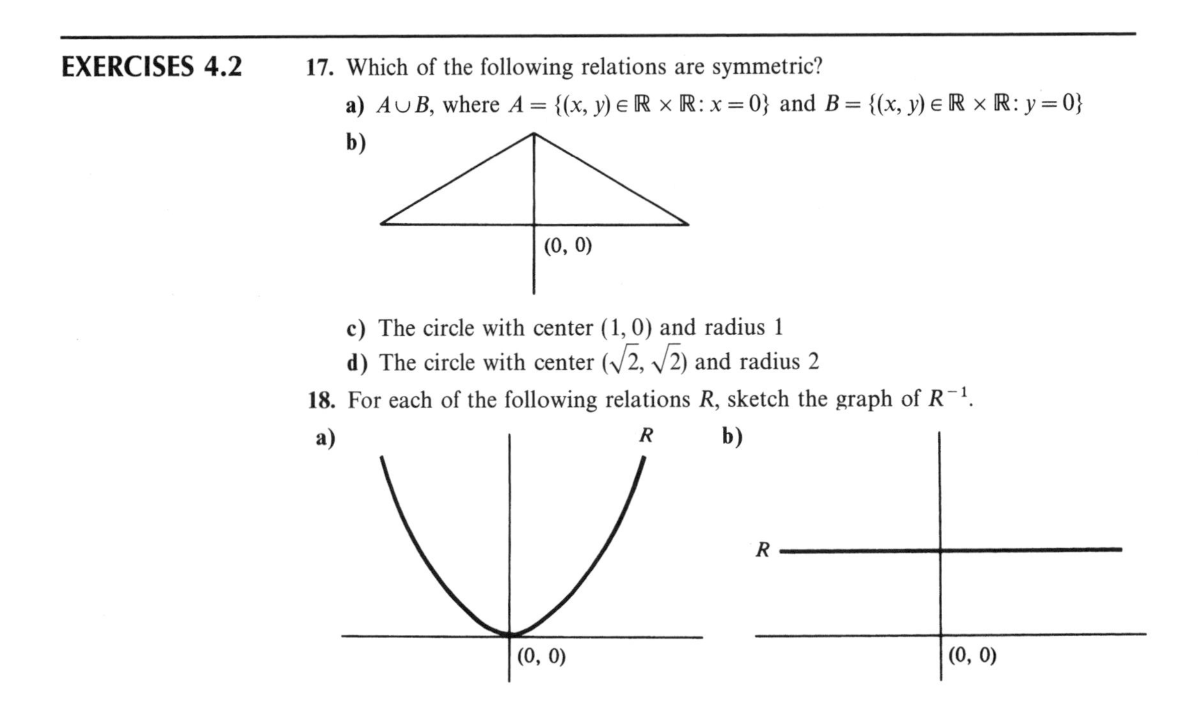


**c)** The circle with center $(1, 0)$ and radius 1

**d)** The circle with center $(\sqrt{2}, \sqrt{2})$ and radius 2

**18.** For each of the following relations $R$, sketch the graph of $R^{-1}$.

**a)** **b)**

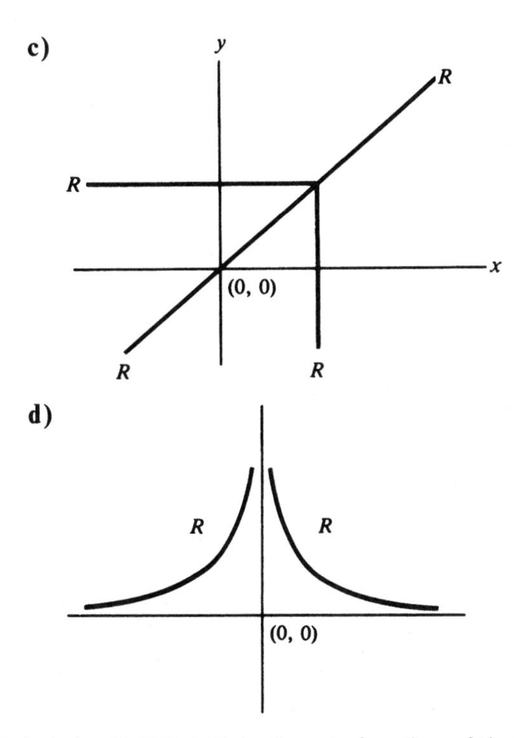


**19.** Let $A = \{1, 2, 3, 4, 5\}$ be the set of vertices of the accompanying directed graph. Draw the Cartesian graph of the relation on $A$ associated with this directed graph.

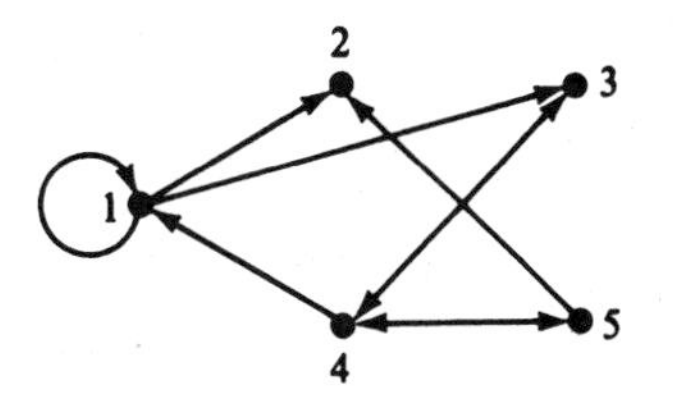


**20.** Let $A = \{1, 2, 3, 4, 5\}$, as in Exercise 19. Draw the directed graphs associated with each of the following relations on $A$.

**a)** $R = \{(a, b) \in A \times A: a \text{ divides } b\}$ **b)** $U = \{(a, b) \in A \times A: a \neq b\}$

**c)** $E = \{(a, b) \in A \times A: a + b \text{ is even}\}$ **d)** $O = \{(a, b) \in A \times A: a + b \text{ is odd}\}$

**21.** Draw the Cartesian graphs of the relations $R$, $U$, $E$, and $O$ of Exercise 20. Which of these relations is symmetric?

**22.** Does there exist a tournament among four players in which all of the following are true?

**a)** Each player plays every *other* player exactly once.

**b)** There are no ties.

**c)** Every player is a Dean. (See Exercise 72, Chapter 3.)

If so, draw a directed graph of such a tournament.

**23.** Consider Exercise 22 in the case that the tournament has exactly five players.

**24.** **a)** Prove Proposition 4.3(a). **b)** Prove Proposition 4.3(b).
**c)** Prove Proposition 4.3(d). **d)** Prove Proposition 4.3(e).

**25.** Let $R$ be a relation such that $R^{-1} \subseteq R$. Must $R$ be symmetric? Prove your answer.

## 4.3 *Equivalence Relations*

Often in mathematics, as in real life, it is convenient to think of two things that are different as being essentially the same. Consider the triangles $\triangle ABC$ and $\triangle DEF$ in Figure 4.8. Certainly as sets these two triangles are not equal; indeed, they are disjoint. But they are congruent triangles, from which it follows that they share geometric properties such as area, length of hypotenuse, and measures of corresponding angles. It is possible to think of properties that the two triangles do not share. For example, the triangle on the right cannot be fitted exactly onto $\triangle ABC$ without lifting it out of the plane and flipping it over. Still, to anyone interested in plane geometry these two triangles, being congruent, have no essential differences. The definition of congruence is somewhat involved, but we believe you have a notion of congruence and will agree that congruence shares with equality the following properties:

1. If $A$ is a subset of the plane, then $A$ is congruent to $A$ (reflexivity).
2. If $A$ and $B$ are subsets of the plane and $A$ is congruent to $B$, then $B$ is congruent to $A$ (symmetry).
3. If $A$, $B$, and $C$ are subsets of the plane, $A$ is congruent to $B$, and $B$ is congruent to $C$, then $A$ is congruent to $C$ (transitivity).

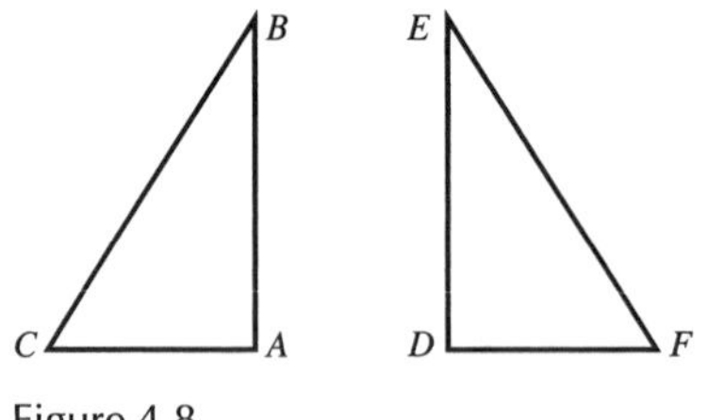

Figure 4.8

Turning from geometry to trigonometry, we might agree that two real numbers $a$ and $b$ are essentially the same provided that there is some integer $k$ such that $a - b = 2k\pi$. Let us write $a \simeq b$ provided this condition holds. We again have the following properties:

**1′.** For any real number $a$, $a \simeq a$.

**2′.** If $a \simeq b$, then $b \simeq a$.

**3′.** If $a \simeq b$ and $b \simeq c$, then $a \simeq c$.

(See Exercise 26.)

Moreover, if $a \simeq b$, then there is an integer $k$ such that $a - b = 2k\pi$; thus,

$$\sin(a) = \sin(b + k\pi) = \sin(b) \quad \text{and} \quad \cos(a) = \cos(b + k\pi) = \cos(b)$$

Therefore, the trigonometric functions cannot distinguish between two numbers $a$ and $b$ such that $a \simeq b$.

Finally, let us consider the set of all rational numbers and agree that two rational numbers are related provided that they can be represented by fractions written in lowest terms that have the same denominator. For example, 14/12 and 25/30 are related because each can be represented by a fraction in lowest terms that has a 6 in its denominator. In Exercise 27, you are asked to verify that

1. Each rational number is related to itself.
2. If $p$ and $q$ are rational numbers and $p$ is related to $q$, then $q$ is related to $p$.
3. If $p$, $q$, and $r$ are rational numbers, $p$ is related to $q$, and $q$ is related to $r$, then $p$ is related to $r$.

Each of the preceding examples can be thought of as a relation (with special properties) that is defined on a set. In the first example, the set is the set of all subsets of the plane and the relation is

$$\{(A, B) \in \mathscr{P}(\mathbb{R} \times \mathbb{R}) \times \mathscr{P}(\mathbb{R} \times \mathbb{R}) \colon A \text{ is congruent to } B\}$$

In the second example, the set is the set $\mathbb{R}$ and the relation is

$$\{(x, y) \in \mathbb{R} \times \mathbb{R} \colon \text{there is an even integer } k \text{ such that } x - y = k\pi\}$$

**Definition**

Let $S$ be a nonempty set and let $R$ be a relation.

**a)** The relation $R$ is **reflexive on** $S$ provided that for each $x \in S$, $(x, x) \in R$.

**b)** The relation $R$ is **symmetric** provided that $(y, x) \in R$ whenever $(x, y) \in R$.

**c)** The relation $R$ is **transitive** provided that whenever $(x, y) \in R$ and $(y, z) \in R$, $(x, z) \in R$.

Let $R$ be a relation on a set $S$. If the relation $R$ is a reflexive relation on $S$ that is symmetric and transitive, then $R$ is said to be an **equivalence relation** on $S$. Note that symmetry and transitivity are internal properties of a relation in the sense that we can determine whether a relation is symmetric or transitive without considering anything other than the given relation. But the relation $R = \{(x, y) \in \mathbb{R} \times \mathbb{R}: xy > 0\}$, which is reflexive on $\mathbb{R} - \{0\}$, fails to be reflexive on $\mathbb{R}$. Thus, for reflexivity we must consider both the relation $R$ and a set $S$. Since we have already encountered symmetry in the preceding section, we note now only that we have given a second definition of symmetry here, which agrees with the definition of symmetry given previously (see Exercise 41). As for transitivity, the definition is somewhat subtler than it looks. It is of the form $P \to Q$, and so a relation $R$ is automatically transitive if there do not exist $x$, $y$, and $z$ such that $(x, y) \in R$ and $(y, z) \in R$.

Note that a relation $R$ on a set $S$ is an equivalence relation on $S$ exactly when the following hold:

1. For each $x \in S$, $x \in [x]$.
2. If $x \in [y]$, then $y \in [x]$.
3. If $y \in [x]$ and $z \in [y]$, then $z \in [x]$.

The set $[x]$ is called the **equivalence class** of $x$ (but the term *box-x* is shorter and suggests the right way to think about $[x]$; it is a set after all, a box filled with $x$ and all its relatives).

We end this section by considering a few simple examples of relations. When these relations are equivalence relations, we consider their corresponding equivalence classes.

**EXAMPLE 4** Let $S = \{1, 2, 3, 4\}$.

**a)** The empty set $\varnothing$ is a symmetric and transitive relation that is not reflexive *on* any nonempty set.

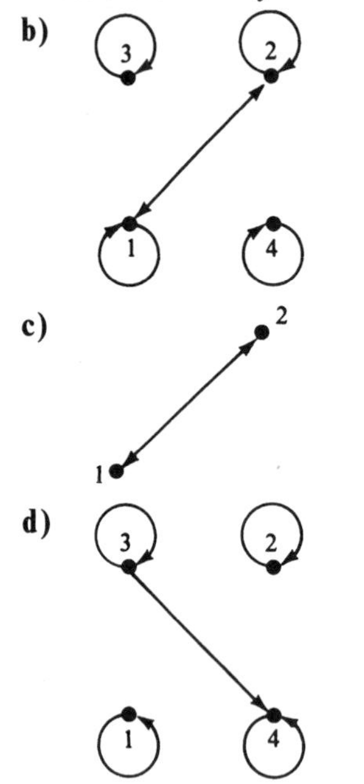


**b)** The relation indicated by the directed graph at the left is an equivalence relation on $S$. The three equivalence classes are $[1] = [2] = \{1, 2\}$, $[3] = \{3\}$, and $[4] = \{4\}$.

**c)** The relation indicated by the directed graph at the left is symmetric but it is *not* transitive.

**d)** The relation indicated by the directed graph at the left is transitive and is reflexive on $S$, but is not symmetric.

**e)**

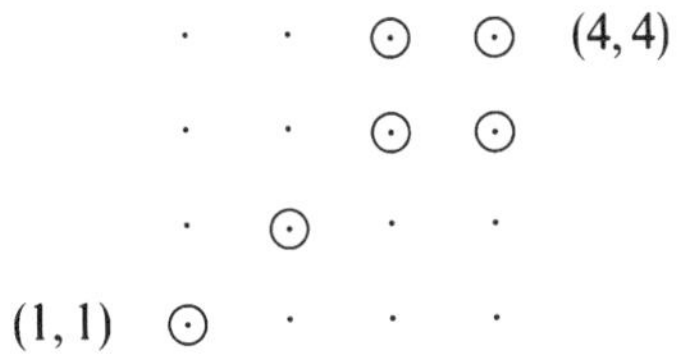

```
      ·  ·  ⊙  ⊙  (4,4)
      ·  ·  ⊙  ⊙
      ·  ⊙  ·  ·
(1,1) ⊙  ·  ·  ·
```

The Cartesian graph at the left indicates an equivalence relation on $S$. The three equivalence classes are $[1] = \{1\}$, $[2] = \{2\}$, and $[3] = [4] = \{3, 4\}$.

**f)** $S \times S$ is an equivalence relation on $S$. The only equivalence class is $\{1, 2, 3, 4\}$.

**g)** $\{(1, 1), (2, 2), (3, 3), (4, 4)\}$ is an equivalence relation on $S$. The four equivalence classes are $[1] = \{1\}$, $[2] = \{2\}$, $[3] = \{3\}$, and $[4] = \{4\}$. ❑

**EXAMPLE 5** Let $S = \{1, 2, 3, 4, 5\}$ and let

$$R = \{(1, 1), (2, 2), (3, 3), (4, 4), (5, 5), (2, 1), (1, 2), (1, 4), (4, 1), (2, 4), (4, 2)\}$$

(See the following Cartesian graph.)

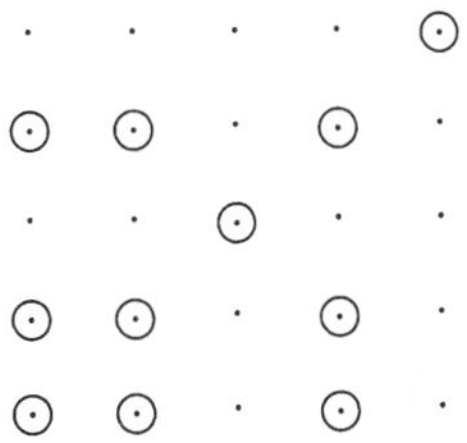

```
·  ·  ·  ·  ⊙
⊙  ⊙  ·  ⊙  ·
·  ·  ⊙  ·  ·
⊙  ⊙  ·  ⊙  ·
⊙  ⊙  ·  ⊙  ·
```

It is clear from the graph that $R$ is reflexive on $S$ and symmetric and it is easily verified that $R$ is transitive. Therefore, $R$ is an equivalence relation on $S$. There are three equivalence classes: $[1] = [2] = [4] = \{1, 2, 4\}$, $[3] = \{3\}$, and $[5] = \{5\}$. Note that every member of $S$ belongs to some equivalence class and that no member of $S$ belongs to two equivalence classes. ❑

**EXAMPLE 6** Let $R = \{(x, y) \in \mathbb{R} \times \mathbb{R}: |x| + |y| = 1\}$. Then $R$ is a symmetric relation that is not reflexive on the closed interval $[-1, 1]$. Furthermore, this relation is not transitive because $(1, 0) \in R$ and $(0, 1) \in R$ but $(1, 1) \notin R$. ❑

**EXAMPLE 7** Let $S = \{1, 2, 3, 4\}$, and for $A, B \in \mathscr{P}(S)$, define $A \sim B$ provided that $A$ and $B$ have the same number of members. Evidently for each subset $A$ of $S$, $A \sim A$. Moreover, if $A \sim B$, then $B \sim A$, and if $A \sim B$ and $B \sim C$, then $A \sim C$. Thus $\sim$ defines an equivalence relation on $\mathscr{P}(S)$. It would be both unnecessary and tiresome to list the ordered pairs belonging to $\sim$. There are five equivalence classes:

$$\{\varnothing\}$$

$$\{\{1\}, \{2\}, \{3\}, \{4\}\}$$

$$\{\{1, 2\}, \{1, 3\}, \{1, 4\}, \{2, 3\}, \{2, 4\}, \{3, 4\}\}$$

$$\{\{1, 2, 3\}, \{2, 3, 4\}, \{1, 2, 4\}, \{1, 3, 4\}\}$$

and

$$\{\{1, 2, 3, 4\}\}$$ ❑

## EXERCISES 4.3

**26.** Let $R = \{(a, b) \in \mathbb{R} \times \mathbb{R}$: there is an integer $k$ such that $a - b = 2k\pi\}$.

**a)** Prove that $R$ is an equivalence relation on $\mathbb{R}$.

**b)** List three members of $[\pi/4]$.

**c)** List three members of $[1]$.

**d)** Which numbers, if any, belong to $[\pi/4] \cap [1]$?

**27.** Let $\mathbb{Q}$ be the set of all rational numbers and let

$R = \{(x, y) \in \mathbb{Q} \times \mathbb{Q}$: when $x$ and $y$ are represented by fractions in lowest terms these fractions have the same denominator$\}$

**a)** Prove that $R$ is an equivalence relation on $\mathbb{Q}$.

**b)** Prove that $[1/6] = [5/6]$.

**c)** Are $[4/6]$ and $[5/6]$ disjoint sets? Prove your answer.

**28.** Let $R = \{(x, y) \in \mathbb{R} \times \mathbb{R}: x^2 - y^2 = 0\}$.

**a)** Prove that $R$ is an equivalence relation on $\mathbb{R}$.

**b)** List all the members of $[3]$.

**29.** Let $R = \{(3, 5)\}$. Is $R$ symmetric? Is $R$ transitive?

**30.** Let $A = \{1, 2, 3, 4\}$.

**a)** Either find a relation that is reflexive on $A$ but neither symmetric nor transitive or prove that no such relation exists.

**b)** Either find a relation on $A$ that is symmetric but neither transitive nor reflexive on $A$ or prove that no such relation exists.

**c)** Either find a relation on $A$ that is transitive but neither symmetric nor reflexive on $A$ or prove that no such relation exists.

**d)** Either find a relation that is reflexive on $A$ and symmetric but not transitive or prove that no such relation exists.

**e)** Either find a relation that is reflexive on $A$ and transitive but not symmetric or prove that no such relation exists.

**f)** Either find a relation on $A$ that is symmetric and transitive but not reflexive on $A$ or prove that no such relation exists.

**31.** Let $R = \{(a, b) \in \mathbb{N} \times \mathbb{N}: a \text{ divides } b\}$. Is $R$ reflexive on $\mathbb{N}$? symmetric? transitive? Prove your answers.

**32.** Let $X$ be a nonempty set and let $A$ be a nonempty proper subset of $X$. Set $R = X \times X - (A \times A)$. Is $R$ reflexive on $X$? symmetric? transitive? Prove your answers.

**33.** Let $S = \{1, 2, 3\}$. Draw a directed graph that represents a nonempty relation on $S$ that is each of the following.

**a)** Reflexive on $S$ and symmetric, but not transitive

**b)** Reflexive on $S$ and transitive, but not symmetric

**c)** Symmetric and transitive but not reflexive on $S$

**d)** An equivalence relation on $S$

**e)** Not reflexive on $S$, not symmetric, and not transitive

**34.** Let $\mathbb{R}$ be the set of all real numbers and consider the following subsets of $\mathbb{R} \times \mathbb{R}$:

**a)** $R_1 = \{(x, y) \in \mathbb{R} \times \mathbb{R}: xy = 0\}$

**b)** $R_2 = \{(x, y) \in \mathbb{R} \times \mathbb{R}: |x - y| < 5\}$

**c)** $R_3 = \{(x, y) \in \mathbb{R} \times \mathbb{R}: xy \neq 0\}$

**d)** $R_4 = \{(x, y) \in \mathbb{R} \times \mathbb{R}: x \geqslant y\}$

**e)** $R_5 = \{(x, y) \in \mathbb{R} \times \mathbb{R}: x^2 + y^2 = 1\}$

Which of the above relations is

**i)** Reflexive on $\mathbb{R}$? **ii)** Symmetric? **iii)** Transitive?

**35.** For each real number $x$, let $f(x) = x^2$. For any two real numbers $a$ and $b$, define $a \simeq b$ provided that $f(a) = f(b)$. Prove that $\simeq$ is an equivalence relation on $\mathbb{R}$ and list all the members of $[-7]$.

**36.** For any two points $(a, b)$ and $(c, d)$ of the plane, define $(a, b) \simeq (c, d)$ provided that $a^2 + b^2 = c^2 + d^2$.

**a)** Prove that $\simeq$ is an equivalence relation on $\mathbb{R} \times \mathbb{R}$.

**b)** List all the members of $[(0, 0)]$.

**c)** Give a geometric description of $[(5, 11)]$.

**37.** Set $S = \{(x, y) \in \mathbb{R} \times \mathbb{R}: y - x \text{ is an integer}\}$.

**a)** Prove that $S$ is an equivalence relation on $\mathbb{R}$.

**b)** List four members of $[\pi]$.

**c)** Which real numbers belong to $[-17]$?

**38.** Let $\mathscr{A} = \{\{1, 2\}, \{3, 4\}, \{5, 6, 7\}, \{8\}\}$.

**a)** List the members of $\cup\mathscr{A}$.

**b)** Set $S = \{(x, y) \in (\cup\mathscr{A}) \times (\cup\mathscr{A}): x \text{ and } y \text{ belong to the same member of } \mathscr{A}\}$. Prove that $S$ is an equivalence relation on $\cup\mathscr{A}$.

**c)** List all the members of $[5]$.

**d)** List all the equivalence classes.

**39.** Set $S = \mathbb{N} \times \mathbb{N}$, and for any two members $(a, b), (c, d)$ of $S$, define $(a, b) \simeq (c, d)$ provided that $ad = bc$. Prove that $\simeq$ is an equivalence relation on $S$ and list four members of $[(6, 8)]$.

**40.** Let $S$ be the collection of all differentiable functions with domain $\mathbb{R}$. Define $f \simeq g$ provided that there is a nonzero number $k$ such that $f'(x) = kg'(x)$ for all $x$ in $\mathbb{R}$.

**a)** Prove that $\simeq$ is an equivalence relation on $S$.

**b)** List four members of $[f]$, where $f(x) = x^2 + 17x + 11$.

**41.** Prove that the definitions of symmetry given on pages 111 and 117 coincide.

**42.** Let $R$ and $S$ be equivalence relations on a set $X$.

**a)** Prove that $R \cap S$ is an equivalence relation on $X$.

**b)** Prove that for each $x \in X$, $(R \cap S)[x] = R[x] \cap S[x]$.

**43.** Let $R$ and $S$ be equivalence relations on a set $X$.

**a)** Is $R \cup S$ reflexive on $X$? Justify your answer.

**b)** Is $R \cup S$ symmetric? Justify your answer.

**c)** Is $R \cup S$ transitive? Justify your answer.

**44.** Let $S = \{(x, y) \in \mathbb{R} \times \mathbb{R}: \sin^2 x + \cos^2 y = 1\}$. Is $S$ reflexive on $\mathbb{R}$? Is $S$ symmetric? Is $S$ transitive? In each case, justify your answer.

The following three problems assume a knowledge of linear algebra. Throughout the problems, $S$ denotes the set of all $n \times n$ matrices with real coefficients for some $n > 1$ and $T$ denotes the subset of all invertible $n \times n$ matrices.

**45.** Define a relation $\sim$ on $S$ by $A \sim B$ provided there is a matrix $M \in T$ such that $A = MBM^{-1}$. Prove that $\sim$ is an equivalence relation on $S$.

**46.** Argue that the relation $\sim$ on $S$ defined by $A \sim B$ provided $A$ is row-equivalent to $B$ is an equivalence relation on $S$.

**47.** Let $\approx$ be an equivalence relation on $\mathbb{R}$. Argue that the relation $\sim$ defined on $S$ by $A \sim B$ provided $|A| \approx |B|$ is an equivalence relation on $S$.

## 4.4 Partitions and Identifications

An equivalence relation $R$ on a given set $S$ determines a collection of equivalence classes that divide $S$ into nonoverlapping subsets. In this section, we show that a division of a given set into nonoverlapping subsets determines an equivalence relation, and we examine the correspondence between such divisions and equivalence relations.

**Definition**

Let $S$ be a nonempty set. A pairwise disjoint family $\mathscr{A}$ of nonempty subsets of $S$ is a **partition** of $S$ provided that $\cup\mathscr{A} = S$. In other words, a family $\mathscr{A}$ of nonempty subsets of $S$ is a partition of $S$ provided that every member of $S$ belongs to one member of $\mathscr{A}$ and no member of $S$ belongs to two members of $\mathscr{A}$.

In the drawing on the left in Figure 4.9, we are assuming that $A_1 \cup A_3 \cup A_4$ is $S$.

**EXAMPLE 8**

Let $S = \{1, 2, 3, 4, 5, 6, 7, 8\}$. Then $\mathscr{A} = \{\{1, 2\}, \{3\}, \{4, 5, 6\}, \{7, 8\}\}$ is a partition of $S$ because each member of $\mathscr{A}$ is nonempty and each member of $S$ belongs to exactly one member of $\mathscr{A}$. ❑

Partitions have a very pleasant feature. Suppose that you are promised that $\mathscr{A}$ is a partition of some set $S$ and that $A$ and $B$ are members of $\mathscr{A}$. Now suppose that you are asked to prove that $A = B$. "Oh," you say to yourself, "that means I'm going to have to show that $A \subseteq B$ and that $B \subseteq A$." But if $\mathscr{A}$ is a partition, no member of $S$ belongs to two different members of $\mathscr{A}$, so in order to show that $A = B$ all you need do is show that there is at least one $x \in S$ such that $x \in A$ and $x \in B$. That is, if $A$ and $B$ are not disjoint, then they are equal.

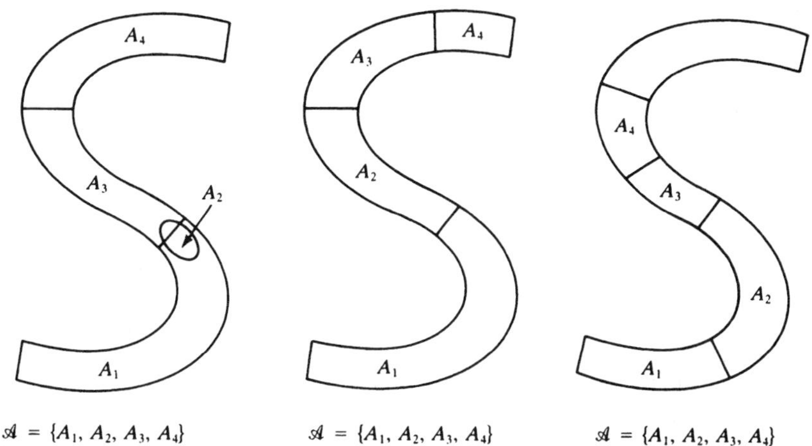
$A_4$
$A_3$
$A_2$
$A_1$
$A_4$
$A_3$
$A_2$
$A_1$
$A_4$
$A_3$
$A_2$
$A_1$
$\mathscr{A} = \{A_1, A_2, A_3, A_4\}$
is not a partition of $S$.
Some member of $S$ belongs
to two members of $\mathscr{A}$.
$\mathscr{A} = \{A_1, A_2, A_3, A_4\}$
is a partition of $S$.
$\mathscr{A} = \{A_1, A_2, A_3, A_4\}$
is not a partition of
$S$ because some
member of $S$ belongs
to no member of $\mathscr{A}$.

Figure 4.9

The remaining theorems of this section establish that partitions and equivalence relations are closely related.

**Theorem 4.4**

> Let $\simeq$ be an equivalence relation on a nonempty set $S$, and let $S/\simeq$ denote the family of all equivalence classes on $S$. Then $S/\simeq$ is a partition of $S$.

**PROOF** Let $x \in S$. Then since $\simeq$ is reflexive on $S$, $x \in [x]$. Thus, every member of $S$ belongs to a member of $S/\simeq$ and no member of $S/\simeq$ is empty. Now suppose that $p$, $x$, and $y$ are members of $S$ and that $p \in [x] \cap [y]$. It suffices to prove that $[x] = [y]$; this equality is established in the standard way. Let $q \in [x]$. Then $x \simeq q$, and because $p \in [x]$, $x \simeq p$. By symmetry we obtain $p \simeq x$, so by transitivity we obtain $p \simeq q$. Since $p \in [y]$, $y \simeq p$ and by transitivity $y \simeq q$. By definition, $q \in [y]$. Therefore, $[x] \subseteq [y]$. The proof that $[y] \subseteq [x]$, which follows similarly, is left for Exercise 48. □

**Theorem 4.5**

> Let $\mathscr{A}$ be a partition of a nonempty set $S$. If $x$ and $y$ are members of $S$, define $x \simeq y$ provided that some member of $\mathscr{A}$ contains both $x$ and $y$. Then $\simeq$ is an equivalence relation on $S$.

**PROOF** See Exercise 49. □

Taken together, Theorems 4.4 and 4.5 show that there is a natural way to retrieve a partition from an equivalence relation and that there is a natural way to retrieve an equivalence relation from a partition. It is conceivable that we could start with an equivalence relation $R$ on a set $S$, build the natural partition $\mathscr{A}(= S/R)$, and then build the equivalence relation $R'$ obtained from $\mathscr{A}$ in Theorem 4.5, and not retrieve the equivalence relation $R$ with which we started. It would be equally conceivable that we could start with a partition $\mathscr{A}$ on a set $S$, build the corresponding equivalence relation, build the partition from this equivalence relation given by Theorem 4.4, and not get back the partition with which we started. Fortunately, as the next theorem shows, both the route from relation to partition to relation and the route from partition to relation to partition are round trips. As a consequence, equivalence relations and partitions can be thought of as the same mathematical concept considered from different points of view. In this respect, the situation is comparable to the relationship between relations and directed graphs, which we have considered previously.

**Theorem 4.6**

Let $S$ be a nonempty set.

**a)** Let $R$ be an equivalence relation on $S$. For any two members $x$ and $y$ of $S$, define $x \simeq y$ provided that $x$ and $y$ belong to the same member of $S/R$. Then $x \simeq y$ if and only if $x \, R \, y$.

**b)** Let $\mathscr{A}$ be a partition of $S$. For any two members $x$ and $y$ of $S$, define $x \simeq y$ provided that $x$ and $y$ belong to the same member of $\mathscr{A}$. Then $\mathscr{A} = \{[s]: s \in S\}$.

**PROOF** See Exercises 55 and 56. □

We turn now to the idea of identification, which is of considerable use in algebraic and geometric topology. As we will see, the concept of identification is best studied from the perspective of equivalence relations and partitions. We begin, however, by considering identifications informally and so we concede that our first intuitive explanations are not mathematically sound. Imagine that you have a strip of paper such as the rectangle in Figure 4.10 labeled $ABCD$. If we were to identify edges $AB$ and $CD$ by rolling up our rectangle, the result would be a small tube.

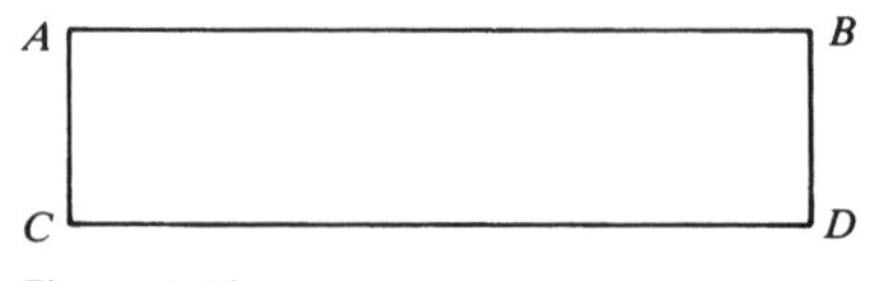


Figure 4.10

Now imagine that we were also to identify edges $AC$ and $BD$. Then what would have been a tube would now become a torus (like the outside of a donut), as shown in Figure 4.11.

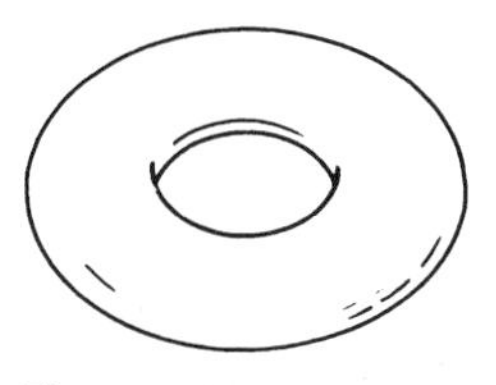

Figure 4.11

In "constructing" a torus by means of these two identifications, we have asked you to connive at several obvious flaws in our construction. First, a torus, like any other two-dimensional geometric object, ought to have no thickness, so at best we have built only a model of a torus. Second, we all know that while we can imagine constructing a torus from the strip of paper, in real life we would wind up constructing a crinkly mess. The question arises whether we do not already have a model of a torus once we have made our identifications, whether or not we actually start constructing an object with paper, scissors, and cellophane tape. For example, suppose we took our rectangle and identified $AC$ with $DB$, matching a point on the directed edge $\overrightarrow{CA}$ that is $x\%$ of the way *up* from $C$ to $A$ with the point on $\overrightarrow{BD}$ that is $x\%$ of the way *down* from $B$ to $D$ (Figure 4.12). In our mind's eye, point $A$ coincides with point $D$ and point $B$ coincides with point $C$. An imaginary trip from $A$ to $B$ (alias $C$) and back to $A$ (alias $D$) is indicated in Figure 4.12. Our trip has been along the only edge of our object, which is a circle. It turns out that this one-edged object is, in fact, quite easy to construct with a piece of paper, as you are asked to do in Exercise 53, but we stress that we have been able to determine that the object has only one edge without cutting up any paper.

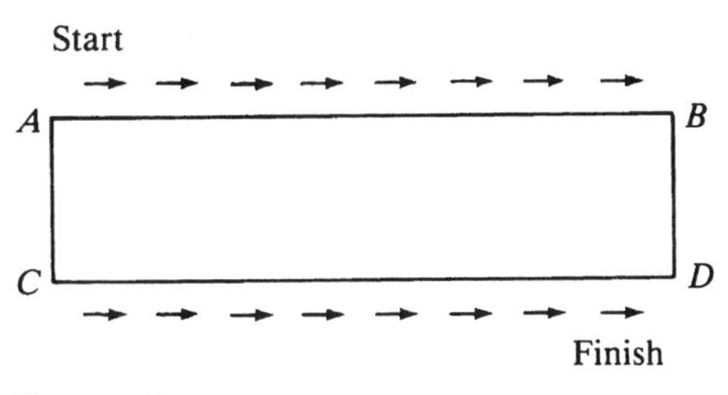


Figure 4.12

In many ways, the piece-of-paper model is a bit old-fashioned; a computer monitor makes a much better model these days. Imagine you are playing some computer game in which you are being chased or shot at by the enemy and in which you must reach certain points in order to gain safety or strength. If the inventor of the game has identified the left and right edges of the monitor in a straightforward manner, your game board is a cylinder. If the programmer identified these edges with a twist, as above, your game board is a Möbius strip, named in honor of Auguste F. Möbius (1790–1868). We are unaware of any computer game played on a Möbius strip, but you may have encountered a game called Pacman whose game board is a cylinder. If top and bottom are identified in a straightforward way and left and right in a straightforward way, the game board is a torus. Can the programmer identify all four edges? Of course. The four corners then coincide and the game board is a sphere. Still other games await. The programmer can identify top and bottom and twist left and right to obtain a Klein bottle or identify both left and right with twist and then top and bottom with twist to obtain a projective plane.

One problem remains. We have been using the term *identify*, which the dictionary defines as meaning "to make or consider the same." Alas, if two sets are different, there is nothing we can do to make them the same, and surely we want mathematics to be founded on something more substantial than talking about theorems that would hold if we considered two objects that we know to be different as if we thought they were the same. It is here that equivalence relations and partitions provide a rigorous way in which to treat two different objects as if they were indistinguishable. Let us consider the rectangle $S$ in the plane (Figure 4.13) and use an equivalence relation on $S$ to build a cylinder:

$$S = \{(x, y) \in \mathbb{R} \times \mathbb{R}: 0 \leqslant x \leqslant 2 \text{ and } 0 \leqslant y \leqslant 1\}$$

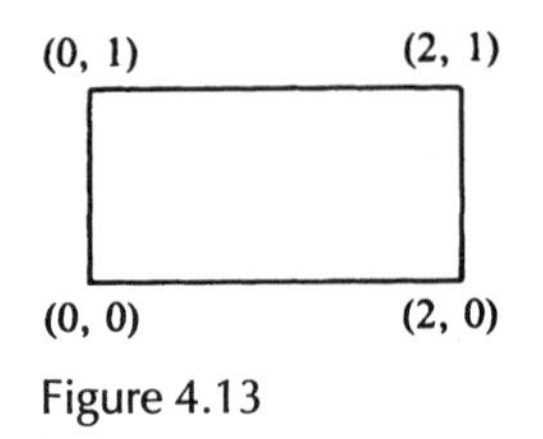

Figure 4.13

We are happy with all points of $S$ except a point on the left or right edge, which we somehow want to think of as being essentially the same as the point at the same height on the opposite edge. Very well; we define the following equivalence relation $\simeq$ on $S$. For $(a, b)$ and $(c, d)$ in $S$, we say that $(a, b) \simeq (c, d)$, provided that

1. $a = c$ and $b = d$, or
2. $a = 0$, $c = 2$, and $b = d$, or
3. $a = 2$, $c = 0$, and $b = d$.

We pause for a moment to allow you to verify that $\simeq$ is, indeed, an equivalence relation on $S$. Note that $[(1/2, 1/3)]$ is the set whose only member is $(1/2, 1/3)$ whereas $[(0, 1/3)] = \{(0, 1/3), (2, 1/3)\}$. Does $(0, 1/3) = (2, 1/3)$? No; of course not. These two ordered pairs have different first terms. Does

$[(0, 1/3)] = [(2, 1/3)]$? Yes. If we take a new set, denoted by $S/\simeq$, consisting of all the equivalence classes under $\simeq$, then some of the members of this set have only one name whereas other members have both a given name and an alias. Admittedly some care must be taken in deciding how far two "points" of $S/\simeq$ are apart. For example, $[(0, 1/2)]$ and $[(1/2, 1/2)]$ are 1/2 apart, but $[(0, 1/2)]$ and $[(2, 3/4)]$ are only 1/4 apart. In this example, it is clear how to determine how far apart any two equivalence classes $[x]$ and $[y]$ are. What is the distance between $[(7/4, 3/4)]$ and $[(0, 3/4)]$? First, observe that $[(7/4, 3/4)] = \{(7/4, 3/4)\}$, whereas $[(0, 3/4)] = \{(0, 3/4), (2, 3/4)\}$. Since the distance between $(7/4, 3/4)$ and $(0, 3/4)$ is 7/4, and the distance between $(7/4, 3/4)$ and $(2, 3/4)$ is 1/4, the distance between $[(7/4, 3/4)]$ and $[(0, 3/4)]$ is 1/4. That is, we must check the distance between each pair of names of $[x]$ and $[y]$ and select the smallest number obtained in this way as the distance between $[x]$ and $[y]$.

## EXERCISES 4.4

**48.** Complete the proof of Theorem 4.4.

**49.** Prove Theorem 4.5.

**50.** Let $S = \{1, 2, 3, 4, 5\}$ and let $\mathscr{A} = \{\{1, 2\}, \{3\}, \{4, 5\}\}$. List all the ordered pairs of the equivalence relation promised by Theorem 4.5.

**51.** The accompanying digraph represents an equivalence relation on $S = \{1, 2, 3, 4, 5, 6\}$. List the equivalence classes of the indicated relation.

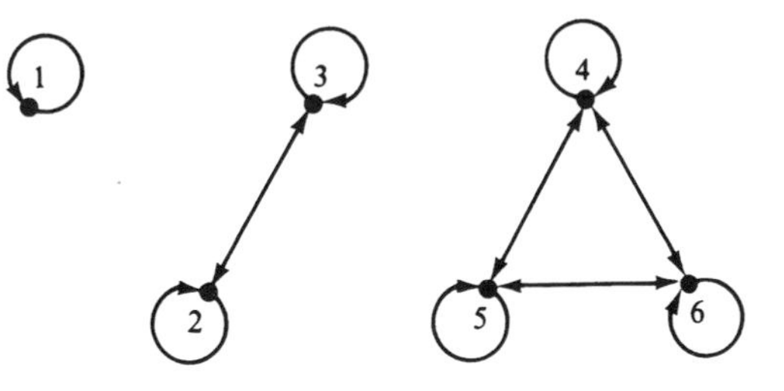


**52.** In the text, we defined an equivalence relation on

$$S = \{(x, y) \in \mathbb{R} \times \mathbb{R}: 0 \leqslant x \leqslant 2, 0 \leqslant y \leqslant 1\}$$

such that the equivalence classes formed a cylinder. Define an equivalence relation on this same set so as to form a Möbius strip.

**53.** Take a thin strip of paper and make a Möbius strip. Count the number of edges and sides. Cut the strip all the way down the middle as if you were gutting a fish. Is the resulting strip a Möbius strip?

**54.** **a)** Let $X = \mathbb{R} \times \mathbb{R}$. For each $(a, b) \in X$, let

$$D_{(a,b)} = \{(x, y) \in X: x = a\} \cup \{(x, y) \in X: x^2 + y^2 = a^2 + b^2\}$$

Is $\{D_{(a,b)}: (a, b) \in X\}$ a partition of $X$? Prove your answer.

**b)** Define a relation $R$ on $X$ by $(x, y)\, R\, (u, v)$ if and only if there exists a member $D_{(a,b)}$ of $\{D_{a,b)}: (a, b) \in X\}$ such that $(x, y)$ and $(u, v)$ both belong to $D_{(a,b)}$. Is $R$ reflexive on $X$? symmetric? transitive? In each case, prove your answer.

c) Suppose that $S$ is a set and $\mathscr{A}$ is a nonempty collection of nonempty subsets of $S$. Is it true that $\mathscr{A}$ is a partition of $S$ if and only if the relation $\simeq$ defined in Theorem 4.5 is an equivalence relation on $S$? Explain.

**55.** Prove Theorem 4.6(a).

**56.** Prove Theorem 4.6(b).

**57.** Let $X = \{(x, y) \in \mathbb{R} \times \mathbb{R}: xy > 0\}$. That is, $X$ is the union of the first and third quadrants of the plane. For each positive real number $\alpha$, let

$$A_\alpha = \{(x, y) \in X: xy = \alpha\}$$

and let

$$\mathscr{A} = \{A_\alpha: \alpha \text{ is a positive real number}\}$$

a) Is the indexed family $\mathscr{A}$ a partition of $X$? (Explain.)

b) Let

$$R = \{((x_1, y_1), (x_2, y_2)): \text{some member of } \mathscr{A} \text{ has both } (x_1, y_1) \text{ and } (x_2, y_2) \text{ as members}\}$$

Is $R$ an equivalence relation on $X$? (Prove your answer.)

c) Draw the graph of $R[(1, 3)]$.

**58.** For each real number $b$, let $A_b = \{(x, y) \in \mathbb{R} \times \mathbb{R}: y = 2x + b\}$, and let $\mathscr{A} = \{A_b: b \in \mathbb{R}\}$. Is $\mathscr{A}$ a partition of $\mathbb{R} \times \mathbb{R}$? Justify your answer.

**59.** For each real number $b$, let $A_b = \{(x, y) \in \mathbb{R} \times \mathbb{R}: y = |x + b|\}$, and let $\mathscr{A} = \{A_b: b \in \mathbb{R}\}$. Is $\mathscr{A}$ a partition of $\mathbb{R} \times \mathbb{R}$? Justify your answer.

**60.** For each real number $b$, let $A_b = \{(x, y) \in \mathbb{R} \times \mathbb{R}: y = |x| + b\}$, and let $\mathscr{A} = \{A_b: b \in \mathbb{R}\}$. Is $\mathscr{A}$ a partition of $\mathbb{R} \times \mathbb{R}$? Justify your answer.

**61.** Economists consider two (usually imaginary) bundles of goods or services to be equivalent provided that a prudent consumer would as soon have the first bundle as the second bundle. An indifference curve indicates the set of all bundles of goods of equal value. Suppose that gold is selling at \$300/oz and silver is selling at \$10/oz.

a) Using the graph below, sketch the indifference curve to which the bundle consisting of 3 oz of gold and 30 oz of silver belongs.

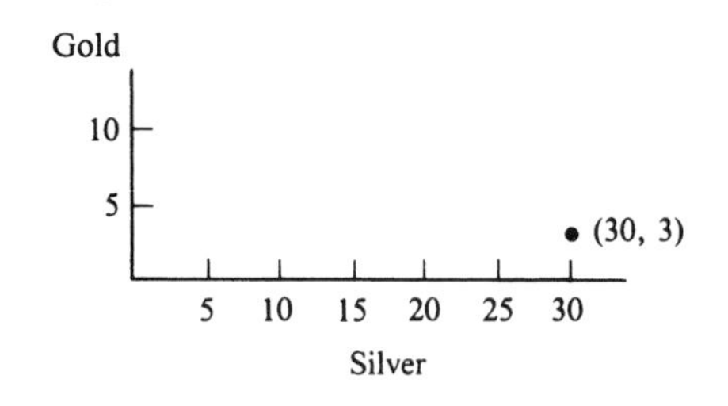


b) Explain briefly how the concepts of indifference and indifference curves relate to partitions and equivalence classes.

## 4.5 Congruence

It is said that the blind can hear the difference between the summer and winter and tell from which direction a passing car has come and how many of its cylinders are firing properly. It is not that the blind hear better than the sighted, but that they listen better. A second way of considering equivalence relations is to agree that any two equivalent members of a set are indistinguishable, as if a film had suddenly obscured all but one essential property of the members of a set. In this way, by blotting out irrelevant differences, we are able for the first time to concentrate on essential distinctions. The idea was first used to full advantage by Carl Friedrich Gauss.

**Definition**

Let $n \in \mathbb{N}$. If $a, b \in \mathbb{Z}$, we say that ***a* is congruent to *b* modulo *n***, written $a \equiv b \pmod{n}$, provided that $a - b$ is divisible by $n$.

As you may guess, everyone says "$a$ is congruent to $b$ mod $n$," rather than "$a$ is congruent to $b$ modulo $n$."

**Theorem 4.7**

Let $n \in \mathbb{N}$, and let $a, b \in \mathbb{Z}$. Then $a \equiv b \pmod{n}$ if and only if there exists an integer $k$ such that $a = b + kn$.

**PROOF** Suppose that $a \equiv b \pmod{n}$. Then there is an integer $k$ such that $a - b = kn$. Thus, $a = b + kn$.

Suppose that there is an integer $k$ such that $a = b + kn$. Then $a - b = kn$, and so $a \equiv b \pmod{n}$. □

**Theorem 4.8**

If $n \in \mathbb{N}$, congruence modulo $n$ is an equivalence relation on the set of integers.

**PROOF** See Exercise 62. □

**Definition**

The equivalence classes for this equivalence relation are called **congruence classes mod *n***, and $n$ is referred to as the **modulus**.

If $n = 1$, this concept is of no interest because any two integers are congruent, modulo 1, so we have only one congruence class.

If $n = 2$, there are two congruence classes—one is the set of all even integers and the other is the set of all odd integers.

If $n = 3$, there are three congruence classes:

$$\{\ldots, -6, -3, 0, 3, 6, \ldots\}$$

$$\{\ldots, -5, -2, 1, 4, 7, \ldots\}$$

$$\{\ldots, -7, -4, -1, 2, 5, \ldots\}$$

In general, for each $n \in \mathbb{N}$, there are $n$ congruence classes, modulo $n$. We prove this by proving the following theorem.

**Theorem 4.9**

If $n \in \mathbb{N}$, then each integer is congruent, modulo $n$, to precisely one of the integers $0, 1, 2, \ldots, n-1$.

**PROOF** Let $n \in \mathbb{N}$, and let $a$ be an integer. By the Division Algorithm for Integers (Theorem 3.3), there are unique integers $q$ and $r$ such that $a = nq + r$ and $0 \leqslant r < n$. Therefore, $a - r = nq$, and hence $n$ divides $a - r$. So $a$ is congruent, modulo $n$, to $r$, and $r$ is one of the integers $0, 1, 2, \ldots, n-1$.

In order to complete the proof of the theorem, we need to show that if $a$ is congruent, modulo $n$, to $s$ and $s$ is one of the integers $0, 1, 2, \ldots, n-1$, then $r = s$. Suppose that $s \in \mathbb{Z}$, $a \equiv s \pmod{n}$, and $0 \leqslant s < n$. Then there exists an integer $t$ such that $a - s = nt$. Hence $a = nt + s$, and $r = s$ by the uniqueness of $r$ in the Division Algorithm. □

We now give a simple but amusing example of the usefulness of Gauss's idea of congruence modulo $n$. We emphasize that this example does not begin to reveal the fundamental importance of congruence, but it does give some hint as to how congruence leads to the construction of new algebraic structures, which we discuss in Chapter 8. Further, our example provides ample evidence that we can perceive new insights by allowing equivalence relations to focus our attention on one particular property of integers.

We choose to consider the nine equivalence classes, $[0], [1], [2], \ldots, [8]$, of the integers, mod 9, and in so doing we fix our attention upon the "nineness" of a number. We no longer care whether an integer is even or odd, positive or negative, large or small, prime or composite. We have partitioned $\mathbb{Z}$ into nine equivalence classes, and, when we are given an integer, our primary interest in it is to determine the congruence class to which it belongs. Given any integer, it is easy to determine by a process of elimination exactly which congruence class contains it. For example, $49 \in [4]$, $-44 \in [1]$, and $-27 \in [0]$. Given an integer, how can we find which of the congruence classes $[0], [1], \ldots, [8]$, contains it? The method of proof of Theorem 4.9 suggests a general method of attack (see Exercise 68), but, as we now show, for congruence mod 9 there is an even better way.

**Definition**

> Let $n$ be a natural number, and let $a$ and $b$ be integers. W
> addition and multiplication of the congruence classes $[a$
> mod $n$, as follows:
>
> $$[a] \oplus [b] = [a + b]$$
> $$[a] \odot [b] = [ab]$$

The definitions of addition and multiplication of congruence classes are not so simple as they may appear. In fact, it is not even clear, a priori, that these definitions make any sense. For example, with respect to congruence mod 9, $[34] \odot [22] = [748]$, but remember we have agreed that we cannot distinguish between 34 and 7 (since they are congruent) nor between 22 and 4. That is $[34]$ *is* $[7]$ and $[22]$ *is* $[4]$, so if our definition of multiplication of congruence classes mod 9 is to make any sense, it had better be true that $[748]$ is $[28]$. The following theorem establishes that our definitions really are independent of the choice of names for our equivalence classes.

**Theorem 4.10**

> Let $n \in \mathbb{N}$, and let $a, b, c, d \in \mathbb{Z}$. If $a \equiv b \pmod{n}$ and $c \equiv d \pmod{n}$, then
>
> **a)** $a + c \equiv b + d \pmod{n}$, and
>
> **b)** $a \cdot c \equiv b \cdot d \pmod{n}$.

**PROOF** We prove part (a) and leave part (b) as Exercise 64. Assume that $a \equiv b \pmod{n}$ and that $c \equiv d \pmod{n}$. Then $n$ divides both $a - b$ and $c - d$, and there are integers $j$ and $k$ such that $nj = a - b$ and $nk = c - d$. Since $a + c = nj + b + nk + d = n(j + k) + (b + d)$, then $(a + c) - (b + d) = n(j + k)$. Therefore, $n$ divides $(a + c) - (b + d)$ and by definition, $a + c \equiv b + d \pmod{n}$. □

Note that Theorem 4.10 can be restated in terms of congruence classes mod $n$ as follows: Let $n \in \mathbb{N}$ and let $a, b, c, d \in \mathbb{Z}$. If $[a] = [b]$ and $[c] = [d]$, then

$$[a + c] = [b + d] \quad \text{and} \quad [ac] = [bd]$$

We have stated Theorem 4.10 for an arbitrary natural number $n$, but we now return to the special case of congruence mod 9.

**Theorem 4.11**

> For the equivalence relation $a \equiv b \pmod{9}$ we have that for each natural number $n$, $[10^n] = [1]$.

**PROOF** The simple proof by induction is left as Exercise 65. □

Theorems 4.10 and 4.11 provide a simple way of determining that $[748]=[28]$.

$$\begin{aligned}[748] &= [700]\oplus[40]\oplus[8] \\ &= [7]\odot[100]\oplus[4]\odot[10]\oplus[8]\odot[1] \\ &= [7]\odot[1]\oplus[4]\odot[1]\oplus[8]\odot[1] \\ &= [7]\oplus[4]\oplus[8] \\ &= [19]\end{aligned}$$

$$[19]=[10]\oplus[9]=[1]\oplus[0]=[1]$$

Also

$$\begin{aligned}[28] &= [20]\oplus[8] \\ &= [2]\odot[10]\oplus[8] \\ &= [2]\odot[1]\oplus[8] \\ &= [2]\oplus[8] \\ &= [10] \\ &= [1]\end{aligned}$$

In general, each natural number belongs to the same congruence class to which the sum of its digits belongs.

Thus, $[123456789101112\ldots99100]=[5050]=[10]=[1]$, and $[142857]=[27]=[9]=[0]$. What does it mean to say that $[142857]=[0]$? It means that $142857\equiv 0 \bmod 9$ or that 9 divides $142857-0$. We have found a simple test for divisibility by 9. But there is more. What does it mean to say that $[n]=[1]$? It means that 9 divides $n-1$. Well, if 9 divides $n-1$, what is the remainder when we divide 9 into $n$? The remainder is 1.

We caution that Theorem 4.11 provides the key for our tricks concerning congruence mod 9. We cannot expect that this theorem will hold for all moduli. Modulus 7 provides a particularly bad modulus in this respect (see Exercise 73).

Theorem 4.10 shows that an addition, subtraction, or multiplication to both sides of a congruence preserves the congruence. However, $7\cdot 2\equiv 4\cdot 2 \pmod 6$ but $7\not\equiv 4 \pmod 6$, so it is not necessarily true that a congruence is preserved when we divide both sides by an integer. The following theorem provides a condition under which a congruence is preserved by division.

**Theorem 4.12**

Let $n\in\mathbb{N}$, let $a, b, c\in\mathbb{Z}$, and let $d=\gcd(c, n)$, and suppose that $ac\equiv bc \pmod n$. Then $a\equiv b \pmod{n/d}$.

**PROOF** Since $ac \equiv bc \pmod{n}$, $n \mid (ac - bc)$. Therefore, there is an integer $k$ such that $c(a-b) = ac - bc = kn$. Thus, $(c/d)(a-b) = k(n/d)$. Since $d = \gcd(c, n)$, by Exercise 103 in Section 3.5, $\gcd(n/d, c/d) = 1$. Therefore, by Exercise 98 in Section 3.5, $(n/d) \mid (a-b)$. Hence $a \equiv b \pmod{n/d}$. □

The following theorem tells us when a linear congruence in one variable has solutions and, if the equation has solutions, the number of incongruent solutions.

**Theorem 4.13**

Let $n \in \mathbb{N}$, let $a, b \in \mathbb{Z}$, and let $d = \gcd(a, n)$.

**a)** If $d$ does not divide $b$, then $ax \equiv b \pmod{n}$ has no solutions.

**b)** If $d \mid b$, then $ax \equiv b \pmod{n}$ has exactly $d$ incongruent solutions modulo $n$.

**PROOF**

**a)** Suppose that $d$ does not divide $b$. By Theorem 4.7, $ax \equiv b \pmod{n}$ if and only if there is an integer $y$ such that $ax = b + yn$. Therefore, the integer $x$ is a solution of $ax \equiv b \pmod{n}$ if and only if there is an integer $y$ such that $ax - ny = b$. By Theorem 3.16, there are no integral solutions to the equation $ax - ny = b$. Thus, $ax \equiv b \pmod{n}$ has no solutions.

**b)** Suppose that $d \mid b$. Then by Theorem 3.16, the equation $ax - ny = b$ has an infinite number of integral solutions, and the solutions are given by $x = x_0 + (n/d)p$ and $y = y_0 + (a/d)p$, where $x = x_0$ and $y = y_0$ is a particular solution and $p \in \mathbb{Z}$. By Theorem 4.7, for each $p \in \mathbb{Z}$, $x = x_0 + (n/d)p$ is a solution of $ax \equiv b \pmod{n}$.

To determine the number of incongruent solutions, we determine when two of the solutions are congruent modulo $n$. Suppose that two solutions $x_1 = x_0 + (n/d)p_1$ and $x_2 = x_0 + (n/d)p_2$ are congruent modulo $n$. Then $x_0 + (n/d)p_1 \equiv x_0 + (n/d)p_2 \pmod{n}$, and so by Theorem 4.10, $(n/d)p_1 \equiv (n/d)p_2 \pmod{n}$. Since $(n/d) \mid n$, $\gcd(n, n/d) = n/d$. Therefore, by Theorem 4.12, $p_1 \equiv p_2 \pmod{d}$. Hence, a complete set of incongruent solutions of $ax \equiv b \pmod{n}$ is given by $x = x_0 + (n/d)p$, where $p = 0, 1, 2, \ldots, d-1$. □

Now we consider systems of congruence that involve one variable but different moduli. Such systems arose in ancient Chinese puzzles such as "Find a number that leaves a remainder of 1 when divided by 3, a remainder of 3 when divided by 7, and a remainder of 5 when divided by 11." This problem leads to the following system of congruences:

$$x \equiv 1 \pmod{3} \qquad x \equiv 3 \pmod{7} \qquad x \equiv 5 \pmod{11}$$

The theory behind the method of solving this system is known as the Chinese Remainder Theorem.

**Theorem 4.14**

**The Chinese Remainder Theorem**

Given a system of congruences

$$x \equiv a_1 \bmod n_1$$

$$x \equiv a_2 \bmod n_2$$

$$\vdots$$

$$x \equiv a_r \bmod n_r$$

for which every pair of moduli are relatively prime, there exists a value for $x$ that satisfies all the given congruences. One such value of $x$ is given by

$$x_0 = a_1 y_1 N_1 + a_2 y_2 N_2 + \cdots a_r y_r N_r$$

where for each $j = 1, 2, \ldots, r$, $y_j$ and $N_j$ are defined as follows:

$N_j$ is the product of all the $n_i$'s except $n_j$ and $y_j$ is an integer, found using the Euclidean algorithm, such that $y_j N_j \equiv 1 \bmod n_j$. Moreover, if $N$ is the product $n_1 n_2 \cdots n_r$, then any number congruent to $x_0 \bmod N$ is a value of $x$ that satisfies all the given congruences, and any value of $x$ that satisfies all the given congruences is congruent to $x_0 \bmod N$.

Before proving the Chinese Remainder Theorem, we use it to solve the system of congruences mentioned prior to the statement of the theorem.

**EXAMPLE 9**

We solve the system of congruences

$$x \equiv 1 \bmod 3 \qquad x \equiv 3 \bmod 7 \qquad x \equiv 5 \bmod 11$$

Clearly, any two of the moduli 3, 7, and 11 are relatively prime. Thus by the Chinese Remainder Theorem $1y_1N_1 + 3y_2N_2 + 5y_3N_3$ is a common solution to the given system of congruences

$$N_1 = (7)(11) = 77 \qquad N_2 = (3)(11) = 33 \qquad N_3 = (3)(7) = 21$$

To find $y_1$, $y_2$, and $y_3$ we must solve

**a)** $77y_1 \equiv 1 \bmod 3$,

**b)** $33y_2 \equiv 1 \bmod 7$, and

**c)** $21y_3 \equiv 1 \bmod 11$.

Since gcd $(3, 77) = 1$, we use the Euclidean algorithm to find integers $A$ and $y_1$ such that $3A + 77y_1 = 1$:

$$77 = 25(3) + 2$$

$$3 = 1(2) + 1$$

Thus, $1 = 3 - 2 = 3 - (77 - 25(3)) = 26(3) - 1(77)$ and so $y_1 = -1$.

Similarly, we find $y_2 = 3$ and $y_3 = -1$ (see Exercise 74.)

Finally, $x_0 = 77y_1 + 99y_2 + 105y_3 = 115$, and the set of all solutions to the system of congruences is the congruence class $[115]$ mod 231. ❑

**PROOF OF THEOREM 4.14** First we construct a solution to the system of congruences. For each $i = 1, 2, \ldots, r$, let $N_i = N/n_i$. Since $\gcd(n_i, n_j) = 1$ for each $i, j = 1, 2, \ldots, r$ such that $i \neq j$, by Exercise 107 in Section 3.5, $\gcd(N_i, n_i) = 1$ for each $i = 1, 2, \ldots, r$. Therefore, by Theorem 4.13, for each $i = 1, 2, \ldots, r$, there exists $y_i$ such that $N_i y_i \equiv 1 \pmod{n_i}$. Let $x = \Sigma_{j=1}^{r} a_j N_j y_j$. To show that $x$ is a solution to the system of congruences, we must show that $x \equiv a_i \pmod{n_i}$ for each $i = 1, 2, \ldots, r$. Since $n_i | N_j$ whenever $i \neq j$, we have $N_j \equiv 0 \pmod{n_i}$. Therefore, in $\Sigma_{j=1}^{r} a_j N_j y_j$, all terms except the $i$th term are congruent to 0 $\pmod{n_i}$. Thus, since $N_i y_i \equiv 1 \pmod{n_i}$, $x \equiv a_i N_i y_i \equiv a_i \pmod{n_i}$.

Now we show that any two solutions of the system of congruences are congruent modulo $N$. Suppose that $x_0$ and $x_1$ are solutions to the system of congruences. Then for each $i = 1, 2, \ldots, r$, $x_0 \equiv x_1 \equiv a_i \pmod{n_i}$, and so $n_i | (x_0 - x_1)$. By Theorem 4.13, $N | (x_0 - x_1)$. Therefore, $x_0 \equiv x_1 \pmod{N}$.

Finally suppose that $x^* \equiv x_0 \bmod N$. Then for each $i = 1, 2, \ldots, r$, $x^* \equiv x_0 \bmod n_i$ and so $x^* \equiv a_i \bmod n_i$. That is, $x^*$ is also a solution to the system of congruences. □

We conclude this section with a congruence for factorials and a congruence for exponents, but we first establish the following theorem.

**Theorem 4.15**

Let $p$ be a prime, and let $a \in \mathbb{N}$. Then $a^2 \equiv 1 \pmod{p}$ if and only if $a \equiv 1 \pmod{p}$ or $a \equiv -1 \pmod{p}$.

**PROOF** Suppose that $a^2 \equiv 1 \pmod{p}$. Then $p | (a^2 - 1)$. Since $a^2 - 1 = (a+1)(a-1)$, by Corollary 3.11, either $p | (a+1)$ or $p | (a-1)$. Therefore, either $a \equiv -1 \pmod{p}$ or $a \equiv 1 \pmod{p}$.

Suppose that $a \equiv 1 \pmod{p}$ or $a \equiv -1 \pmod{p}$. Then $p | (a-1)$ or $p | (a+1)$, and in either case, $p | (a^2 - 1)$. Therefore, $a^2 \equiv 1 \pmod{p}$. □

**THEOREM 4.16**

**Wilson's Theorem**

Let $p$ be a prime. Then $(p-1)! \equiv -1 \pmod{p}$.

**PROOF** Suppose that $p = 2$. Then $(p-1)! = 1$ and $1 \equiv -1 \pmod{p}$. Therefore, the theorem is true for $p = 2$.

Suppose that $p > 2$. By Theorem 4.13, for each natural number $a \leqslant p - 1$, there is an integer $\bar{a}$ such that $a\bar{a} \equiv 1 \pmod{p}$. By Theorem 4.15, $\bar{a} = a$ if and only if $a \equiv 1 \pmod{p}$ or $a \equiv p-1 \pmod{p}$. Therefore, we can group the natural numbers from 2 to $p-2$ into $(p-3)/2$ pairs of integers with the product of each pair congruent to 1 modulo $p$. Thus the product $2 \cdot 3 \cdots (p-3) \cdot (p-2) \equiv 1 \pmod{p}$. Hence by Theorem 4.10,

$$(p-1)! = 2 \cdot 3 \cdots (p-3)(p-2)(p-1) \equiv 1 \cdot (p-1) \equiv -1 \pmod{p} \qquad \square$$

Wilson's Theorem is named after John Wilson, who conjectured but did not prove it. It was proved by Joseph Lagrange in 1770.

**Theorem 4.17**

**Fermat's Little Theorem**

Let $p$ be a prime, and let $a \in \mathbb{N}$ such that $p$ does not divide $a$. Then $a^{p-1} \equiv 1 \pmod{p}$.

**PROOF** None of the $p-1$ integers, $a, 2a, \ldots, (p-1)a$, is divisible by $p$ because if $p \mid ja$ for some $j = 1, 2, \ldots, p-1$, then by Exercise 98 in Section 3.5, $p \mid j$. We show that no two of the integers $a, 2a, \ldots, (p-1)a$ are congruent modulu $p$. To see this, assume that there exist $i, j$ $(i \neq j)$ such that $ia \equiv ja \pmod{p}$. Then by Theorem 4.12, $i \equiv j \pmod{p}$. This is impossible because both $i$ and $j$ are less than $p$. Therefore, we have established that the set of integers $a, 2a, \ldots, (p-1)a$ is congruent (in some order) to the set of integers $1, 2, \ldots, p-1$. Hence $1a \cdot 2a \cdot 3a \cdots (p-1)a \equiv 1 \cdot 2 \cdot 3 \cdots (p-1) \pmod{p}$, and so $a^{p-1}(p-1)! \equiv (p-1)! \pmod{p}$. Since $\gcd((p-1)!, p) = 1$, by Theorem 4.12, $a^{p-1} \equiv 1 \pmod{p}$. $\square$

Fermat's Little Theorem is useful in finding the smallest natural number that is congruent to a given number. For example, by Fermat's Little Theorem, $3^6 \equiv 1 \pmod{7}$. Therefore, $3^{50} = (3^6)^8 \cdot 3^2 \equiv 3^2 \equiv 2 \pmod{7}$.

*Personal Note*

## Pierre de Fermat (1601–1665)

Surely, Pierre de Fermat wrote the world's most famous mathematical apostil. In the margin of his copy of Diophantus' *Arithmeticae*, Fermat wrote, "On the other hand, it is impossible to separate a cube into two cubes ... or in general to separate any power except a square into two powers with the same exponent. I have discovered a truly marvellous proof of this, which however the margin is not large enough to contain." Fermat's assertion has become known as Fermat's Last Theorem, and as this edition of your text goes to print, it appears that Fermat's Last Theorem is at last a theorem. A lively and readable account of the history of this problem is given by David Cox ["Introduction to Fermat's Last Theorem," *Amer. Math. Monthly*, Vol. 101, 1994, pages 3–14].

## EXERCISES 4.5

**62.** Prove Theorem 4.8.

**63.** Prove that if $a \equiv b \pmod{n}$ and $c \equiv d \pmod{n}$, then $a - c \equiv b - d \pmod{n}$.

**64.** Prove Theorem 4.10(b).

**65.** Prove Theorem 4.11.

**66.** Let $p$ be an odd prime and let $S = \{[1], [2], \ldots, [p-1]\}$ be the set of nonzero equivalence classes mod $p$.

**a)** Prove that for any integer $a$ such that $[a] \in S$, $[-a] \in S$ and $[-a] \neq [a]$.

**b)** Prove that if $[a] \in S$ and $[b] \in S$, then $[a] \odot [b] \in S$.

**c)** Prove that if $[a]^2 = [b]^2$ then $[a] = [b]$ or $[a] = [-b]$.

***d)** Prove that exactly $(p-1)/2$ members of $S$ have square roots ($[a]$ has a square root provided there is an integer $x$ such that $[x]^2 = [a]$).

**67.** In the previous problem, instead of taking $p$ to be an odd prime number, take $p = 16$. Show by example that all four statements (a), (b), (c), and (d) are then false.

**68.** Let $n \in \mathbb{N}$, and let $a, b \in \mathbb{Z}$. Prove that $a$ is congruent to $b$, mod $n$, if and only if $a$ and $b$ have the same remainder when divided by $n$.

**69.** Find a natural number $n$ and integers $a$ and $b$ such that $a^2 \equiv b^2 \pmod{n}$ but $a \not\equiv b \pmod{n}$.

**70.** Prove, by induction, that if $n \in \mathbb{N}$, $a, b \in \mathbb{Z}$, and $a \equiv b \pmod{n}$, then $a^m \equiv b^m \pmod{n}$ for each natural number $m$.

**71.** Prove that if $n \in \mathbb{N}$ and $a, b \in \mathbb{Z}$, then there is an integer $c$ such that $a + c \equiv b$ mod $n$.

**72.** Find a natural number $n$ and integers $a$ and $b$ such that $a \cdot c \not\equiv b \pmod{n}$ for any integer $c$.

**73.** Show that

$$\{[10], [100], [1000], [10^4], [10^5], [10^6]\} = \{[1], [2], [3], [4], [5], [6]\}$$

for congruence classes mod 7. (If you are not performing the necessary calculations in your head, you are going about this problem in an unnecessarily difficult way.) [*Remark*: The extremely patient reader may determine that $[10], \ldots, [10^{16}]$ yield 16 different congruence classes mod 17. Gauss asked if there were infinitely many moduli $n$ such that $[10], \ldots, [10^{n-1}]$ yield $n-1$ different congruence classes mod $n$; his question remains unanswered.]

**74.** Find $y_2$ and $y_3$ in Example 9.

**75.** Solve the system of congruences: $x \equiv 1$ mod 3, $x \equiv 2$ mod 5, $x \equiv 3$ mod 7.

**76.** Solve the system of congruences: $x \equiv 4$ mod 11, $x \equiv 3$ mod 17.

**77.** **a)** Prove that for each natural number $n$, $8^n \equiv 1$ mod 7.

***b)** Assume that a number is written in base 8. Explain how to determine whether or not the number is divisible by 7. Illustrate your test on the following base 8 number: 114265.

**78.** Let $S$ be the set of congruence classes mod 5, and for $[a]$ and $[b]$ in $S$ define $[a]^{[b]}$ to be $[a^b]$. Evaluate $[9]^{[4]}$ and evaluate $[4]^{[9]}$.

**79.** Use the results of Exercise 78 to illustrate the most important aspect of the definition of exponentiation of congruence classes as defined in Exercise 78.

**80.** Show that if $n$ is a composite with $n \neq 4$, then $(n-1)! \equiv 0 \pmod{n}$.

**81.** Let $p$ be a prime, and let $a \in \mathbb{N}$. Prove that $a^p \equiv a \pmod{p}$.

**82.** Let $n \in \mathbb{N}$ such that $n > 1$ and $(n-1)! \equiv -1 \pmod{n}$. Prove that $n$ is prime.

**83.** Show that if $p$ is an odd prime, then $2(p-3)! \equiv -1 \pmod{p}$.

**84.** **a)** Show that it is a consequence of Fermat's Little Theorem that if $p$ is a prime, then $p$ divides $2^p - 2$.

**b)** Fermat conjectured that if $2^n - 2$ is divisible by $n$, then $n$ is prime. Verify that this conjecture is false by showing that when $n = 341$, $2^n - 2$ is divisible by $n$. [*Note:* The number $n = 341$ was supplied by Leonhard Euler.]

## 4.6 *Composition of Relations*

The mathematical community's interest in relations has been like a blossoming friendship; in the beginning mathematicians were interested in relations for what they did, and only in modern times have mathematicians become interested in relations for what they are. Hitherto, our presentation of relations has been from the old-time point of view. We have been interested in equivalence relations, for example, because they have done something for us: They have partitioned a given set into equivalence classes. In this section, we reconsider relations, thinking of them this time as objects that are of interest in their own right.

**Definition**

Let $A$, $B$, and $C$ be sets, let $S \subseteq A \times B$, and let $R \subseteq B \times C$. Then

$$R \circ S = \{(a, c) \in A \times C\text{: there is a } b \in B \text{ such that } (a, b) \in S \text{ and } (b, c) \in R\}$$

The relation $R \circ S$ is called $R$ **composite** $S$.

Intuitively, $R \circ S$ is the relation obtained by *first* "doing" $S$ and then "doing" $R$. It has been said (in jest) that a mathematician always does things backward, and the definition of $R \circ S$ is a glaring piece of evidence to support this contention. We will see in Chapter 5 that there is some motivation for having $R \circ S$ mean first $S$ then $R$, but we admit that it takes practice to get used to having $(a, b) \in R$ and $(b, c) \in S$ imply that $(a, c) \in S \circ R$.

**EXAMPLE 10** Let $A = \{1, -1, 2\}$, $B = \{3, 4, 5\}$, $C = \{-2, 0, 6\}$, $S = \{(1, 3), (1, 5), (-1, 5)\}$, and $R = \{(3, 0), (4, 0), (4, 6)\}$. Then $R \circ S = \{(1, 0)\}$. ❑

**EXAMPLE 11** Let $A$, $B$, and $C$ be as in Example 10. Then let

$$S = \{(1, 3), (1, 5), (-1, 5), (2, 4)\} \quad \text{and} \quad R = \{(3, -2), (4, 0), (5, 0), (5, 6)\}$$

Then $R \circ S = \{(1, -2), (1, 0), (1, 6), (-1, 0), (-1, 6), (2, 0)\}$. ❑

**EXAMPLE 12** Let $S = \{(x, y) \in \mathbb{R} \times \mathbb{R}: x = y^2\}$, and let

$$R = \{(x, y) \in \mathbb{R} \times \mathbb{R}: y = 2x + 1\}$$

Then

$$\begin{aligned} R \circ S &= \{(x, y) \in \mathbb{R} \times \mathbb{R}: \text{there exists } z \text{ such that } (x, z) \in S \text{ and } (z, y) \in R\} \\ &= \{(x, y) \in \mathbb{R} \times \mathbb{R}: \text{there exists } z \text{ such that } x = z^2 \text{ and } y = 2z + 1\} \\ &= \{(x, y) \in \mathbb{R} \times \mathbb{R}: y = 2(\pm\sqrt{x}) + 1\} \end{aligned}$$

Also

$$\begin{aligned} S \circ R &= \{(x, y) \in \mathbb{R} \times \mathbb{R}: \text{there exists } z \text{ such that } (x, z) \in R \text{ and } (z, y) \in S\} \\ &= \{(x, y) \in \mathbb{R} \times \mathbb{R}: \text{there exists } z \text{ such that } z = 2x + 1 \text{ and } z = y^2\} \\ &= \{(x, y) \in \mathbb{R} \times \mathbb{R}: y^2 = 2x + 1\} \end{aligned}$$

❑

Let $X$ be a set. Then $i_X$ denotes $\{(x, x): x \in X\}$. Recall that if $R$ is a relation, then $R^{-1}$ (called *R-inverse*) denotes $\{(x, y): (y, x) \in R\}$.

**Theorem 4.18** Let $A$ and $B$ be sets, and let $R$ be a relation with domain $A$ and range $B$. Then the following statements hold:

**a)** $R \circ i_A = i_B \circ R = R$
**b)** $i_B \subseteq R \circ R^{-1}$
**c)** $i_A \subseteq R^{-1} \circ R$

**PROOF** We prove only that $R = i_B \circ R$ and that $i_B \subseteq R \circ R^{-1}$. The remaining parts of this theorem are left as Exercise 85.

Let $(x, y) \in R$. By hypothesis, $R \subseteq A \times B$ and so $(y, y) \in i_B$. It follows that $(x, y) \in i_B \circ R$. Therefore, $R \subseteq i_B \circ R$. Now let $(a, c) \in i_B \circ R$. There exists $b \in B$ such that $(a, b) \in R$ and $(b, c) \in i_B$. Since $b = c$, $(a, c) \in R$ and so $i_B \circ R \subseteq R$. Thus, $R = i_B \circ R$.

To show that $i_B \subseteq R \circ R^{-1}$, let $(x, y) \in i_B$. Then $x = y$ and $y \in B$. Since the range of $R$ is $B$, there exists $z$, necessarily a member of $A$, such that $(z, y) \in R$. Then $(y, z) \in R^{-1}$ and so $(x, y) = (y, y) \in R \circ R^{-1}$. We have shown that $i_B \subseteq R \circ R^{-1}$. □

**Theorem 4.19**

Let $R$ be a relation on a set $A$. Then the following statements hold:

**a)** $R$ is reflexive on $A$ if and only if $i_A \subseteq R$.

**b)** $R$ is transitive if and only if $R \circ R \subseteq R$.

**c)** $R$ is both reflexive on $A$ and transitive if and only if $i_A \subseteq R = R \circ R$.

**PROOF** We prove part (b) and leave the remaining parts as Exercises 86 and 89.

Suppose that $R$ is transitive. Let $(x, z) \in R \circ R$. There exists $y \in A$ such that $(x, y) \in R$ and $(y, z) \in R$. Since $R$ is transitive, $(x, z) \in R$. Thus, $R \circ R \subseteq R$.

Suppose that $R \circ R \subseteq R$. Let $(x, y)$ and $(y, z)$ be members of $R$. Then $(x, z) \in R \circ R$. Since $R \circ R \subseteq R$, $(x, z) \in R$. Therefore, $R$ is transitive. □

**Theorem 4.20**

Let $A$, $B$, and $C$ be sets, and let $R \subseteq B \times C$ and $S \subseteq A \times B$. Then $(R \circ S)^{-1} = S^{-1} \circ R^{-1}$.

**PROOF** Let $(x, z) \in (R \circ S)^{-1}$. Then $(z, x) \in R \circ S$, so there exists $b \in B$ such that $(z, b) \in S$ and $(b, x) \in R$. Therefore, it follows that $(b, z) \in S^{-1}$ and $(x, b) \in R^{-1}$, and so $(x, z) \in S^{-1} \circ R^{-1}$. Thus, $(R \circ S)^{-1} \subseteq S^{-1} \circ R^{-1}$. The containment $S^{-1} \circ R^{-1} \subseteq (R \circ S)^{-1}$ is left as Exercise 87. □

**Theorem 4.21**

Let $A$, $B$, $C$, and $D$ be sets. Let $T \subseteq A \times B$, $S \subseteq B \times C$, and $R \subseteq C \times D$. Then $R \circ (S \circ T) = (R \circ S) \circ T$. In other words, composition is associative.

**PROOF** Let $(x, z) \in R \circ (S \circ T)$. Then there exists a $y \in C$ such that $(x, y) \in S \circ T$ and $(y, z) \in R$. Since $(x, y) \in S \circ T$, there exists $b \in B$ such that $(x, b) \in T$ and $(b, y) \in S$. Because $(y, z) \in R$ and $(b, y) \in S$, $(b, z) \in R \circ S$. As $(x, b) \in T$ and $(b, z) \in R \circ S$, $(x, z) \in (R \circ S) \circ T$ and so $R \circ (S \circ T) \subseteq (R \circ S) \circ T$. The proof of the other containment is equally straightforward. □

## EXERCISES 4.6

**85.** Prove the remaining parts of Theorem 4.18.

**86.** Prove Theorem 4.19(a).

**87.** Complete the proof of Theorem 4.20.

**88.** Complete the proof of Theorem 4.21.

**89.** Prove Theorem 4.19(c).

**90.** Let $R = \{(1, 2), (3, 5), (2, 2)\}$, and let $S = \{(2, 1), (5, 3), (5, 1)\}$. Find each of the following.

**a)** $R^{-1}, S^{-1}$ **b)** $S \circ R$ **c)** $R \circ S$

**d)** $R^{-1} \circ S^{-1}$ **e)** $S^{-1} \circ R^{-1}$

**91.** For each of the following statements, provide a proof or a counterexample.

**a)** If $R$ and $S$ are relations that are reflexive on a set $X$, then $R \circ S$ is reflexive on $X$.

**b)** If $R$ and $S$ are symmetric relations, so is $R \circ S$.

**c)** If $R$ and $S$ are transitive relations, so is $R \circ S$.

**d)** If $R \circ S = S \circ R$, then $S = R$.

**e)** If $R$, $S$, and $T$ are relations from $A$ to $A$, then $R \circ (S \cup T) = (R \circ S) \cup (R \circ T)$.

**f)** If $R$, $S$, and $T$ are relations from $A$ to $A$, then $R \circ (S \cap T) = (R \circ S) \cap (R \circ T)$.

**g)** If $R$ and $S$ are symmetric relations, then $(R \circ S)^{-1} = R \circ S$.

**92.** Give an example to show that there are relations $R$ and $S$ such that $(R \circ S)^{-1} \neq R^{-1} \circ S^{-1}$.

**93.** Let $R = \{(n, p) \in \mathbb{N} \times \mathbb{N} : n < p\}$, and let $S = \{(n, p) \in \mathbb{N} \times \mathbb{N} : n \text{ divides } p\}$. Show that $R \circ S = R$.

***94.** Let $R$ be a relation that is reflexive on a set $A$. For each $n \in \mathbb{N}$, define $R^n$ inductively as follows: $R^1 = R$, and $R^{n+1} = R^n \circ R$. Define $R^\infty$ to be $\cup\{R^n : n \in \mathbb{N}\}$. Prove that $R^\infty$ is a transitive relation that is reflexive on $A$. Further prove that if $T$ is a transitive relation containing $R$, then $R^\infty \subseteq T$.

## 4.7 Types of Orders

Many people, in a rather simplistic way, impose a sense of order on the world in which they live. Their year begins with an orgy of bowl games, which allegedly determine, in order, the top twenty college football teams in the country. By spring their attention has turned to the tortuous "road to the final four," which ranks the top four college basketball teams, and in between these gala events there are such pageants of order as The Miss America Contest and The Miss Universe Contest. In truth, the ordering of both beauty and basketball is more complex and interesting than we like to pretend, and so it should come as no surprise that orders in mathematics are often more involved than $1, 2, 3, \ldots, 19, 20$.

**Definition**

Let $R$ be a relation on a set $S$. The relation $R$ is **irreflexive on $S$** provided that for each $x \in S$, $(x, x) \notin R$. The relation $R$ is **antisymmetric** provided that if $(x, y) \in R$ and $(y, x) \in R$, then $y = x$.

It is clear that if $R$ is a reflexive relation on a nonempty set, then $R$ is not irreflexive.

**EXAMPLE 13** Let $R = \{(x, y) \in \mathbb{R} \times \mathbb{R}: y = 3x^2 + 2\}$. Since the equation $x = 3x^2 + 2$ does not have any real solutions, for each $x \in \mathbb{R}$, $(x, x) \notin R$. Therefore, $R$ is irreflexive. ❑

As Figure 4.14 illustrates, $R$ is irreflexive because sets $R$ and $\{(x, y) \in \mathbb{R} \times \mathbb{R}: x = y\}$ are disjoint.

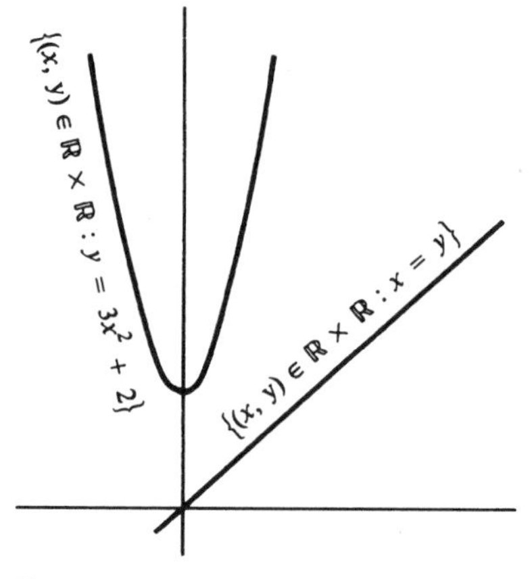

Figure 4.14

It is clear that if $R$ is a symmetric relation having the property that there exist $x$ and $y$ (with $x \neq y$) such that $(x, y) \in R$, then $R$ is not antisymmetric.

**EXAMPLE 14** Let $R = \{(m, n) \in \mathbb{N} \times \mathbb{N}: m \text{ is a divisor of } n\}$. Then $R$ is antisymmetric. ❑

The most basic notion of preference on a set $S$ requires only that if we prefer $a$ to $b$ and $b$ to $c$, then we should prefer $a$ to $c$; thus, any transitive relation can be thought of as indicating a preference. If the notion of order on a set $S$ is "just as good as," then it is reasonable to add that the order relation should be reflexive on $S$. Such an order is called a **preorder**. If the notion is strict preference, then it is reasonable to add that the order relation should be irreflexive on $S$. In either case, assuming that we do not simultaneously prefer $a$ to $b$ and $b$ to $a$, we should insist that the order relation be antisymmetric.

**Definition** Let $S$ be a set. An antisymmetric and transitive relation that is reflexive on $S$ is called a **partial order on $X$**; a transitive relation that is irreflexive on $S$ is called a **strict partial order on $S$**. Finally, if $R$ is a (strict) partial order on a set $S$, then the ordered pair $(S, R)$ is called a **(strict) partially ordered set**.

**EXAMPLE 15** Let $S$ be the set of nonzero real numbers, and let $R = \{(x, y) \in S \times S: xy \text{ is positive}\}$. Then, as you are asked to verify in Exercise 95, although $R$ is a transitive relation that is reflexive on $S$, $R$ is not a partial order. ❑

**EXAMPLE 16** Now let $S$ be the set of all subsets of the natural numbers. Then $\subseteq$ is a partial order on $S$; and if we let $\subset$ mean "is a proper subset of," then $\subset$ is a strict partial order on $S$. ❑

The usual order "less than or equal to" ("less than"), which is denoted $\leqslant$ ($<$), is a partial order (strict partial order). As we will see, these relations are too well behaved to be thought of as typical. Even the very easy partial orders given in the next example are more typical of partial orders than the less-than-or-equal-to order.

**EXAMPLE 17**

**a)** Let $S = \{a, b, c, d, e, f\}$, and for $x, y \in S$, let $x \leqslant y$ mean that $x$ is at least as high up on the page as $y$. Then $(S, \leqslant)$ is a partially ordered set.

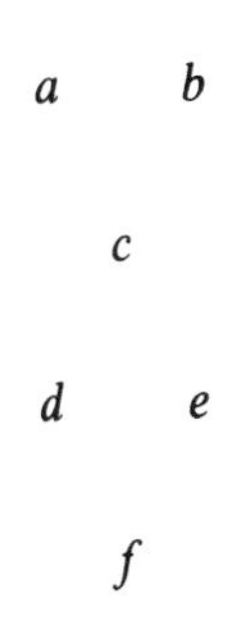


If $S$ were a set of football teams, we would be saying that team $c$ could beat both teams $d$ and $e$ whereas team $f$ would lose to both $d$ and $e$. Perhaps team $d$ would beat team $e$ in good weather but lose to $e$ in the rain. At any rate, no judgment of superiority has been indicated between $d$ and $e$.

**b)** Let $R$ by the relation determined by the directed graph on $\mathscr{P}\{1, 2, 3\}$, and let $R^T$ be the smallest transitive relation that contains $R$. Then $R^T$ is the strict partial order, "is a proper subset of." If we had also drawn loops at all eight corners, then $R^T$ would have been the partial order, "is a subset of."

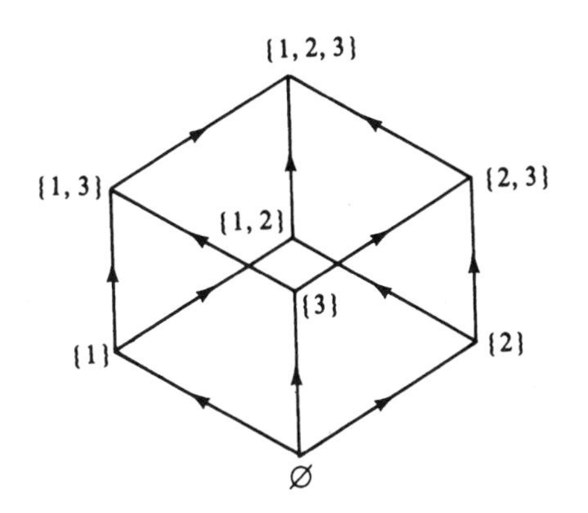


❑

**EXAMPLE 18**

Let $S$ be the set consisting of the first 1000 natural numbers and for each $m, n \in S$ define $m \leqslant n$ provided that $m$ divides $n$. Then $(S, \leqslant)$ is a partially ordered set (Exercise 96). This partially ordered set, like the partially ordered sets of the preceding example, has members that are not compared. Evidently $3 \leqslant 6$, but it is not true that $3 \leqslant 7$ nor is it true that $7 \leqslant 3$. ❑

It is intuitively obvious that partial orders and strict partial orders are allied concepts (in a way that Exercises 100 and 107 make precise). The following fundamental theorem reveals that preorders and partial orders are also kin. The theorem is a result of Ernst Schroeder (1841–1902), a mathematician whose work we will encounter again in Chapter 7.

**Theorem 4.22**

Let $R$ be a transitive relation that is reflexive on a set $S$, and let $E = R \cap R^{-1}$. Then $E$ is an equivalence relation on $S$, and if for any two equivalence classes $[a]$ and $[b]$ we define $[a] \leqslant [b]$ provided that for each $x \in [a]$ and each $y \in [b]$, $(x, y) \in R$, then $(S/E, \leqslant)$ is a partially ordered set.

**PROOF** The proof that $E$ is an equivalence relation is left as Exercise 97. Let $[a] \in S/E$, and let $x$ and $y$ be members of $[a]$. Then $x \, R \, a$ and $a \, R \, y$ hold, and because $R$ is transitive, $(x, y) \in R$. Thus, by definition $[a] \leqslant [a]$ and $\leqslant$ is reflexive on $S/E$.

Now suppose that $[a] \leqslant [b]$ and that $[b] \leqslant [a]$. In order to show that $[a] = [b]$, it suffices to show that $(a, b) \in E$. (Why?) Since $[a] \leqslant [b]$, $(a, b) \in R$ and because $[b] \leqslant [a]$, $(b, a) \in R$. Thus, $(a, b) \in R \cap R^{-1} = E$, and we have shown that $\leqslant$ is antisymmetric.

The proof that $\leqslant$ is transitive is left as an exercise (Exercise 98). □

**Definition**

Let $(S, \leqslant)$ be a partially ordered set. Two members $x$ and $y$ of $S$ are said to be **comparable** provided that $x \leqslant y$ or $y \leqslant x$. A subset $C$ of $S$ with the property that any two members of $C$ are comparable is called a **chain**. If the subset $S$ itself is a chain, we say that $\leqslant$ is a **linear order** and that $(S, \leqslant)$ is a **linearly ordered set.**

**EXAMPLE 19**

In Example 18, $\{2, 4, 8, 16, 32, 64\}$ is a chain, as is $\{2, 8, 32\}$, but $\{2, 4, 6, 8\}$ is not a chain because 6 and 8 are not comparable. ❑

**EXAMPLE 20**

As the name **linear** order suggests, the real line $\mathbb{R}$ with its usual order $\leqslant$ is a linearly ordered set. ❑

It should be noted that linearly ordered sets need not be "lined up" so nicely as the members of $(\mathbb{R}, \leqslant)$ are, and in Exercise 105, we will consider one maverick linearly ordered set, which should suffice to illustrate this point.

Finally, let us briefly consider the concept of a well-ordered set.

**Definition**

If $(S, \leqslant)$ is a partially ordered set with the property that every nonempty subset has a least member, we say that $\leqslant$ is a **well order on $S$** and that $(S, \leqslant)$ is a **well-ordered set.**

No doubt these definitions have a familiar ring. Sure enough, the natural numbers with their usual order provide a basic example of a well-ordered set. Of course, $(\mathbb{R}, \leqslant)$ is *not* a well-ordered set. The open interval $(6, 7)$ is a nonempty subset of $\mathbb{R}$ that has no least member.

Perhaps the concept of a well-ordered set would have remained a relatively obscure idea if it were not for a famous result of Zermelo. We have already seen that the Principle of Mathematical Induction is equivalent to the assumption that the usual order on $\mathbb{N}$ is a well order. What Zermelo proved was that our Rule 10 from page 61 (the Axiom of Choice) is equivalent to the assumption that for each set $S$ there is a linear order $\leqslant$ on $S$ such that $(S, \leqslant)$ is a well-ordered set. It is beyond the scope of this text to prove Zermelo's Theorem. It suffices to remark here that most mathematicians now accept the Axiom of Choice and hence the principle that every set can be well ordered. This principle is known as the *Well-Ordering Principle.*

## EXERCISES 4.7

**95.** Let $S$ be the set of all nonzero real numbers, and let

$$R = \{(x, y) \in S \times S: xy \text{ is positive}\}$$

**a)** Show that $R$ is a preorder on $S$ that is not a partial order on $S$.

**b)** Describe the equivalence classes under the equivalence relation $E$ of Theorem 4.22.

**96.** Let $S$ be the set of all natural numbers less than or equal to 1000. Let

$$\leqslant\, = \{(x, y) \in S \times S: x \text{ divides } y\}$$

Prove that $(S, \leqslant)$ is a partially ordered set.

**97.** Let $R$ be a preorder on a set $S$. Prove that $R \cap R^{-1}$ is an equivalence relation on $S$.

**98.** Complete the proof of Theorem 4.22 by showing that $\leqslant$ is a transitive relation.

***99.** Let $S$ be the collection of all real-valued functions that are continuous on $[0, 1]$. For each $f, g \in S$, define $f \leqslant g$ provided that $\int_0^1 f(x)\,dx \leqslant \int_0^1 g(x)\,dx$.

**a)** Prove that $\leqslant$ is a preorder on $S$ that is not a partial order on $S$.

**b)** Explain what it means for two functions $f$ and $g$ to be equivalent with respect to the equivalence relation $E$ of Theorem 4.22.

**100.** Let $S$ be a nonempty set.

**a)** Prove that if $R$ is a strict partial order on $S$, then $R \cup i_S$ is a partial order on $S$.

**b)** Prove that if $R$ is a partial order on $S$ and for each $x, y \in S$ we define $x < y$ provided $(x, y) \in R$ and $x \neq y$, then $<$ is a strict partial order on $S$.

**101.** Let $S$ be a nonempty finite set. Prove or give a counterexample to each of the following statements.

**a)** If $R$ is a preorder on $S$, then $R$ is a partial order on $S$.

**b)** If $R$ is a partial order on $S$, then $R$ is a linear order on $S$.

**c)** If $R$ is a linear order on $S$, then $R$ is a well order on $S$.

**102.** Let $S = \{1, 2, 3\}$, and let $X = \mathscr{P}(S)$. How many chains with exactly two elements are there in the partially ordered set $(X, \subseteq)$?

**103.** Let $S$ be a nonempty set, and let $R$ be a relation on $S$. Prove that $R$ is a partial order on $S$ if and only if $R^{-1}$ is a partial order on $S$.

**104.** Let $(S, R)$ be a linearly ordered set, and let $X \subseteq S$. Set $R^* = R \cap (X \times X)$. Prove that $(X, R^*)$ is a linearly ordered set.

**105.** In any linearly ordered set $(S, \leqslant)$, we can define open intervals and closed intervals in the natural way. For example, if $a \leqslant b$ and $a \neq b$, the open interval

$$(a, b) = \{x \in S \colon a \leqslant x \leqslant b \text{ and } a \neq x \text{ and } b \neq x\}$$

**a)** Give an example of an open interval $(a, b)$ in a linearly ordered set $(S, \leqslant)$ (with $a \leqslant b$ and $a \neq b$) such that $(a, b) = \varnothing$.

**b)** For any two points $(x_1, y_1)$ and $(x_2, y_2)$ of the plane, define $(x_1, y_1)\, R\, (x_2, y_2)$ provided that $x_1 < x_2$ or $x_1 = x_2$ and $y_1 < y_2$. Define $(x_1, y_1)\, R^*\, (x_2, y_2)$ provided that $(x_1, y_1) = (x_2, y_2)$ or $(x_1, y_1)\, R\, (x_2, y_2)$. Prove that $R^*$ is a linear order on the plane.

**c)** Sketch the open interval $(a, b)$ with respect to the linear order $R^*$ when

**i)** $a = (1/2, 3/4)$ and $b = (1/2, 1)$

**ii)** $a = (0, 1)$ and $b = (1, 0)$

[*Remark*: The linear order considered in this problem is called the "dictionary order" because deciding which of $(x_1, y_1)$ and $(x_2, y_2)$ comes first is like deciding which word of two two-letter words appears first in the dictionary. (Consider "at" versus "so," and "be" versus "by.") We hope this example serves to alert you that linearly ordered sets are not necessarily very line-like. If it has not served this purpose, we invite you to consider some intervals in the dictionary of all three-letter words (= 3-space).]

**106.** Prove that every well order is a linear order.

**107.** Prove or find a counterexample to the following statement: Every strict partial order is a partial order.

## WRITING EXERCISES

1. The text gives two methods of describing finite relations: Cartesian graphs and directed graphs. The incidence matrix of a relation $R$ on a finite set $\{a_1, a_2, a_3, \ldots, a_n\}$ is an $n \times n$ zero-one matrix whose $(i, j)$th entry is 0 if $(a_i, a_j) \notin R$ and is 1 if $(a_i, a_j) \in R$. Explain how to tell from its incidence matrix whether a given relation on $S$ is reflexive on $S$, symmetric, or transitive.

2. Compare the use of incidence matrices, Cartesian graphs, and directed graphs; discuss the advantages and disadvantages of each method.

3. Prove that for congruence mod 11, $[10^n] = [1]$ if $n$ is an even natural number and $[10^n] = [-1]$ if $n$ is an odd natural number. Explain how this result is the basis for a method of finding the remainder when a number is divided by 11. Illustrate your explanation with the number 10,472.

# CHAPTER 5

# FUNCTIONS

You have already encountered functions in the study of calculus and trigonometry. In this chapter, we consider these important relations from a more sophisticated point of view. We develop notation for certain functions called permutations, which play a central role in modern algebra, and we also consider set-valued functions, which are used in analysis and topology.

## 5.1 *Functions as Relations*

In elementary mathematics, we are taught to think of a function as a machine that assigns to each member of one set a member of another set. Then in algebra, trigonometry, and calculus, we study graphs of functions and begin to think of functions in a more precise manner. One way of thinking about functions is in terms of matching the members of one set with the members of another set; that is, we think about the members of the function as being ordered pairs. When we think of a function in this manner, we know that it must be a relation, and this motivates the following definition.

**Definition**

A **function** is a relation $f$ such that no distinct members of $f$ have the same first term. We denote the domain of $f$ by Dom $(f)$. If the function $f$ is a relation from $X$ to $Y$, we write $f: X \to Y$ and say that $f$ **maps** $X$ **into** $Y$. (Recall that $f$ is a relation from $X$ to $Y$ provided that $X$ is the domain of $f$ and $f \subseteq X \times Y$). Thus the notation $f: X \to Y$ promises that $f$ is a function whose domain is $X$ and whose range is contained in $Y$. Although $\varnothing$ is a function, we assume in this text that all functions have nonempty domain and range.

Personal Note

## Functional Notation

Although we now take the functional notation $f(x)$ for granted, this notation evolved from the writings of Jean Bernoulli (1667–1748) and his famous student Leonhard Euler (1707–1783). In a letter to Leibniz, Bernoulli wrote, "For denoting any function of a variable quantity $x$, I prefer to use the capital letter having the same name $X$ or the Greek $\xi$, for it appears at once of what variable it is a function; this relieves the memory" [2], page 267. Leibniz suggested an elaboration on this notation, which mercifully both he and Bernoulli abandoned. For a while, however, Bernoulli's notation was widely adopted, especially on the European continent. In 1734, Euler chose the letter $f$ to denote an arbitrary function, and in 1753 he denoted a function of two variables $x$ and $t$ by $\Phi$: $(x, t)$. This appears to be the first instance of enclosing the variables of a function in parentheses, but the notation was accepted quickly and within a year D'Alembert (1717–1783) was writing, "Let $\phi(z)$ be a function of the variable $z$...."

**EXAMPLE 1** Each of the following is a function $f$ that maps $X$ into $Y$.

**a)** $X = \{1, 2, 3, 4\},\ Y = \{5, 6, 7\},\ f = \{(1, 6), (2, 7), (3, 7), (4, 6)\}$

**b)** $X = \{1, 2, 3, 4\},\ Y = \{5, 6, 7\},\ f = \{(1, 7), (2, 5), (3, 7), (4, 6)\}$

**c)** $X = \{1, 2, 3\},\ Y = \{4, 5, 6, 7\},\ f = \{(1, 5), (2, 7), (3, 4)\}$

**d)** $X = \mathbb{N},\ Y = \mathbb{R},\ f = \{(n, x) \in \mathbb{N} \times \mathbb{R} : x = \sqrt{n}\}$

**e)** $X = \mathbb{R},\ Y = \mathbb{R},\ f = \{(x, y) \in \mathbb{R} \times \mathbb{R} : y = x^2\}$ ❑

If $f$ is a function and $(x, y) \in f$, we write $y = f(x)$ in place of $\{y\} = f[x]$. It is also customary to describe functions by using equations. For example, the terminology "Define $f: \mathbb{N} \to \mathbb{R}$ by $f(n) = \sqrt{n}$" describes the function in Example 1(d).

Suppose that we are given functions $f$ and $g$ and we wish to determine whether $f = g$. By definition, we must show that, as collections of ordered pairs, $f \subseteq g$ and $g \subseteq f$. The following important theorem allows us to show that the functions are equal without going all the way back to the primitive notion of a function as a set of ordered pairs.

**Theorem 5.1**

Let $f$ and $g$ be functions. Then $f = g$ if and only if

**a)** $\text{Dom}(f) = \text{Dom}(g)$, and

**b)** For each $x$ in the domain of $f$, $f(x) = g(x)$.

**PROOF** If $f = g$, we see by substitution that conditions (a) and (b) hold. Now suppose that conditions (a) and (b) hold. Let $(x, y) \in f$. Then $x \in \text{Dom}(f) = \text{Dom}(g)$, and $g(x) = f(x) = y$. By definition, $(x, y) \in g$. Thus, $f \subseteq g$. A similar argument establishes that $g \subseteq f$, and so $f = g$. □

**EXAMPLE 2** Define $f: \mathbb{Z} \to \mathbb{Z}$ by $f(x) = x^2$ and define $g: \mathbb{Z} \to \mathbb{N} \cup \{0\}$ by $g(x) = x^2$. Then by Theorem 5.1, $f = g$. ❑

**Definition** If $f: X \to Y$ and $Y$ is the range of $f$, then $f$ is said to **map $X$ onto $Y$**. A function $f$ is said to be a **one-to-one function** provided that no two distinct members of $f$ have the same second term.

Note that $f: X \to Y$ is a one-to-one function provided that $x_1 = x_2$ whenever $f(x_1) = f(x_2)$.

**OPINION** The English language uses nouns as adjectives, so we have school days, Guy Fawkes days, dog days, and even Shakespeare's salad days. But the promotion of a preposition to an adjective is a bit much. The phrase one sometimes hears that $f$ is "an onto" function is poor English, but it is ubiquitous.

**Definition** The term **surjective** is also used to describe a function $f: X \to Y$ that maps $X$ onto $Y$, and such a function is called a **surjection**. A one-to-one function $f: X \to Y$ is called an **injection** and is said to be **injective**.

**EXAMPLE 3** Let $X = \{1, 2, 3\}$, $Y = \{4, 5, 6, 7\}$, $A = \{8, 9, 10, 11\}$, and $B = \{12, 13, 14\}$. Then $f = \{(1, 4), (2, 5), (3, 6)\}$ is a one-to-one function from $X$ into $Y$ that does not map $X$ onto $Y$, and $g = \{(8, 12), (9, 13), (10, 14), (11, 14)\}$ is a function from $A$ onto $B$ that is not one-to-one. The functions $f$ and $g$ are illustrated in Figure 5.1.

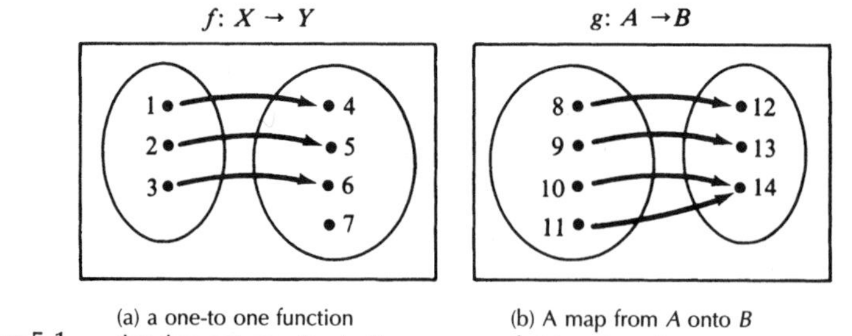


Figure 5.1 (a) a one-to one function that does not map $X$ onto $Y$ (b) A map from $A$ onto $B$ that is not one-to-one ❑

The terminology *one-to-one* is meant to suggest that if $x_1$ and $x_2$ are in the domain of the given function $f$ and $x_1 \neq x_2$, then $f(x_1) \neq f(x_2)$. That is, no two points are mapped to the same place. Nonetheless, our definition of a one-to-one function is not the statement that if $x_1 \neq x_2$, then $f(x_1) \neq f(x_2)$, but, instead, is its contrapositive: If $f(x_1) = f(x_2)$, then $x_1 = x_2$. The decision to choose the contrapositive is not pure perverseness, because the contrapositive is the easier of the two equivalent statements with which to work. The proof of Theorem 5.2, which follows, illustrates this point.

**Theorem 5.2**

> Let $f: X \to Y$. Then $f^{-1}: f(x) \to x$ is a one-to-one function if and only if $f^{-1}$ is a function.

**PROOF** Suppose that $x_1 = x_2$ whenever $f(x_1) = f(x_2)$. Let $(a, b)$ and $(c, d)$ be members of $f^{-1}$ and suppose that these ordered pairs have the same first term. Then $a = c$ and $(b, a)$ and $(d, a)$ are members of $f$. Since $f(b) = a = f(d)$, $b = d$. We have shown that no *two distinct* members of $f^{-1}$ have the same first term and so, by definition, $f^{-1}$ is a function.

Now suppose that $f^{-1}$ is a function and that $f(x_1) = f(x_2)$. Then $(x_1, f(x_1)) \in f$ and $(x_2, f(x_2)) \in f$ and so $(f(x_1), x_1) \in f^{-1}$ and $(f(x_2), x_2) \in f^{-1}$. Since $f(x_1) = f(x_2)$, we have that $x_1 = x_2$. □

**Definition**

> If $f: X \to Y$ and $A \subseteq X$, we define the *restriction* of $f$ to $A$ to be the function $\{(a, f(a)): a \in A\}$. This function is denoted by $f|A$.

Note that $f|A: A \to Y$.

**EXAMPLE 4**

Let $f: \mathbb{R} \to \mathbb{R}$ be defined by $f(x) = x^3$. Then $f$ is a one-to-one function mapping $\mathbb{R}$ onto $\mathbb{R}$. Figure 5.2 illustrates $f|[0, 1]$, $f|[-1, 1]$, and $f|\{-1, 1\}$.

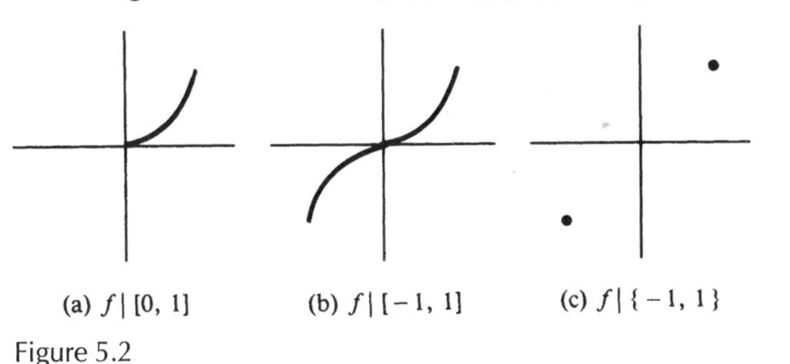

(a) $f|[0, 1]$ (b) $f|[-1, 1]$ (c) $f|\{-1, 1\}$

Figure 5.2 ❑

In general, the restriction of any one-to-one function to any subset of its domain is a one-to-one function.

**EXAMPLE 5**

Consider the five relations that follow.

**I.**

| | | | | |
|---|---|---|---|---|
| $d$ | · | ⊙ | · | · |
| $c$ | · | · | · | ⊙ |
| $b$ | · | ⊙ | ⊙ | · |
| $a$ | ⊙ | ⊙ | · | · |
| | $a$ | $b$ | $c$ | $d$ |

$f_1 = \{(a, a), (b, a), (b, b), (b, d), (c, b), (d, c)\}$

II.

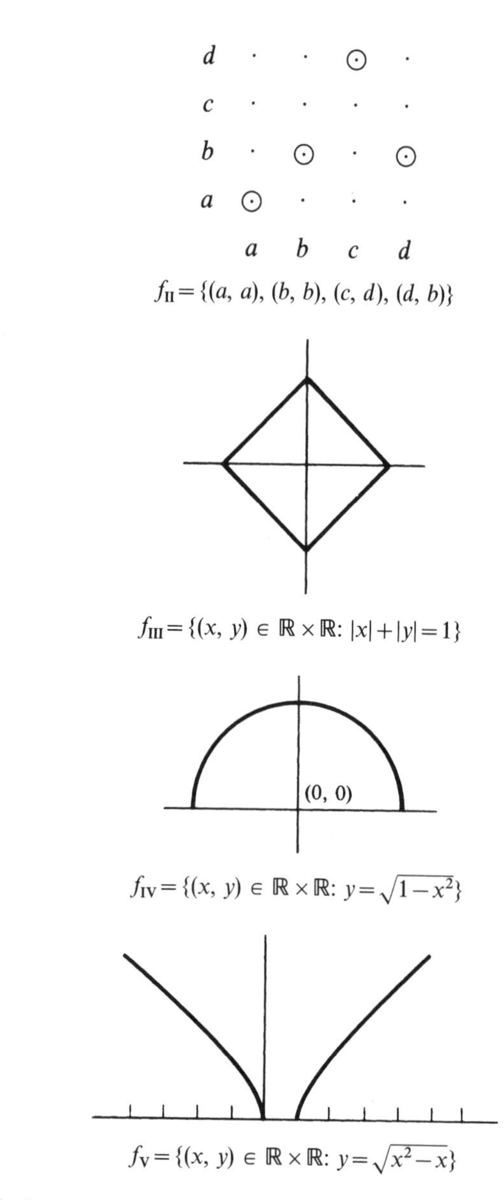


$f_{\text{II}} = \{(a, a), (b, b), (c, d), (d, b)\}$

III.

$f_{\text{III}} = \{(x, y) \in \mathbb{R} \times \mathbb{R}\colon |x| + |y| = 1\}$

IV.

$f_{\text{IV}} = \{(x, y) \in \mathbb{R} \times \mathbb{R}\colon y = \sqrt{1 - x^2}\}$

V.

$f_{\text{V}} = \{(x, y) \in \mathbb{R} \times \mathbb{R}\colon y = \sqrt{x^2 - x}\}$

The relations $f_{\text{I}}$ and $f_{\text{III}}$ are not functions (in each of the graphs, there is a vertical line that hits the graph in two or more places). The relations $f_{\text{II}}$, $f_{\text{IV}}$, and $f_{\text{V}}$ are functions, but none is a one-to-one function (in the graphs of each of these functions, there is a horizontal line that hits the graph in two or more places). The domain of $f_{\text{V}}$ is

$$\{x \in \mathbb{R}\colon x \leqslant 0\} \cup \{x \in \mathbb{R}\colon x \geqslant 1\}$$

Note that the restriction of $f_{\text{V}}$ to $[1, \infty)$ is a one-to-one function. ❏

Those of you who have not omitted Section 4.6, which is optional, will realize that, since functions are relations, we can define the composition $f \circ g$ of two functions $f$ and $g$ using the definition of composition of relations given on page 138. Indeed, we are about to repeat this definition; however, Theorem 5.1 allows us to state the definition in another form, which is often more pleasant to use. You will find, for example, that this other definition is useful in Exercise 10.

**Definition**

Let $f\colon Y \to Z$ and $g\colon X \to W$ be functions. Then

$$f \circ g = \{(x, z) \in X \times Z\colon \text{there is a } y \in Y \cap W \text{ such that } (x, y) \in g \text{ and } (y, z) \in f\}$$

Alternatively, if $D = \{x \in X\colon g(x) \in \text{Dom}(f)\}$, then $f \circ g\colon D \to Z$ is defined by $f \circ g(x) = f(g(x))$.

The following example illustrates how to find $f \circ g(x)$ and $g \circ f(x)$ given $x \in \mathbb{R}$ and the graphs of $f$ and $g$.

**EXAMPLE 6** Figure 5.3 is evidence that the calculations of $f \circ g(x)$ and $g \circ f(x)$ are not trivial.

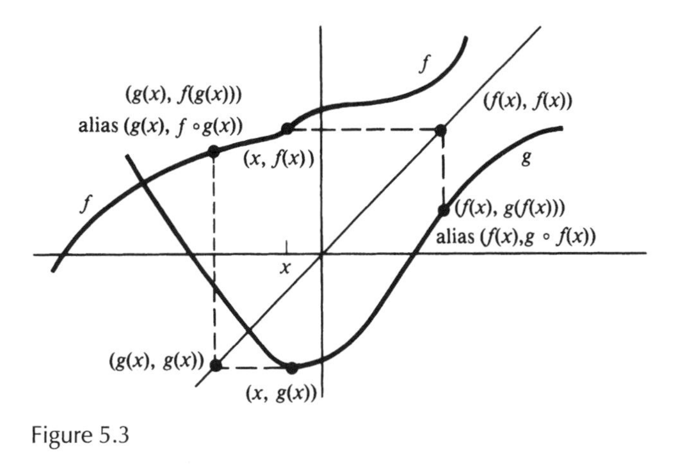


Figure 5.3 ❑

In Example 6, we considered the composition of two functions mapping $\mathbb{R}$ into $\mathbb{R}$. The composition of functions plays a less obvious role in the study of sequences. The intuitive concept of sequence is expressed in the non-mathematical use of the word in phrases such as "the sequence of events that led to the Second World War." There is an implied (well) ordering of the events: first event, second event, third event, and so forth. Just as we are unable to use the set $\{a, b\}$ for the ordered pair $(a, b)$ (since $\{a, b\} = \{b, a\}$), we cannot use the set $\{1, 1/2, 1/3, 1/4, \ldots\}$ to denote the sequence whose $n$th "event" is $1/n$. Sets do not respect our intended order. The solution to our predicament is easy.

**Definition**

A **sequence** is a function whose domain is the set of all natural numbers.

Technically, the sequence $\langle 1/n \rangle$ is the function $f: \mathbb{N} \to \mathbb{R}$ whose $n$th term is $(n, 1/n)$. This leads to the definition of the $n$th term of a sequence.

**Definition**

The ***n*th term** of a sequence $\langle f(n) \rangle$ is $(n, f(n))$.

Often when we are given a sequence such as $\langle 1/n \rangle$, we wish to select a subsequence—say, for example, the subsequence $\langle 1/(3n+1) \rangle$, whose first term is $(1, 1/4)$ and whose fifth term is $(5, 1/16)$. We need a new function whose domain is $\mathbb{N}$, and we see that the function we have in mind is $f \circ g$ where, for each $n \in \mathbb{N}$, $g(n) = 3n + 1$. This leads to the definition of a *subsequence* of a sequence $\langle f(n) \rangle$.

**Definition**

Let $\langle f(n) \rangle$ be a sequence, and let $g: \mathbb{N} \to \mathbb{N}$ be a sequence such that for each $n \in \mathbb{N}$, $g(n+1) > g(n)$. Then $\langle f \circ g(n) \rangle$ is a **subsequence** of $\langle f(n) \rangle$.

**EXAMPLE 7**

Let $f: \mathbb{N} \to \mathbb{R}$ be the sequence $\langle f(n) \rangle = \langle 5 + (-1)^n/n^2 \rangle$. The first four terms of this sequence are $(1, 4)$, $(2, 21/4)$, $(3, 44/9)$, and $(4, 81/16)$. This sequence has a largest possible subsequence whose range is a subset of $\{x \in \mathbb{R}: x > 5\}$. The first four terms of this subsequence are $(1, 21/4)$, $(2, 81/16)$, $(3, 181/36)$, and $(4, 321/64)$. The subsequence is given by $\langle f \circ g(n) \rangle$, where $g(n) = 2n$. ❑

## EXERCISES 5.1

1. Each of the following functions maps $\mathbb{Z}$ into $\mathbb{Z}$. For each function, determine whether it is a one-to-one function and whether it maps $\mathbb{Z}$ onto $\mathbb{Z}$. Prove your answers.

   **a)** $f(n) = 2n + 1$ **b)** $f(n) = 4n + 2$

   **c)** $f(n) = n^2$ **d)** $f(n) = n^3 - n^2$

**2.** Indicate the position of $(a, f \circ g(a))$ and $(a, g \circ f(a))$ on each of the following graphs.

**a)**

**b)**

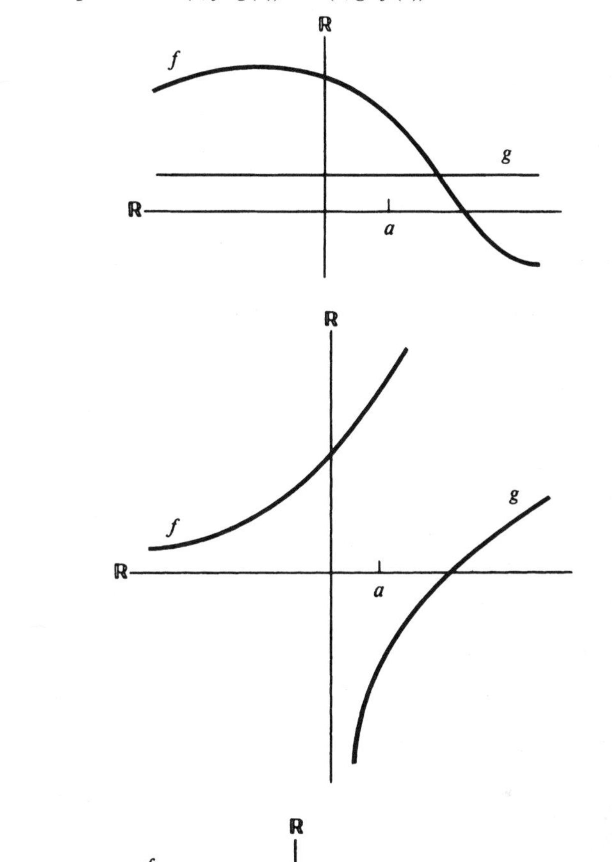


**c)**

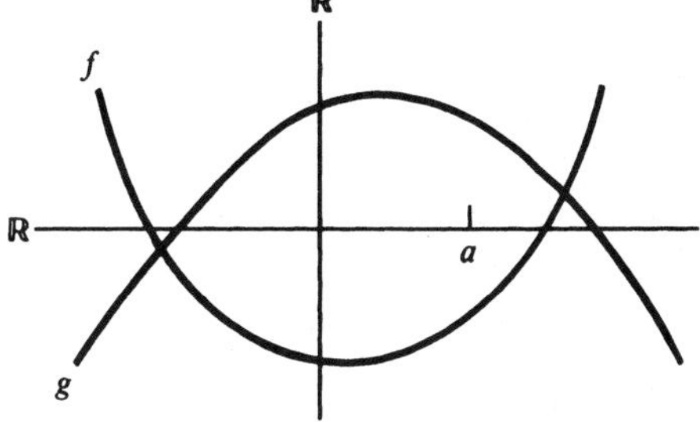


**3.** Let $f$ and $g$ be the following one-to-one functions:

(4, 4) (4, 4)

(1, 1) (1, 1)

$f$ $g$

Draw the graph of

**a)** $f \circ g$ **b)** $g \circ f$ **c)** $(f \circ g)^{-1}$ **d)** $g^{-1} \circ f^{-1}$

**4.** Let $f: \mathbb{R} \to \mathbb{R}$. Then $f$ is **strictly increasing** provided that $f(a) < f(b)$ whenever $a < b$, and $f$ is **strictly decreasing** provided that $f(a) > f(b)$ whenever $a < b$. Prove that if $f$ is strictly increasing, then $f$ is one-to-one and $f^{-1}$ is strictly increasing.

**5.** Let $f: \mathbb{R} \to \mathbb{R}$. Prove that if $f$ is strictly decreasing, then $f$ is one-to-one and $f^{-1}$ is strictly decreasing.

**6.** For each natural number $n$, let $f(n)$ be the number of digits in $n$ (for example, $f(437) = 3$ and $f(19) = 2$). Then $f: \mathbb{N} \to \mathbb{N}$ is a function. Is $f$ a one-to-one map? Does $f$ map $\mathbb{N}$ onto $\mathbb{N}$?

**7. a)** Give an example of a one-to-one function $f: \mathbb{N} \to \mathbb{N}$ that does not map $\mathbb{N}$ onto $\mathbb{N}$.

**b)** Give an example of a function $f$ mapping $\mathbb{N}$ onto $\mathbb{N}$ that is not a one-to-one function.

**c)** Give an example of a function $f: \mathbb{N} \to \mathbb{N}$ that is not one-to-one and does not map onto $\mathbb{N}$.

**d)** Let $X = \{1, 2, 3, 4\}$. Does there exist a one-to-one function $f: X \to X$ that does not map $X$ onto $X$? Does there exist a function $f$ mapping $X$ onto $X$ that is not a one-to-one function?

**8.** Let $f: (-\infty, -1] \to [4, \infty)$ be defined by $f(x) = x^2 + 2x + 5$. Prove that $f$ is a one-to-one function that maps onto $[4, \infty)$.

**9.** Let $A = \{X: X \text{ is a nonempty finite subset of } \mathbb{Z}\}$, and define $f: A \to \mathbb{R}$ by $f(X) =$ the average of the members of $X$ [for example, if $X = \{2, 17, 21, 23\}$, then $f(X) = (2 + 17 + 21 + 23)/4 = 15.75$]. Is $f$ one-to-one? Does $f$ map $A$ onto $\mathbb{R}$? In each case, justify your answer.

**10.** Let $A$, $B$, $C$, and $D$ be sets, and let $h: A \to B$, $g: B \to C$, and $f: C \to D$.

**a)** Prove that Dom $[f \circ (g \circ h)]$ = Dom $[(f \circ g) \circ h]$.

**b)** Use Theorem 5.1 to show that $f \circ (g \circ h) = (f \circ g) \circ h$.

**11.** Let $f$ and $g$ be functions. Either prove or give a counterexample to the conjecture that $f \cup g$ is a function.

**12. a)** Let $A$ be a set, and let $f$ and $g$ be functions from $A$ to $A$. Prove that $(f \circ g)^{-1} = g^{-1} \circ f^{-1}$.

**b)** Give an example to show that there is a set $A$ and functions $f$ and $g$ from $A$ to $A$ such that $(f \circ g)^{-1} \neq f^{-1} \circ g^{-1}$.

**13.** Find the second, ninth, and tenth terms of each of the following sequences.

**a)** $\langle (-1)^n \rangle$ **b)** $\langle (n^2 - 2n)/5n \rangle$ **c)** $\langle \sin (n\pi/2) \rangle$

**14.** Let $\langle f(n) \rangle = \langle (n^2 - 2n)/5n \rangle$, and let $g(n) = 3n - 1$.

**a)** Find the second, ninth, and tenth terms of the subsequence $\langle f \circ g(n) \rangle$.

**b)** What is a formula for the $n$th term of this subsequence?

**15.** Given the sequence $\langle f(n) \rangle$, what function $g: \mathbb{N} \to \mathbb{N}$ has the property that $\langle f \circ g(n) \rangle$ is the subsequence consisting of all the odd terms of the original sequence $\langle f(n) \rangle$?

***16.** A sequence $f: \mathbb{N} \to \mathbb{R}$ converges to $p$ provided that for each $\varepsilon > 0$ there is a natural number $N$ such that if $n \geqslant N$, then $|f(n) - p| < \varepsilon$. Prove that if a sequence $f: \mathbb{N} \to \mathbb{R}$ converges to $p$, then so does each of its subsequences.

## 5.2 Functions Viewed Globally

To this point, we have been considering functions more or less one point at a time and one function at a time. In this section, we deal with the way in which functions interact under composition, and instead of asking questions such as "where does the function $f$ map $x$," we look for the global effects of $f$. We begin by characterizing the one-to-one functions by the way in which these functions interact with other functions. Recall that if $S$ is a set, then $i_S: S \to S$ is the function defined by $i_S(x) = x$ for each $x \in S$.

**Theorem 5.3**

Let $f: A \to B$ be a function. Then $f$ is one-to-one if and only if whenever $C$ is a set and $g: C \to A$ and $h: C \to A$ are functions such that $f \circ g = f \circ h$, it follows that $g = h$.

**PROOF** Suppose that whenever $C$ is a set and $g: C \to A$ and $h: C \to A$ are functions such that $f \circ g = f \circ h$, it follows that $g = h$. Let $p$ and $q$ be elements of $A$ such that $f(p) = f(q)$. To show that $f$ is one-to-one, we must show that $p = q$. Define $g: A \to A$ by $g(x) = x$ if $x \neq p$ and $g(p) = q$. Since $f \circ g = f \circ i_A$, it follows that $g = i_A$. In particular, $q = g(p) = i_A(p) = p$.

Now suppose that $f$ is one-to-one, let $C$ be a set, and let $g: C \to A$ and $h: C \to A$ be functions such that $f \circ g = f \circ h$. Then for each $x \in C$, $f(g(x)) = f(h(x))$, and because $f$ is one-to-one, $g(x) = h(x)$. Since $g$ and $h$ have the same domain, by Theorem 5.2, $g = h$. □

Stated informally, Theorem 5.3 says that the one-to-one functions are those functions that can be canceled from the left. It is natural to ask what property characterizes those functions that can be canceled from the right.

**Theorem 5.4**

Let $f: A \to B$ be a function. Then $f$ maps $A$ onto $B$ if and only if whenever $C$ is a set and $g: B \to C$ and $h: B \to C$ are functions such that $g \circ f = h \circ f$, it follows that $g = h$.

**PROOF** We prove only that if $f$ maps $A$ onto $B$, then $f$ is *right cancelable.* The remainder of the proof is left as Exercise 30. Suppose that $f$ maps $A$ onto $B$, and suppose that $C$ is a set and $g: B \to C$ and $h: B \to C$ are functions such that $g \circ f = h \circ f$. Let $b \in B$. Since $\text{Dom}(g) = B = \text{Dom}(h)$, by Theorem 5.2, it suffices to show that $g(b) = h(b)$. Since $f$ maps $A$ onto $B$, there is an $a \in A$ such that $f(a) = b$. By hypothesis, $g \circ f(a) = h \circ f(a)$, and so $g(b) = h(b)$. □

**Definition**

A one-to-one function from a set $A$ onto a set $B$ is called a **bijection** from $A$ to $B$.

Often in mathematics we define two mathematical objects $G$ and $H$ to be essentially the same if there is a bijection from $G$ to $H$ that behaves "nicely." The meaning of "nicely" varies from situation to situation, but in each case an equivalent relation is defined by agreeing that two objects $G$ and $H$ are equivalence if there is a "nice" bijection from $G$ to $H$. In algebra "nice" bijections from $G$ to $H$ are called *isomorphisms*, and in topology they are called *homeomorphisms*. Because in many branches of mathematics it is important to be able to decide whether a function $f: G \to H$ is a bijection from $G$ to $H$, the consequences of Theorems 5.3 and 5.4 given below are useful in their own right.

**Theorem 5.5**

Let $f: A \to B$ and $g: B \to A$.

**a)** If $f \circ g = i_B$, then $f$ maps $A$ onto $B$.

**b)** If $g \circ f = i_A$, then $f$ is a one-to-one function.

**c)** Moreover, if the hypotheses of conditions (a) and (b) are true, then $f = g^{-1}$ and $g = f^{-1}$.

**PROOF** Suppose that $f \circ g = i_B$. Let $C$ be a set, and let $h: B \to C$ and $k: B \to C$ be functions such that $k \circ f = h \circ f$. In order to show that $f$ maps $A$ onto $B$, we must show that $k = h$. Since $(k \circ f) \circ g = (h \circ f) \circ g$, $k \circ (f \circ g) = h \circ (f \circ g)$, and so $k = k \circ i_B = h \circ i_B = h$.

Now suppose that $g \circ f = i_A$. Let $C$ be a set, and let $h: C \to A$ and $k: C \to A$ be functions such that $f \circ h = f \circ k$. In order to show that $f$ is one-to-one, we must show that $h = k$. Since $g \circ (f \circ h) = g \circ (f \circ k)$, $(g \circ f) \circ h = (g \circ f) \circ k$, and so $h = i_A \circ h = i_A \circ k = k$.

Now suppose that both hypotheses of conditions (a) and (b) hold. We show that $f = g^{-1}$, from which it follows that $g = f^{-1}$. Since $g \circ f = i_A$, $g^{-1} = g^{-1} \circ i_A = g^{-1} \circ (g \circ f) = (g^{-1} \circ g) \circ f = i_B \circ f = f$. □

In Proposition 5.6, we illustrate how Theorem 5.5 can be used.

**Proposition 5.6**

Let $X = (-1, 1)$, and let $f: \mathbb{R} \to X$ be defined by

$$f(x) = \frac{x}{1 + |x|}$$

Then $f$ is a one-to-one function from $\mathbb{R}$ onto $X$.

**PROOF** Let $g: X \to \mathbb{R}$ be defined by

$$g(x) = \frac{x}{1 - |x|}$$

Then

$$f \circ g(x) = f\left(\frac{x}{1-|x|}\right)$$

$$= \frac{x}{1-|x|} \div \left(1 + \frac{|x|}{|1-|x||}\right)$$

$$= \frac{x}{1-|x|} \div \frac{1}{1-|x|}$$

$$= x$$

By Theorem 5.5(a), $f$ maps $\mathbb{R}$ onto $X$.

Moreover, as you should verify, $g \circ f(x) = x$, so by Theorem 5.5(b), $f$ is a one-to-one function. □

As Proposition 5.6 indicates, there are times when Theorem 5.5 is useful. The following example illustrates a standard approach to showing that a given function $f: X \to Y$ is a one-to-one function mapping $X$ onto $Y$. This standard approach is not so powerful as Theorem 5.5. Still, in those situations in which it applies, it is generally easier to use. We have another use for the example as well. It may seem that the function $g$ used in the proof of Proposition 5.6 appeared miraculously out of the clouds. But the equation $f(f^{-1}(x)) = x$ can often be used to determine a formula for $f^{-1}$, and that is how we found the function $g$ used in Proposition 5.6. Rather than burden you with the somewhat involved algebraic manipulations we used to determine $g$, we find a formula for the inverse of the simpler function given in Example 8.

**EXAMPLE 8**

Let $f: [-3, \infty) \to [-8, \infty)$ be defined by $f(x) = x^2 + 6x + 1$. Then $f$ is a one-to-one function that maps onto $[-8, \infty)$. Moreover, $f^{-1}: [-8, \infty) \to [-3, \infty)$ is defined by $f^{-1}(x) = -3 + \sqrt{8 + x}$.

To see that $f$ is one-to-one, suppose that $f(a) = f(b)$. Then $a^2 + 6a + 1 = b^2 + 6b + 1$ so that $(a - b)(a + b) + 6(a - b) = (a - b)(a + b + 6) = 0$. As $a \geqslant -3$ and $b \geqslant -3$, $a + b + 6 = 0$ only when $a = b = -3$. Of course, $a - b = 0$ only when $a = b$. Therefore, $a = b$ whenever $f(a) = f(b)$, and so $f$ is one-to-one.

To see that $f$ maps onto $[-8, \infty)$, let $y \geqslant -8$. We are searching for an $x \geqslant -3$ such that $f(x) = x^2 + 6x + 1 = y$. Using the quadratic formula, we see that $x = -3 + \sqrt{9 - (1 - y)}$ is the number in Dom $(f)$ that maps to $y$.

Since $f$ is one-to-one, $f^{-1}$ exists, and since $f$ maps onto $[-8, \infty)$, we have that Dom $(f^{-1}) = [-8, \infty)$. For each $x \in [-8, \infty)$, $f(f^{-1}(x)) = x$, and so $[f^{-1}(x)]^2 + 6f^{-1}(x) + 1 = x$. Using the quadratic formula, we obtain $f^{-1}(x) = -3 \pm \sqrt{9 - (1 - x)}$ and since the range of $f^{-1}$ is $[-3, \infty)$,

$f^{-1}(x) = -3 + \sqrt{8+x}$. Note that the algebra used to determine that $f^{-1}(x) = -3 + \sqrt{8+x}$ is the same as the algebra used to show that $f$ maps onto $[-8, \infty)$. ❑

We end this section with a theorem that is painfully obvious, but so important that it must be stated in a prominent way.

**Theorem 5.7**

Let $g\colon A \to B$ and $f\colon B \to C$.

**a)** If $g$ maps $A$ onto $B$ and $f$ maps $B$ onto $C$, then $f \circ g$ maps $A$ onto $C$.

**b)** If $g$ and $f$ are one-to-one functions, so are $g^{-1}$ and $f \circ g$.

**c)** If $g$ is a one-to-one function from $A$ onto $B$, then $g^{-1}$ is a one-to-one function from $B$ onto $A$.

**PROOF**

**a)** Suppose that $g$ maps $A$ onto $B$ and $f$ maps $B$ onto $C$. Let $c \in C$. Since $f$ maps $B$ onto $C$, there exists $b \in B$ such that $f(b) = c$. Since $g$ maps $A$ onto $B$, there is an $x \in A$ such that $g(x) = b$. Thus, $f \circ g(x) = f(g(x)) = f(b) = c$, and so $f \circ g$ maps onto $C$.

**b)** Suppose that $f$ and $g$ are one-to-one functions. We first prove that $g^{-1}$ is a one-to-one function. Since $g$ is one-to-one, by Theorem 5.2, $g^{-1}$ is a function. Moreover, $(g^{-1})^{-1} = g$, which is a function. By Theorem 5.2, $g^{-1}$ is one-to-one.

To see that $f \circ g$ is one-to-one, suppose that $f \circ g(x) = f \circ g(y)$. Then $f(g(x)) = f(g(y))$, and since $f$ is one-to-one, $g(x) = g(y)$. Since $g$ is one-to-one, $x = y$. Thus, $f \circ g$ is one-to-one.

**c)** Suppose that $g$ is a one-to-one function from $A$ onto $B$. By part (b), $g^{-1}$ is a one-to-one function, and it is obvious that $g^{-1}$ maps $B$ onto $A$. □

## EXERCISES 5.2

**17.** Let $f\colon \mathbb{R} \to \mathbb{R}$ be defined by $f(x) = 17 - 3x$. Prove that $f$ is a one-to-one function mapping $\mathbb{R}$ onto $\mathbb{R}$, and determine a formula for $f^{-1}\colon \mathbb{R} \to \mathbb{R}$.

**18.** Let $f\colon (-\infty, 17/3] \to \mathbb{R}$ be defined by $f(x) = \sqrt{17 - 3x}$. Prove that $f$ is a one-to-one function mapping onto $[0, \infty)$. Determine a formula for $f^{-1}\colon [0, \infty) \to \mathbb{R}$.

**19.** Let $f\colon (-\infty, 17/3] \to \mathbb{R}$ be defined by $f(x) = (17 - 3x)^2$. Prove that $f$ is a one-to-one function mapping onto $[0, \infty)$. Determine a formula for $f^{-1}\colon [0, \infty) \to (-\infty, 17/3]$.

**20.** Let $f\colon (17/3, \infty) \to (-\infty, 0)$ be defined by $f(x) = 1/(17 - 3x)$. Prove that $f$ is a one-to-one function mapping onto $(-\infty, 0)$, and determine a formula for $f^{-1}\colon (-\infty, 0) \to (17/3, \infty)$.

21. Define $f: [-3, \infty) \to [-8, \infty)$ by $f(x) = x^2 + 6x + 1$, and define $g: \mathbb{R} \to \mathbb{R}$ by $g(x) = 2x + 3$. Prove that $g \circ f$ is one-to-one.
22. Define $f: (-\infty, 0] \to \mathbb{R}$ by $f(x) = x^2$, and define $g: [0, \infty) \to \mathbb{R}$ by $g(x) = x/(1-x)$ if $0 \leqslant x < 1$, and $g(x) = 1 - x$ if $x \geqslant 1$. Does $g \circ f$ map $(-\infty, 0]$ onto $\mathbb{R}$? Justify your answer.
23. Let $g: A \to B$ and $f: B \to C$ be functions. Prove the following result: If $f \circ g: A \to C$ maps $A$ onto $C$, then $f: B \to C$ maps $B$ onto $C$. Show by example that the converse does not hold. How does this exercise relate to Theorem 5.7(a)?
24. Let $g: A \to B$ and $f: B \to C$ be functions. Prove the following result: If $f \circ g: A \to C$ is a one-to-one function, then $g: A \to B$ is a one-to-one function. Show by example that the converse does not hold. How does this exercise relate to Theorem 5.7(b)?
25. Prove Theorem 5.5(a) without using Theorem 5.4.
26. Prove Theorem 5.5(b) without using Theorem 5.3.
27. Use Theorem 5.4 to prove Theorem 5.7(a).
28. Use Theorem 5.3 to prove that if $g: A \to B$ and $f: B \to C$ are one-to-one functions, then $f \circ g: A \to C$ is one-to-one.

*29. Use Theorems 5.5(a) and (b) to prove Theorem 5.7(c).

30. Let $f: A \to B$. Suppose that whenever $C$ is a set and $g: B \to C$ and $h: B \to C$ are functions such that $g \circ f = h \circ f$, it follows that $g = h$. Prove that $f$ maps $A$ onto $B$.

## 5.3 Permutations

**Definition**

Let $S$ be a nonempty set. A one-to-one function mapping $S$ onto $S$ is called a **permutation on $S$.**

The collection of all permutations on $S$ is denoted by Sym $(S)$. The proof of the first proposition, which consists of tracking down results that have already been established, is left as Exercise 31.

**Proposition 5.8**

Let $S$ be a nonempty set. The collection Sym $(S)$ of all permutations on $S$ has the following properties:

**a)** $\text{Sym}(S) \neq \varnothing$.

**b)** If $f, g \in \text{Sym}(S)$, then $f \circ g \in \text{Sym}(S)$.

**c)** If $f \in \text{Sym}(S)$, then $f^{-1} \in \text{Sym}(S)$.

**d)** There exists $i_S \in \text{Sym}(S)$ such that for each $f \in \text{Sym}(S)$, $f \circ i_S = i_S \circ f = f$.

**e)** For each $f \in \text{Sym}(S)$, $f \circ f^{-1} = f^{-1} \circ f = i_S$.

**f)** For $f, g, h \in \text{Sym}(S)$, $f \circ (g \circ h) = (f \circ g) \circ h$.

If S is a set with exactly one member, say $x$, then $\{(x, x)\}$ is the only permutation on $S$. If $S$ has exactly two members, say $x_1, x_2$, then there are two permutations on $S$. They are $\{(x_1, x_1), (x_2, x_2)\} = i_S$ and $\{(x_1, x_2), (x_2, x_1)\}$. Although we have considered only sets with one or two members, we hope you are already beginning to sense that our ordered-pair notation for functions needs to be replaced by less cumbersome notation. First we note that if $S$ is some finite set $\{x_1, x_2, \ldots, x_n\}$ and we are interested in the permutations on $S$ rather than in $S$ itself, then there is no real harm in considering the permutations on $\{1, 2, \ldots, n\}$ instead of the permutations on $\{x_1, x_2, \ldots, x_n\} = S$. Since in this text, we consider only permutations on a finite set, we restrict our attention to sets of the form $\{1, 2, \ldots, n\}$ for some natural number $n$. It is customary to denote the set of all permutations on $\{1, 2, 3, \ldots, n\}$ by $S_n$ rather than by the more cumbersome Sym $(\{1, 2, \ldots, n\})$. Let us consider a permutation $\alpha = \{(1, 3), (2, 2), (3, 4), (4, 1)\}$ belonging to $S_4$. The function $\alpha$ maps 1 to 3, 3 to 4, and 4 to 1, whereas it maps 2 to 2. The associated directed graph is shown in Figure 5.4. The directed graph seems easier to think about, but if you imagine having to type a paper about $S_{24}$, you will soon come to the conclusion that we need some still more convenient way to write permutations. We denote $\alpha$ by $(134)(2)$. This notation, which for obvious reasons is called *cycle notation*, is illustrated in the next examples.

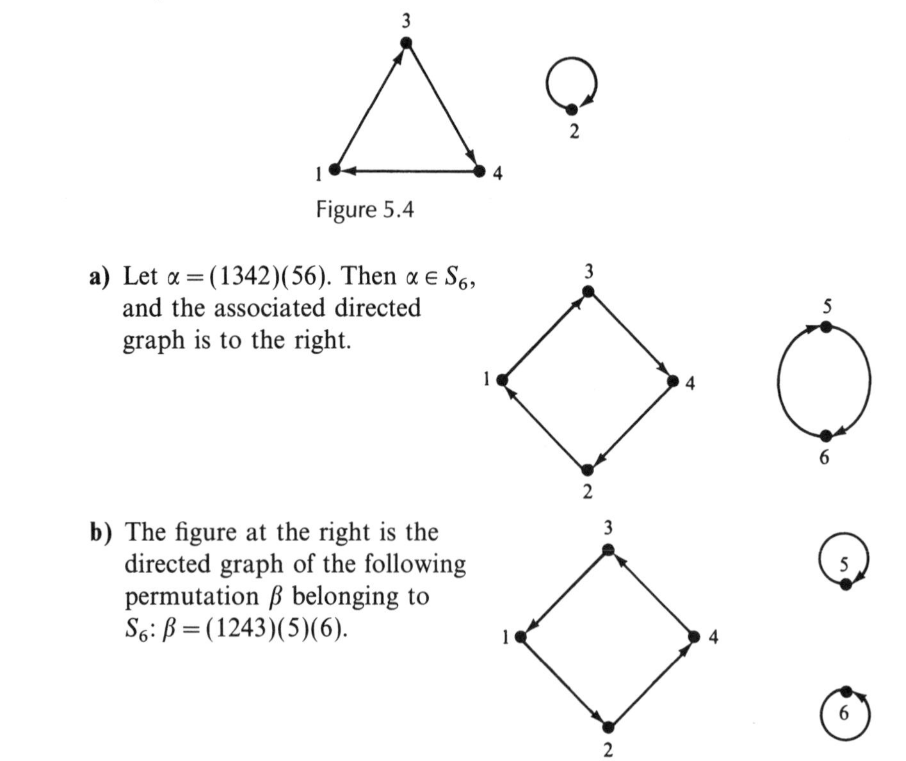

Figure 5.4

**EXAMPLE 9**

**a)** Let $\alpha = (1342)(56)$. Then $\alpha \in S_6$, and the associated directed graph is to the right.

**b)** The figure at the right is the directed graph of the following permutation $\beta$ belonging to $S_6$: $\beta = (1243)(5)(6)$.

Once we recall that the inverse of a relation is obtained from the relation's associated directed graph by reversing all the arrows in the graph, it becomes easy to write the directed graphs of $\alpha^{-1}$ and $\beta^{-1}$.

**c)** $\alpha^{-1} = (1243)(56)$

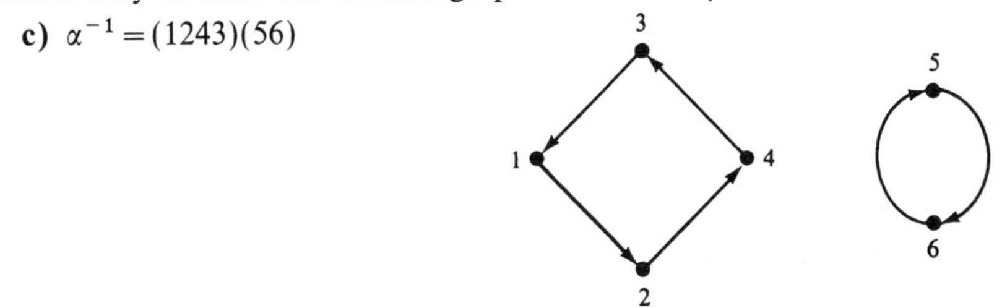


Note that reversing the arrows between 5 and 6 accomplished nothing, so we continued to write $\alpha^{-1} = (1243)(56)$ in place of $\alpha^{-1} = (1243)(65)$. There is nothing wrong with the latter notation; indeed, we could have written $\alpha^{-1} = (4312)(65)$.

**d)** Since $\beta = (1243)(5)(6)$, we obtain $\beta^{-1}$ by writing each parenthetical expression backwards. That is, $\beta^{-1} = (3421)(5)(6)$ or $\beta^{-1} = (1342)(5)(6)$.

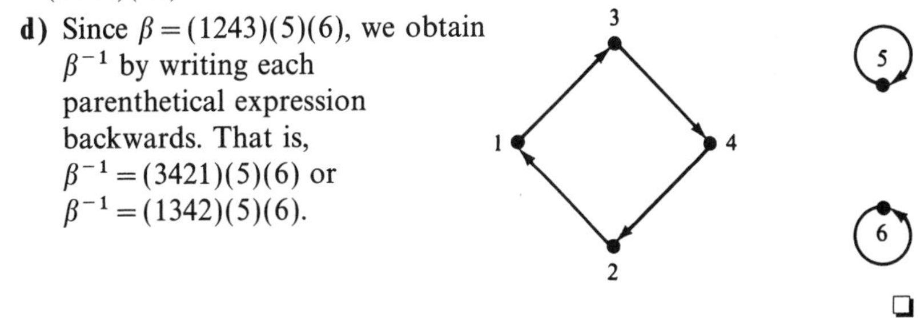


❑

**EXAMPLE 10**

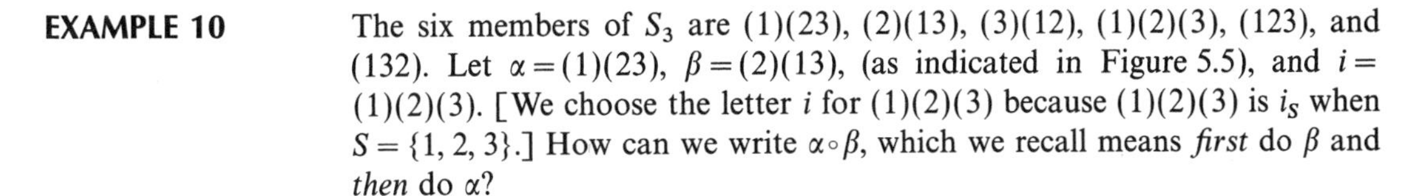

The six members of $S_3$ are (1)(23), (2)(13), (3)(12), (1)(2)(3), (123), and (132). Let $\alpha = (1)(23)$, $\beta = (2)(13)$, (as indicated in Figure 5.5), and $i = (1)(2)(3)$. [We choose the letter $i$ for (1)(2)(3) because (1)(2)(3) is $i_S$ when $S = \{1, 2, 3\}$.] How can we write $\alpha \circ \beta$, which we recall means *first* do $\beta$ and *then* do $\alpha$?

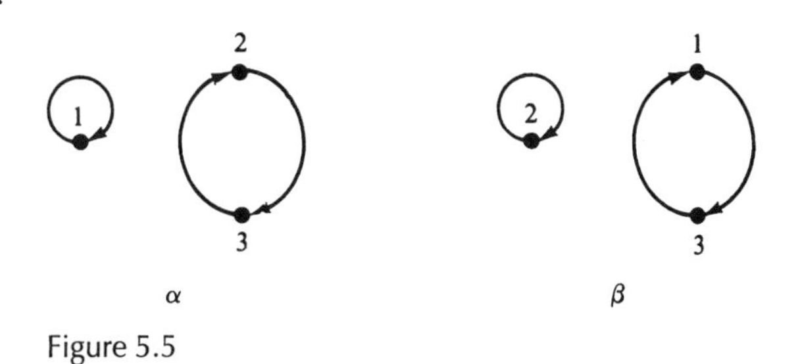


Figure 5.5

We first determine $\alpha \circ \beta(1)$. Well, $\beta$ takes 1 to 3 and $\alpha$ takes 3 to 2. Thus, $\alpha \circ \beta(1) = \alpha(\beta(1)) = \alpha(3) = 2$. Now, $\alpha \circ \beta(2) = \alpha(\beta(2)) = \alpha(2) = 3$. Therefore, we have $\alpha \circ \beta = (123)$. Note that we never checked up on $\alpha \circ \beta(3)$ because, by Proposition 5.8(b), $\alpha \circ \beta$ is a permutation on $\{1, 2, 3\}$, and since 2 and 3 have already appeared in the range of $\alpha \circ \beta$, 3 has nowhere left to go but to 1.

❑

Unless a given permutation is the identity permutation, which is denoted by $i$, it is customary to omit its 1-cycles when writing it in cycle notation. Thus the permutation $\beta$ of Example 9 is written (1243) rather than (1243)(5)(6). For example, the permutation $\alpha=(1342)(56)$ of Example 9 is the composition $\alpha=\delta\circ\varepsilon$, where $\delta=(1342)$ and $\varepsilon=(56)$. In the more cumbersome notation of sets of ordered pairs

$$\alpha=\{(1,3),(3,4),(4,2),(2,1),(5,6),(6,5)\}$$

$$\delta=\{(1,3),(3,4),(4,2),(2,1),(5,5),(6,6)\}$$

and $$\varepsilon=\{(1,1),(2,2),(3,3),(4,4),(5,6),(6,5)\}$$

So $\alpha=\delta\circ\varepsilon$.

## EXERCISES 5.3

**31.** Prove Proposition 5.8 by citing appropriate previous results where possible.

**32.** Write in cycle notation the members of $S_7$ whose directed graphs follow.

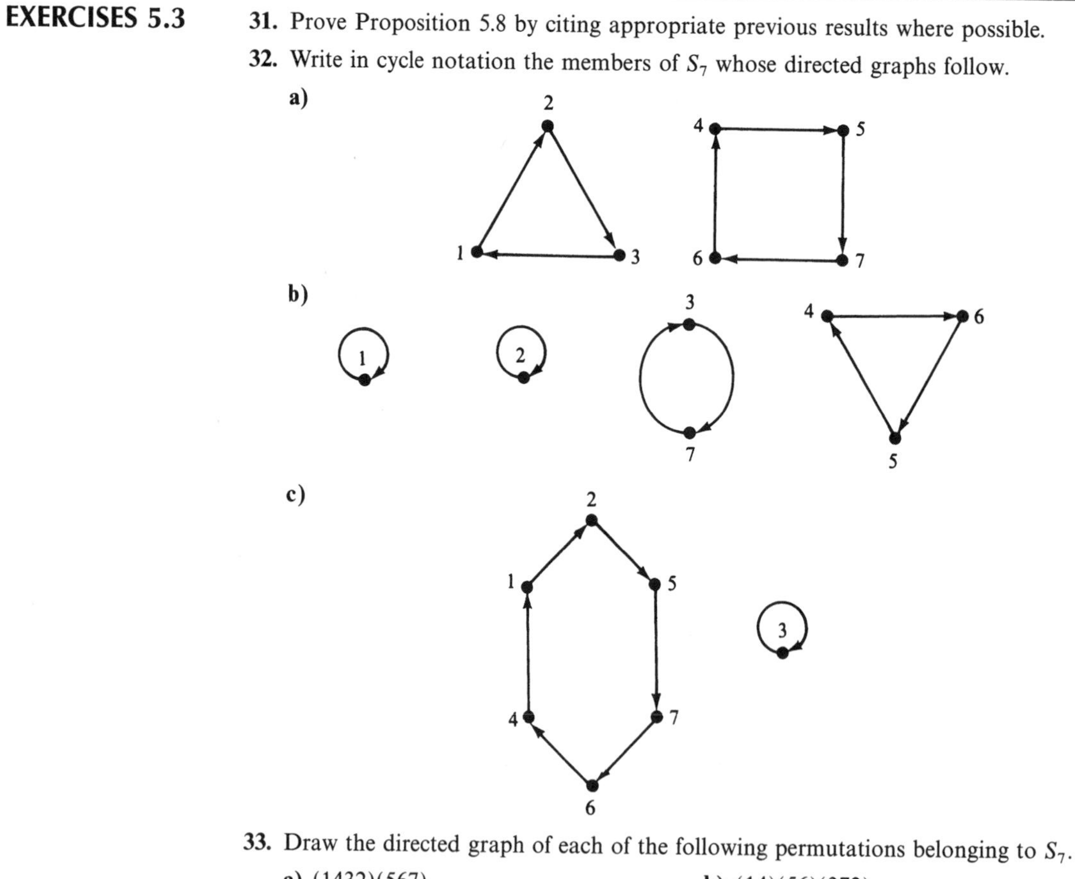


**33.** Draw the directed graph of each of the following permutations belonging to $S_7$.

**a)** (1432)(567)

**b)** (14)(56)(372)

**c)** (1)(2)(3)(4)(5)(6)(7)

**d)** (1234567)

**34.** Which of the following pairs denote the same permutation in $S_6$?

**a)** (365)(241), (536)(412)

**b)** (365)(241), (356)(214)

**c)** (36)(52)(41), (63)(25)(14)

**d)** (1)(2)(3)(4)(5)(6), (6)(5)(2)(3)(4)(1)

**e)** (654321), (123456)

**35.** Let $\alpha = (326451)$. Find each of the individual permutations and draw its associated directed graph.

**a)** $\alpha \circ \alpha$

**b)** $\alpha \circ \alpha \circ \alpha$

**c)** $\alpha \circ \alpha \circ \alpha \circ \alpha$

**d)** $\alpha \circ \alpha \circ \alpha \circ \alpha \circ \alpha$

**e)** $\alpha \circ \alpha \circ \alpha \circ \alpha \circ \alpha \circ \alpha$

**36.** Give an example of $\alpha \in S_6$ such that $\alpha \neq (1)(2)(3)(4)(5)(6)$, but $\alpha \circ \alpha = (1)(2)(3)(4)(5)(6)$.

**37. a)** Give an example of two different members $\alpha$ and $\beta$ of $S_6$ such that $\alpha \circ \beta = \beta \circ \alpha$.

**b)** Give an example of two different members $\alpha$ and $\beta$ of $S_6$ such that $\alpha \circ \beta \neq \beta \circ \alpha$.

**38.** Let $\alpha = (124)(35)(6)$ and $\beta = (145)(326)$.

**a)** Write $\alpha \circ \beta$ in cycle notation and draw its directed graph.

**b)** Write $\beta \circ \alpha$ in cycle notation and draw its directed graph.

**c)** Write $\alpha^{-1}$ in cycle notation and $\beta^{-1}$ in cycle notation.

**d)** Draw the directed graph of $(\alpha \circ \beta)^{-1}$.

**e)** Draw the directed graph of $\alpha^{-1} \circ \beta^{-1}$.

**f)** Write $\beta^{-1} \circ \alpha^{-1}$ in cycle notation.

## 5.4 *Functions and Partitions*

According to Theorem 4.6, equivalence relations and partitions are just two different ways of considering the same mathematical concept. In this section, we see that partitions and functions are also closely allied ideas.

**Proposition 5.9**

> Let $f: X \to Y$ and define a relation $R$ on $X$ by agreeing that $(a, b) \in R$ provided that $f(a) = f(b)$. Then $R$ is an equivalence relation on $X$.

**PROOF** See Exercise 39. □

Proposition 5.9 assures us that whenever we are given a set $X$ and a function $f$ with domain $X$, there is an associated partition of $X$ with the property that two points $a$ and $b$ of $X$ belong to the same member of the partition if and only if $f(a) = f(b)$. Let us reverse the situation. Suppose that we are given a partition $P$ of a nonempty set $X$. Is there some natural function $f$ with domain $X$ such that two points $a$ and $b$ of $X$ belong to the same member of the given partition if and only if $f(a) = f(b)$? Yes. That's easy. We define $f: X \to P$ by letting $f(x)$ be *the* member of $P$ that contains $x$.

Note that our relation $f$ is a function because no member of $X$ can belong to two members of a partition. We already know that the equivalence relation $R$ that yields the partition $P$ is given by

$$R = \{(a, b) \in X \times X: a \text{ and } b \text{ belong to the same member of } P\}$$

Thus, $R = \{(a, b) \in X \times X: f(a) = f(b)\}$, and for each $x \in X$, $f(x) = [x]$ (where, as usual, $[x]$ is the equivalence class that contains $x$). The function $f: X \to P$ is called the *canonical function.*

**EXAMPLE 11** Let $f: \mathbb{N} \to \mathbb{N}$ be defined by letting $f(n)$ be the smallest prime divisor of $n$ if $n \neq 1$ and by letting $f(1) = 1$. With respect to the equivalence relation defined in Proposition 5.9, the equivalence classes are $[1] = \{1\}$ and for each $k \in \mathbb{N}$ such that $k > 1$,

$$[k] = \{n \in \mathbb{N}: \text{the smallest prime divisor of } k \text{ is the smallest prime divisor of } n\}$$

Note that $[18] = [2]$ and $[91] = [7]$. ❑

**EXAMPLE 12** Let $f: \mathbb{R} \to \mathbb{R}$ be given by $f(x) = \sin x$. With respect to the equivalence relation defined in Proposition 5.9, for each real number $y$, the equivalence class

$$[y] = \{x \in \mathbb{R}: \sin x = \sin y\}$$

Note that $[0] = [\pi]$ and $[\pi/4] = [3\pi/4]$. ❑

**EXAMPLE 13** Let $X = \{1, 2, 3, 4, 5, 6\}$, and let $P$ be the following partition of $X$:

$$P = \{\{1, 2, 3\}, \{4, 6\}, \{5\}\}$$

Let $f: X \to P$ be the canonical function. Then $f(4) = f(6) = \{4, 6\}$, $f(1) = f(2) = f(3) = \{1, 2, 3\}$, and $f(5) = \{5\}$. ❑

**EXAMPLE 14** Let $S$ be the set of all lines in the plane and define two lines to be equivalent provided that they are parallel or identical. It is easily verified that we have defined an equivalence relation on $S$, which partitions $S$ into a partition $P$. Let $f: S \to P$ be the canonical function. If $L$ is a horizontal line, then $f(L)$ is the collection of all horizontal lines. If $L$ is the line $y = x$, then $f(L)$ is the collection of all lines with slope 1. ❑

The words *canonical* and *heretical* are antonyms, so the term *canonical function* suggests that the canonical function is a natural and orthodox function to think of whenever we are presented with a partition (or equivalence relation). We do not, however, wish to leave the impression that the canonical map is the only natural function associated with a partition. Often when we are given a partition $P$, it is natural to define a function with domain $P$ (whereas the canonical function has range $P$). In this situation, which comes up again and again in Chapter 8, we must first make sure that the function we are trying to define really makes sense.

Suppose, for example, that we are given the partition $P$ of $\mathbb{Z}$ consisting of the nine congruence classes mod 9, and we wish to define a function $f: P \to P$ by $f([x]) = [5x]$. Thus $f([111]) = [555]$ and $([111], [555]) \in f$. Now $[111] = [3]$ and so $([111], [15]) \in f$. If our "function" $f$ really is a function, it cannot have two ordered pairs with the same first term and so we must know that $[555] = [15]$. Luckily, this equality holds, but we have not yet *proved* that $f$ is a function, because we have considered only one case. If $f$ turns out to be a function, we say that $f$ is *well defined.*

We now show that $f$ is a function. Suppose that $[x] = [x']$. We must show that $f([x]) = f([x'])$ or, in other words, we must show that $[5x] = [5x']$. Since we are considering a partition, it is enough to show that $5x' \in [5x]$. Because $[x] = [x']$, $x \equiv x' \pmod 9$ and there is an integer $q$ such that $x - x' = 9q$. Thus, $9(5q) = 5x - 5x'$ and $5x \equiv 5x' \pmod 9$. It follows that $5x' \in [5x]$, and so $f([x])$ is the same equivalence class no matter which member of $[x]$ is used to calculate it.

**EXAMPLE 15** Let $P$ be the partition of $\mathbb{Z}$ into the ten congruence classes mod 10. We define a relation $f$ on $P$ as follows: Let $x \in \mathbb{Z}$. If $x \geqslant 0$, we define $f([x]) = [b]$, where $b$ is the sum of the digits of $x$. If $x < 0$, we define $f([x]) = [b]$, where $b$ is the negative of the sum of digits of $x$. Thus, $f([25]) = [7]$, $f([123]) = [6]$, and $f([-34]) = [-7]$. Alas, $f$ is not a function, and it is customary to say that $f$ is *not* well defined because $[5] = [125]$, but $f([5]) \neq f([125])$. ❑

---

## EXERCISES 5.4

**39.** Prove Proposition 5.9.

**40.** Suppose that $P$ is a partition of a nonempty set $X$ and that the canonical function $f: X \to P$ is one-to-one. What can you say about the partition $P$?

**41.** For nonempty subsets $A$ and $B$ of $\mathbb{N}$, define $A \simeq B$ provided that there is a one-to-one function mapping $A$ onto $B$.

**a)** Prove that $\simeq$ is an equivalence relation on $\mathscr{P}(\mathbb{N}) - \{\varnothing\}$.

**b)** Let $f$ be the canonical function associated with the equivalence relation $\simeq$. In plain English, what can you say about $A$ and $B$ if $f(A) = f(B)$?

**42.** Define two points $(x_1, y_1)$ and $(x_2, y_2)$ of the plane to be equivalent provided that $x_1 = x_2$. It is easily verified that you have defined an equivalence relation. Let $f$ be the associated canonical function. What set is $f((3, 4))$? In plain English, what does it mean about two points $p$ and $q$ if $f(p) = f(q)$?

**43.** Let $P$ be the partition of $\mathbb{Z}$ into the nine congruence classes mod 9. Define a relation $f$ on $P$ as in Example 15. Is $f$ a function on $P$?

**44.** Let $P$ be the partition of $\mathbb{Z}$ into nine congruence classes mod 9, and define $f: P \to P$ by $f([x]) = [5x]$. We have already shown that $f$ is a well-defined function. Prove that $f$ is a permutation on $P$.

**45.** Let $P$ be the partition of $\mathbb{Z}$ given in Exercise 44. Define a relation for $P$ by $f([x]) = [x^2]$. Is $f$ a function on $P$? Prove your answer.

## 5.5 Real-valued Functions

**Definition** If $D$ is a set and $f: D \to \mathbb{R}$, then $f$ is said to be a **real-valued function.**

From your study of calculus, you are probably familiar with real-valued functions whose domains are subsets of $\mathbb{R}$ (or in the case of vector calculus, real-valued functions whose domains are subsets of $\mathbb{R}^n$). Nonetheless, a real-valued function can have any set as its domain, and two real-valued functions can be added, subtracted, multiplied, or divided, even if their domains are sets that have no algebraic structure.

**Definition** Let $D_1$ and $D_2$ be the two sets such that $D_1 \cap D_2 \neq \varnothing$, and let $f$ and $g$ be real-valued functions with domains $D_1$ and $D_2$. Then

$f+g: D_1 \cap D_2 \to \mathbb{R}$ is defined by $(f+g)(x) = f(x) + g(x)$.
$f-g: D_1 \cap D_2 \to \mathbb{R}$ is defined by $(f-g)(x) = f(x) - g(x)$.
$f \cdot g: D_1 \cap D_2 \to \mathbb{R}$ is defined by $(f \cdot g)(x) = f(x)g(x)$.
$f/g: D_1 \cap D_2 - \{x \in D_1 \cap D_2 : g(x) = 0\} \to \mathbb{R}$ is defined by $(f/g)(x) = f(x)/g(x)$.

**EXAMPLE 16** Let $f: [0, \infty) \to \mathbb{R}$ be defined by $f(x) = \sqrt{x} + x^2 + 4$, and let $g: \mathbb{R} \to \mathbb{R}$ be defined by $g(x) = x - 1$. Then

$f+g: [0, \infty) \to \mathbb{R}$ is defined by $(f+g)(x) = x^2 + x + 3 + \sqrt{x}$.
$f-g: [0, \infty) \to \mathbb{R}$ is defined by $(f-g)(x) = x^2 - x + 5 + \sqrt{x}$.
$f \cdot g: [0, \infty) \to \mathbb{R}$ is defined by $(f \cdot g)(x) = (x-1)(\sqrt{x} + x^2 + 4)$.
$f/g: [0, 1) \cup (1, \infty) \to \mathbb{R}$ is defined by $(f/g)(x) = (\sqrt{x} + x^2 + 4)/(x-1)$.

❑

**EXAMPLE 17** Let $S$ be a nonempty set, and for each subset $A$ of $S$, let $\chi_A$ ("chi sub $A$") be the real-valued function defined by

$$\chi_A(x) = 1 \qquad \text{if } x \in A$$

$$\chi_A(x) = 0 \qquad \text{if } x \notin A$$

The function $\chi_A$ is called the **characteristic function** *of* $A$. Note that $\chi_\varnothing: S \to \mathbb{R}$ is given by $\chi_\varnothing(x) = 0$ for all $x \in S$, and $\chi_S: S \to \mathbb{R}$ is given by $\chi_S(x) = 1$ for all $x \in S$. For any two subsets $A$ and $B$ of $S$, $\chi_A = \chi_B$ if and only if $A = B$ (Exercise 48). Moreover, $\chi_A \cdot \chi_B = \chi_{A \cap B}$ and $\chi_A + \chi_{A^\sim} = \chi_S$ (Exercise 49). ❑

Suppose that we wish to define some function. How much information are we obligated to divulge in order to have our function defined unambiguously? The answer to this question depends on the setting. For example, if $\mathbb{R}^+$ denotes $(0, \infty)$, the present notion of function makes no distinction between $f: \mathbb{R} \to \mathbb{R}^+$ defined by $f(x) = 2^x$ and $g: \mathbb{R} \to \mathbb{R}$ defined by $g(x) = 2^x$. As sets of ordered pairs, $f = g$. It may be important, however, to make clear that the range of $f$ is considered to be only a subset of $\mathbb{R}^+$, whereas the range of $g$ is considered to be a subset of $\mathbb{R}$, and in this case our definition of a function as a relation has not really completely described the idea we have in mind. In some branches of mathematics, therefore, the definition of a function would have been given as follows:

**Definition**

A *function* is an ordered triple $(f, D, C)$ where $f$ is a relation in which no two ordered pairs have the same first term, $D$ is the domain of $f$, and $C$ is a set called the **codomain** of $f$ that contains the range of $f$. If a function $(f, D, C)$ maps $D$ onto its codomain $C$, the function is called a *surjection* (*surjective function*) and if $(f, D, C)$ is one-to-one, it is called an *injection* (*injective function*). A function $(f, D, C)$ that is both surjective and injective is called a *bijection* (*bijective function*).

With this notation, when $f$ is the relation $\{(x, 2^x): x \in \mathbb{R}\}$, $(f, \mathbb{R}, \mathbb{R}^+)$ and $(f, \mathbb{R}, \mathbb{R})$ are two different functions. Indeed, only the first of these functions is a surjection. We have no need for such precision in this text, and we are convinced that those of you who encounter the other definition of function in subsequent work will have no trouble adapting. After all, the ordered-triple definition gives you more information with which to work. At the other extreme, there are phrases from freshman calculus such as "let $f(x) = \sqrt{x-1}$" and

$$\lim_{x \to 1} [(x^2 - 1)/(x - 1)] = \lim_{x \to 1} (x + 1)$$

Here you are expected to guess both the domain and the codomain of the functions being discussed. Presumably the functions are real-valued, so the phrase $f(x) = \sqrt{x-1}$ suggests that there is some subset $D$ of $\mathbb{R}$ such that $f: D \to \mathbb{R}$ and *by convention* we always take $D$ to be the largest subset of $\mathbb{R}$ that makes sense. In this case, $D = [1, \infty)$. Thus, "let $f(x) = \sqrt{x-1}$" is shorthand for "let $D = [1, \infty)$, and let $f: D \to \mathbb{R}$ be defined by $f(x) = \sqrt{x-1}$." The conventional shorthand is indeed convenient, but it can be misleading. If $g(x) = (x^2 - 1)/(x - 1)$ and $h(x) = x + 1$, then as we just noted

$$\lim_{x \to 1} g(x) = \lim_{x \to 1} h(x)$$

but those of you who say to yourself that these limits coincide because $g$ and $h$ are the same functions have been tricked. The functions $g$ and $h$ are different because $\text{Dom}(g) = (-\infty, 1) \cup (1, \infty)$, whereas $\text{Dom}(h) = \mathbb{R}$. A similar problem arises in the following example.

**EXAMPLE 18** Let $f(x) = 3x + 4$ and $g(x) = \sqrt{7x - 9}$.

1. $\text{Dom}(f) = \text{range}(f) = \mathbb{R}$
   $\text{Dom}(g) = [9/7, \infty)$ and $\text{range}(g) = [0, \infty)$
2. Both $f$ and $g$ are one-to-one maps.
   a) Suppose that $f(a) = f(b)$. Then $3a + 4 = 3b + 4$ and so $a = b$.
   b) Suppose that $g(a) = g(b)$. Then $\sqrt{7a - 9} = \sqrt{7b - 9}$ and so $7a - 9 = 7b - 9$ and $a = b$.

Since $f$ and $g$ are one-to-one maps, we can determine an explicit expression for $f^{-1}$ and $g^{-1}$. Intuitively [since $\text{Dom}(g) = [9/7, \infty)$], we think of $g$ as a little machine that accepts only numbers greater than or equal to 9/7. When given such a number, the machine $g$ multiplies it sevenfold, subtracts 9, and takes the nonnegative square root. We find $g^{-1}$ by running the machine backward. The machine $g^{-1}$ accepts only numbers in the range of $g$; that is, it accepts only nonnegative numbers. Given such a number $x$, the machine squares it, adds 9, and divides by 7. Thus, $g^{-1}(x) = (x^2 + 9)/7$, and $\text{Dom}(g^{-1}) = [0, \infty)$. We can determine $f^{-1}$ in a similar manner: $f^{-1}(x) = (x - 4)/3$, and $\text{Dom}(f^{-1}) = \mathbb{R}$.

3. We can find explicit expressions for $f \circ g$ and $(f \circ g)^{-1}$. We first consider $f \circ g$. Working intuitively, we have two machines $f$ and $g$, and we are asked to place an object $x$ that machine $g$ accepts into that machine. If the resulting object $g(x)$ that has been machined by $g$ is acceptable to $f$, we let $f$ operate on the machined part $g(x)$ to form $f \circ g(x)$. Thus,

$$f \circ g(x) = f(g(x)) = f(\sqrt{7x - 9}) = 3\sqrt{7x - 9} + 4$$

The domain of $f \circ g$ is defined to be $\{x \in \text{Dom}(g): g(x) \in \text{Dom}(f)\}$. That is, $x \in \text{Dom}(f \circ g)$ provided that $x \geqslant 9/7$ and $g(x) \in \mathbb{R}$. This set is exactly the domain accepted by convention for $h(x) = 3\sqrt{7x - 9} + 4$. We also note that the range of $f \circ g$ is $[4, \infty)$.

To find $(f \circ g)^{-1}$, we recall from Chapter 4 that $(f \circ g)^{-1} = g^{-1} \circ f^{-1}$. Hence,

$$\begin{aligned}(f \circ g)^{-1}(x) &= (g^{-1} \circ f^{-1})(x) \\ &= g^{-1}(f^{-1}(x)) \\ &= \frac{(f^{-1}(x))^2 + 9}{7} \\ &= \frac{((x - 4)/3)^2 + 9}{7}\end{aligned}$$

Alternatively, we could have determined $(f \circ g)^{-1}$ by running the machine $f \circ g$ backward:

**a)** Take $x$ and subtract 4.

**b)** Divide by 3.

**c)** Square.

**d)** Add 9.

**e)** Divide by 7.

So, $(f \circ g)^{-1}(x) = ([(x-4)/3]^2 + 9)/7$.

**Warning** No matter which way we calculate the explicit expression for $(f \circ g)^{-1}(x)$, we are about to be misled by our convention. We see that $(f \circ g)^{-1}$ does not equal the function $h$ defined by $h(x) = ([(x-4)/3]^2 + 9)/7$ because $\text{Dom}\,(f \circ g)^{-1} = \text{Range}\,(f \circ g) = [4, \infty)$, but $\text{Dom}\,(h) = \mathbb{R}$. ❑

**Proposition 5.10**

Let $f$ and $g$ be functions, and let $A$ denote the domain of $f$, $B$ denote the domain of $g$, and $C$ denote $A \cap B$. Then

**a)** $f \cup g$ is a function if and only if $f \mid C = g \mid C$. (In particular, if $C = \varnothing$, then $f \cup g$ is a function.)

**b)** $f \cap g$ is a function, and $\text{Dom}\,(f \cap g) = C$ if and only if $f \mid C = g \mid C$.

**c)** If $f$ and $g$ are one-to-one functions, so is $f \cap g$.

**d)** If $f$ and $g$ are one-to-one functions whose ranges are disjoint and $f \cup g$ is a function, then $f \cup g$ is a one-to-one function.

**PROOF** The proof of this proposition is left as Exercise 55. □

Parts (a) and (d) of Proposition 5.10 are more useful than one might imagine, because it is quite common to define a function as the union of two other functions. We give a typical example of the union of two real-valued functions. A more useful example of the union of two functions is given in Exercise 80 in Section 5.6.

**EXAMPLE 19**

Let $f(x) = \sqrt{x-1}$ and $g(x) = \sqrt{1-x^2}$. Since $\sqrt{x-1}$ is defined only when $x \geqslant 1$, by convention $\text{Dom}\,(f) = [1, \infty)$. Thus, $\text{Dom}\,(f) \cap \text{Dom}\,(g) = \{1\}$. Since $f(1) = g(1)$, by Proposition 5.10(a), $f \cup g$ is a function, but part (d) does not apply. We have shown the graph of $f \cup g$ in Figure 5.6.

**Remark** You are probably aware that both $f$ and $g$ are continuous functions. Proposition 5.10(a) is important in part because there is a theorem of analysis that promises that whenever $f$ and $g$ are continuous functions satisfying the condition of Proposition 5.10(a) that $f \mid C = g \mid C$ [where $C =$

Dom $(f) \cap$ Dom $(g)$], then $f \cup g$ will be a continuous function if the set $C$ is closed. All finite sets are closed, and so the function $f \cup g$ that we have been considering is continuous (see Figure 5.6).

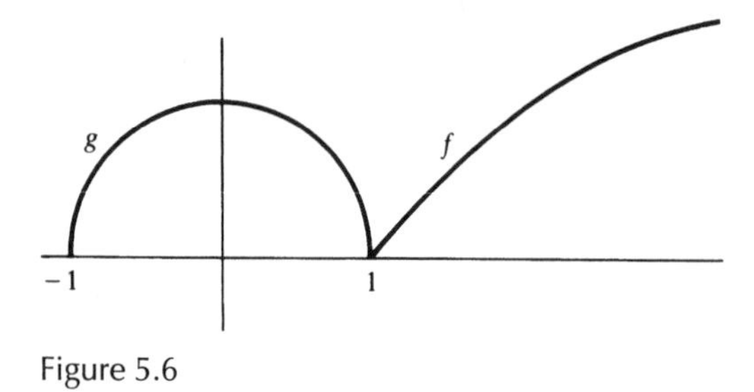


Figure 5.6 □

**EXAMPLE 20** Let $A$ and $B$ be nonempty sets. Then $\pi_1 \colon A \times B \to A$ is defined by $\pi_1((a, b)) = a$, and $\pi_2 \colon A \times B \to B$ is defined by $\pi_2((a, b)) = b$. These functions are called **projections** because each function maps a point of $A \times B$ onto its shadow (see Figure 5.7). Note that $\pi_1$ maps $A \times B$ onto $A$ and $\pi_2$ maps $A \times B$ onto $B$, but, except for the trivial case in which $A$ or $B$ is a singleton set, neither $\pi_1$ nor $\pi_2$ is one-to-one. Although the definitions of $\pi_1$ and $\pi_2$ make sense for the Cartesian product of any two sets, in the case that we are considering, the projections on $\mathbb{R} \times \mathbb{R}$, both $\pi_1$ and $\pi_2$ become real-valued functions. In particular, if $D$ is a set and $f \colon D \to \mathbb{R} \times \mathbb{R}$, then both $\pi_1 \circ f$ and $\pi_2 \circ f$ are real-valued functions with domain $D$. We postpone further consideration of the projection maps until Section 5.6.

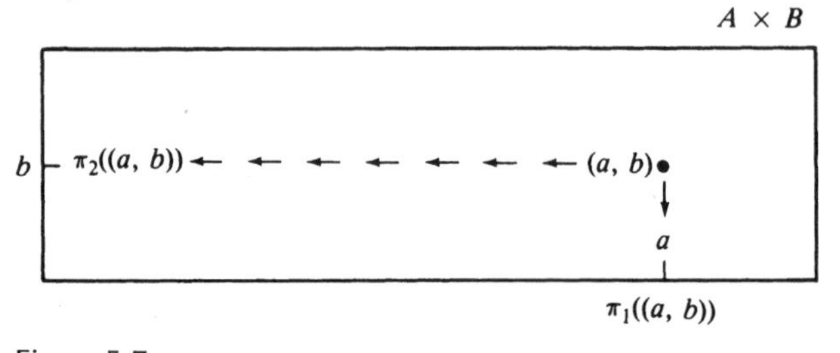

Figure 5.7 □

Our last two examples of real-valued functions use the usual linear order on $\mathbb{R}$.

**EXAMPLE 21** Let $D_1$ and $D_2$ be sets such that $D_1 \cap D_2 \neq \varnothing$, and let $f$ and $g$ be real-valued functions with domains $D_1$ and $D_2$. Then

$$f \wedge g \colon D_1 \cap D_2 \to \mathbb{R} \text{ is defined by } (f \wedge g)(x) = \text{minimum of } f(x) \text{ and } g(x)$$

Similarly,

$$f \vee g \colon D_1 \cap D_2 \to \mathbb{R} \text{ is defined by } (f \vee g)(x) = \text{maximum of } f(x) \text{ and } g(x)$$

The symbols $\wedge$ and $\vee$ are called "meet" and "join." In Exercise 56, you are asked to show that for each $x \in D_1 \cap D_2$,

$$(f \vee g)(x) = [f(x) + g(x)]/2 + |f(x) - g(x)|/2$$

and you are asked to find a similar explicit expression for $(f \wedge g)(x)$. ❑

**Definition**

For each real number $x$, $\lfloor x \rfloor$ is defined to be the greatest integer that is less than or equal to $x$ and $\lceil x \rceil$ is defined to be the least integer that is greater than or equal to $x$. We define the **floor function** $f\colon \mathbb{R} \to \mathbb{Z}$ by $f(x) = \lfloor x \rfloor$ and the **ceiling function** $g\colon \mathbb{R} \to \mathbb{Z}$ by $g(x) = \lceil x \rceil$. The floor function is also called the **greatest-integer function**.

**EXAMPLE 22**

The graphs of $h(x) = \sqrt{x}$ and $f \circ h(x) = \lfloor \sqrt{x} \rfloor$ are shown in Figure 5.8. You are asked to graph $g \circ h(x) = \lceil \sqrt{x} \rceil$ in Exercise 60.

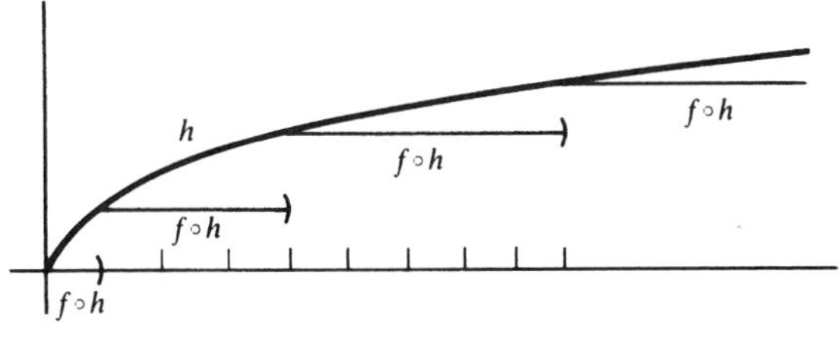


Figure 5.8

When considering $\lfloor x \rfloor$ for a negative real number $x$, we can easily be tripped up. Note that $\lfloor -\pi \rfloor = -4$ (not $-3$). With this warning in mind, you should have no trouble sketching the graph of $h(x) = \lfloor -x \rfloor$. (See Exercise 59c.) ❑

## EXERCISES 5.5

**46.** Let $f(x) = 1/(x-3)$ and $g(x) = \sqrt{x-3}$. Find the domain of each of the following functions.

**a)** $f$ **b)** $g$ **c)** $f^{-1}$ **d)** $g^{-1}$ **e)** $f+g$
**f)** $f/g$ **g)** $f \circ g$ **h)** $g \circ f$ **i)** $(f \circ g)^{-1}$ **j)** $(g \circ f)^{-1}$

(Assume that the domains of $f$ and $g$ are the largest subsets of $\mathbb{R}$ that make sense.)

**47.** Let $f$ and $g$ be the real-valued functions given in Exercise 46. Find explicit expressions for each of the following functions.

**a)** $f^{-1}$ **b)** $g^{-1}$ **c)** $f+g$ **d)** $f \cdot g$
**e)** $f \circ g$ **f)** $g \circ f$ **g)** $(f \circ g)^{-1}$ **h)** $(g \circ f)^{-1}$

**48.** Let $S$ be a nonempty set, and let $A$ and $B$ be subsets of $S$. Prove that $\chi_A = \chi_B$ if and only if $A = B$.

**49.** Let $S$ be a nonempty set, and let $A$ and $B$ be subsets of $S$.

**a)** Prove that $\chi_A + \chi_{A^{\sim}} = \chi_S$. **b)** Prove that $\chi_A \cdot \chi_B = \chi_{A \cap B}$.

**50.** **a)** Show that there is a set $S$ and subsets $A$ and $B$ of $S$ such that $\chi_A + \chi_B \neq \chi_{A \cup B}$.

**b)** State and prove the correct formula for $\chi_A + \chi_B$ that is analogous to the equation given in Exercise 49(b).

**51.** Let $A = [0, 2]$ be considered as a subset of $\mathbb{R}$. Sketch the graph of $\chi_A$.

**52.** Let $f$ and $g$ be the real-valued functions given in Example 18. Find explicit expressions for $g \circ f$ and $(g \circ f)^{-1}$. State the domain and the range of each of these functions.

**53.** State the domain dictated by convention for each of the following real-valued functions. For those functions $f$ that are one-to-one, prove that $f$ is one-to-one and state the domain of $f^{-1}$.

**a)** $f(x) = 7/(x-3)$ **b)** $f(x) = 7/\sqrt{x-3}$

**c)** $f(x) = 7/(x-3)^2$ **d)** $f(x) = 7/\sqrt{x^2+4x-21}$

**54.** Consider the following graphs, each of which is the graph of a function. Identify which of the functions is a one-to-one function, and for each given function $f$ that is one-to-one, sketch the graph of $f^{-1}$.

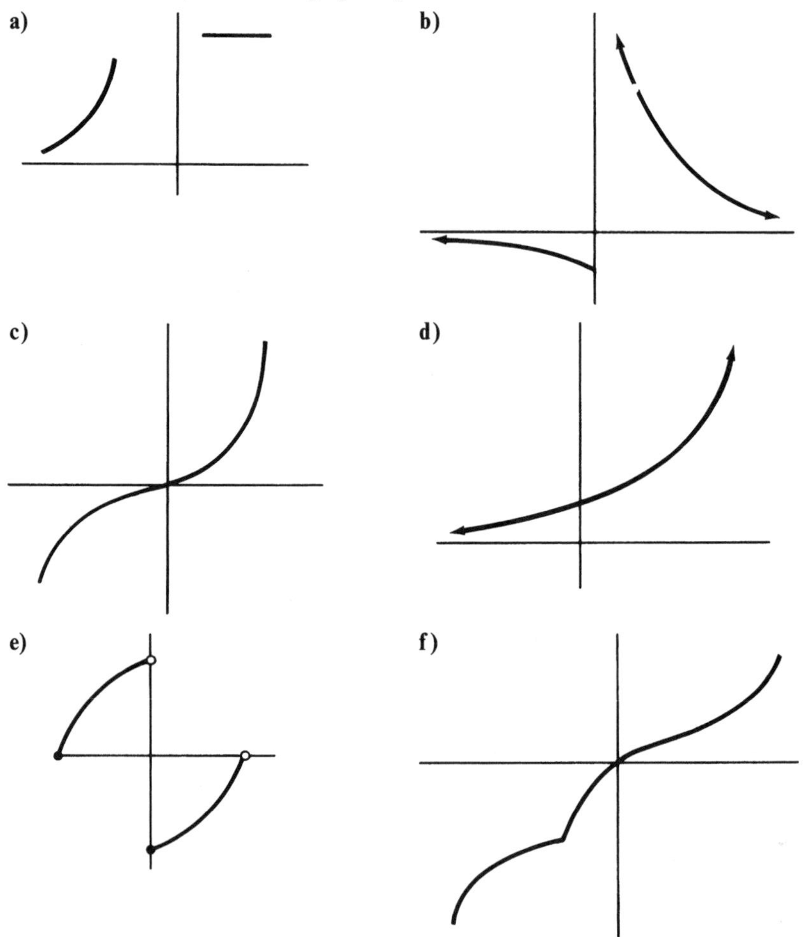

**55. a)** Prove Proposition 5.10(a).

**b)** Prove Proposition 5.10(b).

**c)** Prove Proposition 5.10(c).

**d)** Prove Proposition 5.10(d).

**56. a)** Prove that $(f \vee g)(x) = (f(x) + g(x))/2 + |f(x) - g(x)|/2$.

**b)** Find an explicit expression for $(f \wedge g)(x)$ similar to the one given above for $(f \vee g)(x)$.

**57.** Give an example to show that there are sets $A$, $B$, and $C$ and functions $f\colon A \to C$ and $g\colon B \to C$ such that $f \cup g$ is not a function.

**58.** Give an example to show that there are sets $A$, $B$, and $C$ and one-to-one functions $f\colon A \to C$ and $g\colon B \to C$ such that $f \cup g$ is a function but $f \cup g$ is not a one-to-one function.

**59.** Sketch the graph of each of the following.

**a)** $f(x) = \lfloor 2x \rfloor$ **b)** $f(x) = \lfloor x/3 \rfloor$ **c)** $f(x) = \lfloor -x \rfloor$

**d)** $f(x) = \lceil 2x \rceil$ **e)** $f(x) = \lceil x/3 \rceil$ **f)** $f(x) = \lceil -x \rceil$

**60.** Define $h\colon \mathbb{R} \to \mathbb{R}$ by $h(x) = \sqrt{x}$. Graph $g \circ h\colon \mathbb{R} \to \mathbb{R}$, where $g$ is the ceiling function.

**61.** Let $n$ and $d$ be positive integers. Use the Division Algorithm to show that the number of positive integers less than or equal to $n$ that are divisible by $d$ is $\lfloor n/d \rfloor$.

**62.** Let $f\colon \mathbb{R} \to \mathbb{Z}$ be the floor function and $g\colon \mathbb{R} \to \mathbb{Z}$ be the ceiling function. Prove that for each real number $x$, $f(x) = -g(-x)$ and $g(x) = -f(-x)$.

**63.** Let $f\colon \mathbb{R} \to \mathbb{Z}$ be defined by $f(x) = \lfloor x/2 \rfloor$. Is $f \mid \mathbb{Z}$ a one-to-one function? Does $f \mid \mathbb{Z}$ map onto $\mathbb{Z}$? Explain.

***64.** For a given one-to-one function $f$ whose domain and range are subsets of $\mathbb{R}$, it is a common mistake to confuse $f^{-1}$ with $1/f$. Let $\alpha$ be a permutation of $(0, \infty)$ such that $\alpha^{-1} = 1/\alpha$.

**a)** Show that $\alpha(1) = 1$.

**b)** Show that if $x \neq 1$ and $x > 0$, then $\alpha(x) \neq x$.

**c)** Show that if $x \neq 1$ and $x > 0$, then $\alpha \circ \alpha(x) \neq x$.

**d)** Show that if $x \neq 1$ and $x > 0$, then $\alpha \circ \alpha \circ \alpha(x) \neq x$.

**e)** Show that $\alpha \circ \alpha \circ \alpha \circ \alpha = i_{(0,\infty)}$.

***65.** Legendre's identity states that for each real number $x > 1$, we have that $\lfloor x \rfloor!$ is the product over all primes $p \leqslant x$ of $p^{\alpha(p)}$, where $\alpha(p) = \Sigma_{n=1}^{\infty} \lfloor x/p^n \rfloor$. (For example, $\lfloor 10 \rfloor! = 2^{\alpha(2)} \cdot 3^{\alpha(3)} \cdot 5^{\alpha(5)} \cdot 7^{\alpha(7)}$.)

**a)** Verify Legendre's identity for $x = \pi$.

**b)** Verify Legendre's identity for $x = 6$.

**c)** Verify Legendre's identity for $x = 6.9$.

**66.** Let $f\colon \mathbb{R} \to \mathbb{R}$ be the floor function $f(x) = \lfloor x \rfloor$ and $g\colon \mathbb{R} \to \mathbb{R}$ be the ceiling function $g(x) = \lceil x \rceil$.

**a)** Find the range of $f$ and the range of $g$.

**b)** Is $f$ a one-to-one function?

**c)** Use Theorem 5.1 to prove that $f = g \circ f$ and $g = f \circ g$.

## 5.6 *Images and Inverse Images of Sets*

Let $f: X \to Y$. It is clear that we can associate with the given function $f$ a new function $F: \mathscr{P}(X) \to \mathscr{P}(Y)$, which is defined in the following way: For each subset $A$ of $X$, $F(A) = \{f(a): a \in A\}$. For example, if $f: \mathbb{N} \to \mathbb{N}$ is defined by $f(n) = 2n$, then $F(\{1, 2, 3\}) = \{2, 4, 6\}$, and if $A$ is the set of all odd natural numbers and $E$ is the set of all even natural numbers, then $F(A) = E$. At present, the symbols $f(\{1, 2, 3\})$ have no meaning, but a reasonable person forced to guess what $f(\{1, 2, 3\})$ should mean would surely guess correctly that what is meant is $f(\{1, 2, 3\}) = \{2, 4, 6\}$.

**Definition**

Let $X$ be a set such that $X$ and $\mathscr{P}(X)$ are disjoint sets, let $f: X \to Y$, and let $A \in \mathscr{P}(X)$.Then we define $f(A) = \{f(a): a \in A\}$.

Most authors do not specify that $X$ and $\mathscr{P}(X)$ must be disjoint in the definition of $f(A)$. We give an example to illustrate why we have added this requirement. Let $X = \mathbb{N} \cup \mathscr{P}(\mathbb{N})$. Then $X \cap \mathscr{P}(X) \neq \varnothing$. Define $g: X \to \mathbb{N}$ as follows: For each $n \in \mathbb{N}$, $g(n) = 2n$, and for each nonempty subset $B$ of $\mathbb{N}$, $g(B)$ is the least member of $B$. Set $g(\varnothing) = 17$. If $A = \{3, 5, 7\}$, then $A$ is a nonempty subset of $\mathbb{N}$ and hence $g(A) = 3$. However, $A \subseteq \mathbb{N}$ and, if we did not require that $X$ and $\mathscr{P}(X)$ be disjoint, then, by the definition above, $g(A)$ would be $\{6, 10, 14\}$. One seldom encounters functions such as $g$, so we assume that whenever we write $f: X \to Y$, then $X \cap \mathscr{P}(X) = \varnothing$. In the example just given, ambiguity can be avoided by denoting the set $\{g(a): a \in A\} = \{6, 10, 14\}$ by $\vec{g}(A)$ and the element 3 by $g(A)$.

Given $f: X \to Y$, there is a second function involving $\mathscr{P}(X)$ and $\mathscr{P}(Y)$ that we can obtain from $f$. Define $\hat{F}: \mathscr{P}(Y) \to \mathscr{P}(X)$ by saying that for each subset $A$ of $Y$, $\hat{F}(A) = \{x \in X: f(x) \in A\}$. Note that even when $f$ is not a one-to-one function, $\hat{F}$ is a function. Once again, given a subset of $A$ of $Y$, the symbols $f^{-1}(A)$ have no meaning, so we are forced to give a definition, which is the obvious analogue of the preceding definition (see Exercise 81).

**Definition**

Let $Y$ be a set such that $Y$ and $\mathscr{P}(Y)$ are disjoint, let $f: X \to Y$, and let $A \in \mathscr{P}(Y)$. Then $f^{-1}(A) = \{x \in X: f(x) \in A\}$.

If $y \in Y$, we write $f^{-1}(y)$ to mean $f^{-1}(\{y\})$.

**EXAMPLE 23**

Let $f$ be the function whose Cartesian graph is given on the next page, let $A = \{2, 4, 5\}$, and let $B = \{1, 3, 5\}$. We have indicated $f(A)$ and $f^{-1}(B)$ beside the appropriate axes.

⊙ · · · · (5, 5)
· · ⊙ ⊙ ·
$f(A) = \{3, 4\}$ · ⊙ · · ⊙
· · · · ·
(1, 1) · · · · ·
A A A

B ⊙ · · · · (5, 5)
· · ⊙ ⊙ ·
B · ⊙ · · ⊙
· · · · ·
(1, 1) B · · · · ·

$f^{-1}(B) = \{1, 2, 5\}$

❑

**EXAMPLE 24** Let $f\colon \mathbb{R} \to \mathbb{R}$ be defined by $f(x) = x^2$, let $A = [2, 3]$, and let $B = [4, 9]$. We have indicated, beside the appropriate axis, $f(A)$ in Figure 5.9(a) and $f^{-1}(B)$ in Figure 5.9(b). Note that $f^{-1}(B) = f^{-1}(f(A)) \neq A$. Also note that $f$ is not a one-to-one function.

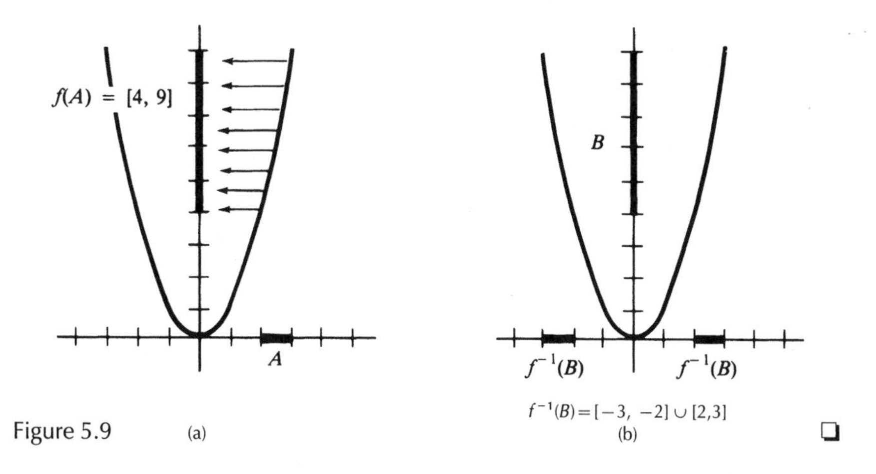


Figure 5.9 (a) (b) $f^{-1}(B) = [-3, -2] \cup [2, 3]$

❑

**EXAMPLE 25** Let $\pi_1$ and $\pi_2$ be the usual projections mapping $\mathbb{R} \times \mathbb{R} \to \mathbb{R}$, and let $A$ be the three-piece subset of the plane sketched in Figure 5.10. We have indicated $\pi_1(A)$ and $\pi_2(A)$ beside the appropriate axes. Note that $\pi_2(A)$ has only two pieces.

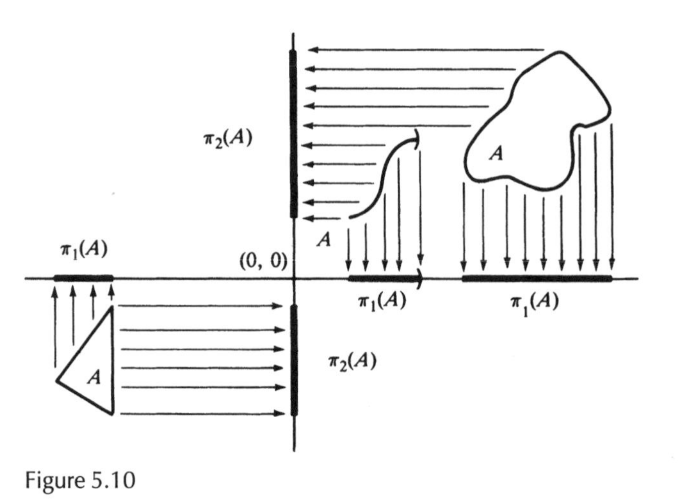


Figure 5.10 ❑

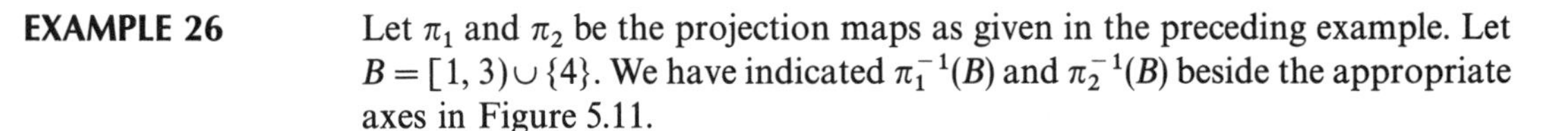

**EXAMPLE 26** Let $\pi_1$ and $\pi_2$ be the projection maps as given in the preceding example. Let $B=[1,3)\cup\{4\}$. We have indicated $\pi_1^{-1}(B)$ and $\pi_2^{-1}(B)$ beside the appropriate axes in Figure 5.11.

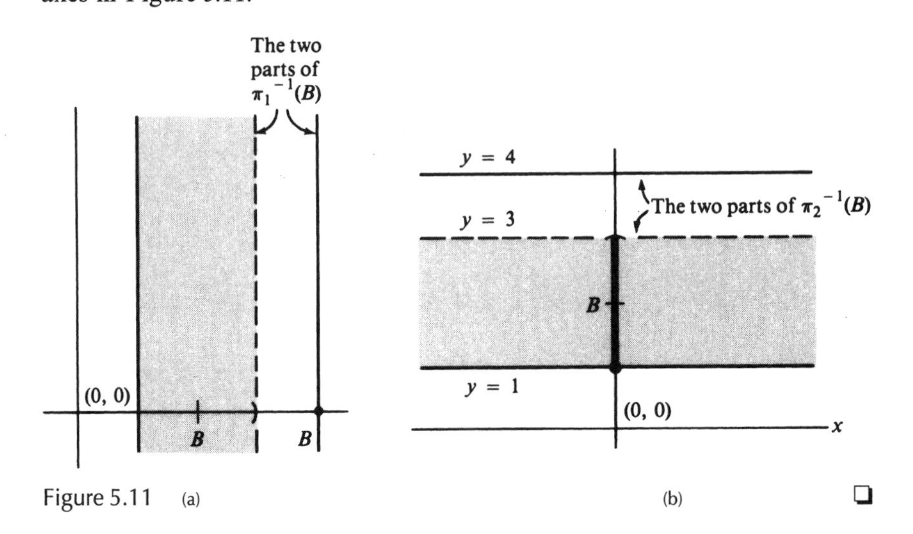


Figure 5.11 (a) (b) ❑

Recall that whenever we have a function $f: X \to Y$, there are two associated set-valued functions $F: \mathscr{P}(X) \to \mathscr{P}(Y)$ and $\hat{F}: \mathscr{P}(Y) \to \mathscr{P}(X)$ defined by $F(A)=\{f(a): a \in A\}$ for each $A \in \mathscr{P}(X)$ and $\hat{F}(B)=\{x \in X: f(x) \in B\}$ for each $B \in \mathscr{P}(Y)$.

It is terribly tempting to guess that $\hat{F}$ is $F^{-1}$, from which it would follow that $f(f^{-1}(B))=B$ and $f^{-1}(f(A))=A$, where $A \in \mathscr{P}(X)$ and $B \in \mathscr{P}(Y)$. Unfortunately, we have already seen in Example 24 that the second of these equations does not always hold, and, as is shown in Theorem 5.11(b) and (d), neither of these pleasant equations holds in general. For the first to hold we need $B$ to be a subset of $f(X)$, which is certain to happen when $f$ maps $X$ onto $Y$, and for the second to hold it is enough that $f$ be a one-to-one function.

Before proving these results, let us draw a Venn-like diagram to see what can go wrong when we consider a function $f: X \to Y$ that is not one-to-one or that does not map $X$ onto $Y$. Let $X = \{1, 2, 3, 4, 5\}$ and $Y = \{6, 7, 8\}$ and define $f: X \to Y$ by $f(x) = 6$ if $x$ is even and $f(x) = 7$ if $x$ is odd.

Let $B = \{7, 8\}$. Then $f^{-1}(B) = \{1, 3, 5\}$ and $f(f^{-1}(B)) = \{7\}$. Thus, $f(f^{-1}(B)) \subseteq B$ but $f(f^{-1}(B)) \neq B$. (See Figure 5.12a.)

Let $A = \{3, 5\}$. Then $f(A) = \{7\}$ and $f^{-1}(f(A)) = \{1, 3, 5\}$. Thus, $A \subseteq f^{-1}(f(A))$ but $f^{-1}(f(A)) \neq A$. (See Figure 5.12b.)

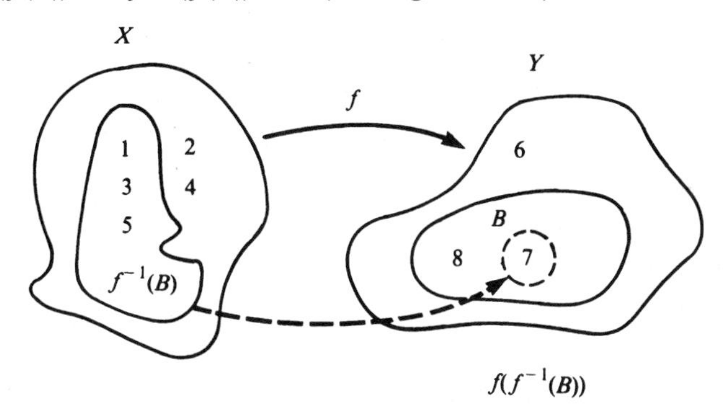

(a)

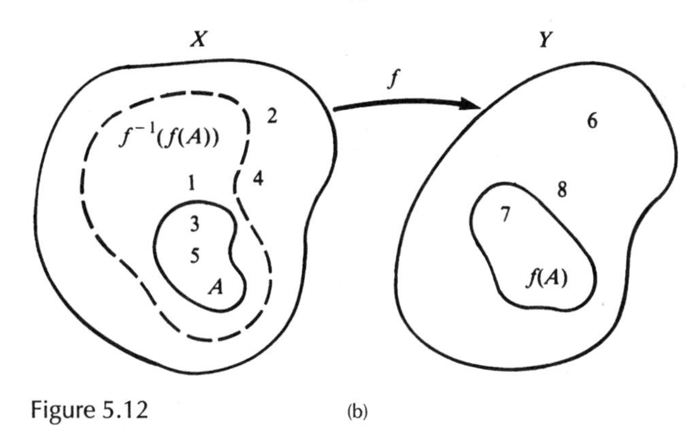


Figure 5.12 (b)

**Theorem 5.11**

> Let $f: X \to Y$, let $A$ be a subset of $X$, and let $B$ be a subset of $Y$. Then
>
> **a)** $f(f^{-1}(B)) \subseteq B$
>
> **b)** $f(f^{-1}(B)) = B$ if and only if $B \subseteq f(X)$.
>
> **c)** $A \subseteq f^{-1}(f(A))$
>
> **d)** $A = f^{-1}(f(A))$ if $f$ is one-to-one.

**PROOF**

**a)** Let $y \in f(f^{-1}(B))$. Then there is an $x \in f^{-1}(B)$ such that $f(x) = y$. Since $x \in f^{-1}(B)$, there is a $b \in B$ such that $f(x) = b$. Since $y = f(x)$, $y \in B$.

**b)** Suppose that $B \subseteq f(X)$. In order to show that $f(f^{-1}(B)) = B$, it suffices to show that $B \subseteq f(f^{-1}(B))$. (Why?) Let $b \in B$. By hypothesis, there is an $x \in X$ such that $f(x) = b$. Thus, $x \in f^{-1}(B)$ and $b = f(x) \in f(f^{-1}(B))$ as required.

Now suppose that $f(f^{-1}(B)) = B$. Let $b \in B$. Then there is an $x \in f^{-1}(B) \subseteq X$ such that $f(x) = b$. By definition $b \in f(X)$, so $B \subseteq f(X)$.

Parts (c) and (d) are left as exercises. □

Before leaving you to venture forth on the exercises to follow, we give a parting warning, which if heeded will serve you well.

The following statement is true, and so is its converse:

$$\text{If } x \in f^{-1}(A), \text{ then } f(x) \in A.$$

The following statement is true:

$$\text{If } x \in A, \text{ then } f(x) \in f(A).$$

*But be forewarned* that the converse

$$\text{If } f(x) \in f(A), \text{ then } x \in A$$

does not necessarily hold unless $f$, perchance, is one-to-one. Indeed in Figure 5.12(b), since $f(1) = 7$, $f(1) \in f(A)$ even though $1 \notin A$. Of course, even for one-to-one functions, the converse cannot be used until it is proved (see Exercise 73).

## EXERCISES 5.6

**67.** Let $f: \mathbb{R} \to \mathbb{R}$ be defined by $f(x) = 3x^2 + 2$. Find each of the following.

**a)** $f([2, 3])$ **b)** $f^{-1}([55, 307])$ **c)** $f^{-1}([1, 2])$

**d)** $f^{-1}(f(3))$ **e)** $f(\{-1, -2, -3\})$ **f)** $f(\{1, 2, 3\})$

**68.** Let $f: \mathbb{R} \to \mathbb{R}$ be defined by $f(x) = \sin x$, and let $D = [0, \pi/2]$. Find each of the following.

**a)** $f(D)$ **b)** $f^{-1}(D)$ **c)** $f(f^{-1}(D))$ **d)** $f^{-1}(f(D))$

**69.** Let $f: \mathbb{N} \times \mathbb{N} \to \mathbb{N}$ be defined by $f((m, n)) = mn$. Let

$$A = \{(a, b) \in \mathbb{N} \times \mathbb{N}: a = 1 \text{ and } b \text{ is even}\}$$

$$B = \{(a, b) \in \mathbb{N} \times \mathbb{N}: a \text{ and } b \text{ are even}\}$$

$$C = \{(a, b) \in \mathbb{N} \times \mathbb{N}: a \text{ is even or } b \text{ is even}\}$$

$$D = \{n \in \mathbb{N}: n \text{ is odd}\}$$

$$E = \{n \in \mathbb{N}: n \text{ is even}\}$$

$$F = \{14\}$$

Find each of the following.

**a)** $f(A)$ **b)** $f(B)$ **c)** $f(C)$

**d)** $f^{-1}(D)$ **e)** $f^{-1}(E)$ **f)** $f^{-1}(F)$

**70.** Let $f\colon \mathbb{R}\to\mathbb{R}$ be defined by $f(x)=\lfloor x\rfloor$, and let

$$A=[3,5]\cup(7,9)\cup(11,15)$$

Find $f(A)$, $f^{-1}(A)$, $f(f^{-1}(A))$, and $f^{-1}(f(A))$.

**71.** For each of the following statements, either cite a theorem that establishes the given statement or draw a picture that shows that the statement is false.

**a)** If $A\subseteq\mathbb{R}\times\mathbb{R}$, then $\pi_1(A)\times\pi_2(A)=A$.

**b)** If $A\subseteq\mathbb{R}$, $\pi_1\circ\pi_1^{-1}(A)=A$.

**c)** If $A\subseteq\mathbb{R}\times\mathbb{R}$, $\pi_1^{-1}\circ\pi_1(A)=A$.

**d)** If $A$ and $B$ are subsets of $\mathbb{R}\times\mathbb{R}$, then

$$\pi_1(A\cup B)=\pi_1(A)\cup\pi_1(B)$$

**e)** If $A$ and $B$ are subsets of $\mathbb{R}\times\mathbb{R}$, then

$$\pi_1(A)\cap\pi_1(B)=\pi_1(A\cap B)$$

**72.** Let $X$ be a nonempty set. Define the complement function $C_X\colon \mathscr{P}(X)\to\mathscr{P}(X)$ by $C_X(A)=X-A$. Prove that $C_X$ is a one-to-one function from $\mathscr{P}(X)$ onto $\mathscr{P}(X)$. What is $C_X^{-1}$?

**73.** Let $f\colon X\to Y$, let $A\subseteq X$, and let $B\subseteq Y$.

**a)** Prove that if $y\in B$, then $f^{-1}(y)\subseteq f^{-1}(B)$.

**b)** Prove that if $f(x)\in f(A)$ and $f$ is a one-to-one map, then $x\in A$.

**74.** Prove parts (c) and (d) of Theorem 5.11.

**75.** Let $f\colon X\to Y$. Prove that $f$ is a one-to-one function if and only if $f(A)\cap f(B)=f(A\cap B)$ for all pairs of subsets of $X$.

**76.** What, if anything, is wrong with the following argument that if $f\colon X\to Y$ is a one-to-one function, then for any subset $A$ of $X$, $f(X-A)=f(X)-f(A)$? [*Argument*: $f(X-A)=f(X\cap A')=f(X)\cap f(A')=f(X)\cap f(A)'=f(X)-f(A)$.]

**77.** Let $f\colon X\to Y$.

**a)** Prove that for any two subsets $A$ and $B$ of $X$,

$$f(A)-f(B)\subseteq f(A-B)$$

**b)** Prove that for any two subsets $A$ and $B$ of $Y$,

$$f^{-1}(A)-f^{-1}(B)\subseteq f^{-1}(A-B)$$

**c)** Can containment be replaced by equality in either (a) or (b)? Either prove or give a counterexample.

**d)** If $f$ is a one-to-one function, can containment be replaced by equality in both (a) and (b)? Either prove or give a counterexample.

***78.** Let $f\colon X\to Y$, and let $\hat{F}\colon \mathscr{P}(Y)\to\mathscr{P}(X)$ be defined by $\hat{F}(B)=\{x\in X\colon f(x)\in B\}$. Prove that $\hat{F}$ is one-to-one if and only if $f$ maps onto $Y$.

**79.** Let $X$ and $Y$ be nonempty sets, let $G$ be a nonempty subset of $X$, and let $H$ be a nonempty subset of $Y$.

**a)** Prove that $G \times H = (X \times H) \cap (G \times Y)$.

**b)** Prove that $\pi_2^{-1}(H) = X \times H$ and $\pi_1^{-1}(G) = G \times Y$.

**80.** Let $X_1$ and $X_2$ be nonempty sets and suppose that there are subsets $A_1$ of $X_1$ and $A_2$ of $X_2$ and there are one-to-one functions $f_1: A_1 \to X_2$ and $f_2: A_2 \to X_1$ such that $f_1(A_1) = X_2 - A_2$ and $f_2(A_2) = X_1 - A_1$. Prove that $f_1 \cup f_2^{-1}$ is a one-to-one function mapping $X_1$ onto $X_2$.

**81.** Define $f: \mathbb{N} \to \mathbb{N} \cup \mathscr{P}(\mathbb{N})$ as follows:

$$f(n) = n, \qquad \text{if } n \text{ is odd}$$
$$f(n) = \{k \in \mathbb{N}: k \text{ is odd and } k < n\}, \qquad \text{if } n \text{ is even}$$

**a)** Is $f$ one-to-one?

**b)** Evaluate $f^{-1}(A)$, where $A = \{1, 3, 5\}$.

**c)** Explain briefly the bearing this problem has on the first two definitions of this section.

## 5.7 Functions and Indexed Families

We begin this section with an example of the inverse image of the union of sets. Then we prove a theorem that says that a function is well behaved only with respect to unions, whereas the inverse of a function is well behaved with respect to both unions and intersections.

**EXAMPLE 27** Let $X = \{n \in \mathbb{N}: n > 1\}$, let $Y = \{q \in \mathbb{Q}: q = 1/n \text{ for some } n \in \mathbb{N}\}$, and define $f: X \to Y$ by $f(n) = 1/p$, where $p$ is the smallest prime that divides $n$. Let $B_1 = \{q \in Y: 1/4 \leqslant q \leqslant 1/2\}$, let $B_2 = \{q \in Y: 1/9 \leqslant q \leqslant 1/3\}$, and let $B_3 = \{q \in Y: 1/16 \leqslant q \leqslant 1/4\}$. Then $f^{-1}(B_1) = \{n \in \mathbb{N}:$ the smallest prime that divides $n$ is either 2 or 3$\}$, $f^{-1}(B_2) = \{n \in \mathbb{N}:$ the smallest prime that divides $n$ is either 3, 5, or 7$\}$, and $f^{-1}(B_3) = \{n \in \mathbb{N}:$ the smallest prime that divides $n$ is either 5, 7, 11, or 13$\}$. Thus $\cup_{i=1}^3 f^{-1}(B_i) = \{n \in \mathbb{N}:$ the smallest prime that divides $n$ is either 2, 3, 5, 7, 11, or 13$\}$. Also, $\cup_{i=1}^3 B_i = \{q \in Y: 1/16 \leqslant q \leqslant 1/2\}$, so $f^{-1}(\cup_{i=1}^3 B_i) = \cup_{i=1}^3 f^{-1}(B_i)$. ❑

**Theorem 5.12** Let $f: X \to Y$, let $\mathscr{A} = \{A_\alpha: \alpha \in \Lambda)$ be a nonempty family of subsets of $X$, and let $\mathscr{B} = \{B_\beta: \beta \in \Gamma\}$ be a nonempty family of subsets of $Y$. Then:

**a)** $$f\left(\bigcup_{\alpha \in \Lambda} A_\alpha\right) = \bigcup_{\alpha \in \Lambda} f(A_\alpha)$$

**b)** $$f\left(\bigcap_{\alpha \in \Lambda} A_\alpha\right) \subseteq \bigcap_{\alpha \in \Lambda} f(A_\alpha)$$

c) $f^{-1}\left(\bigcap_{\beta\in\Gamma} B_\beta\right) = \bigcap_{\beta\in\Gamma} f^{-1}(B_\beta)$

d) $f^{-1}\left(\bigcup_{\beta\in\Gamma} B_\beta\right) = \bigcup_{\beta\in\Gamma} f^{-1}(B_\beta)$

**PROOF** We prove (d) and leave the remaining parts as exercises.

Let

$$x \in f^{-1}\left(\bigcup_{\beta\in\Gamma} B_\beta\right)$$

Then

$$f(x) \in \bigcup_{\beta\in\Gamma} B_\beta$$

so there is a $\beta \in \Gamma$ such that $f(x) \in B_\beta$.For this $\beta$, $x \in f^{-1}(B_\beta)$. Therefore,

$$x \in \bigcup_{\beta\in\Gamma} f^{-1}(B_\beta)$$

Now let

$$x \in \bigcup_{\beta\in\Gamma} f^{-1}(B_\beta)$$

There exists $\beta \in \Gamma$ such that $x$ belongs to $f^{-1}(B_\beta)$, so $f(x) \in B_\beta \subseteq \cup_{\beta\in\Gamma} B_\beta$. Thus, $x \in f^{-1}(\cup_{\beta\in\Gamma} B_\beta)$ as required. □

It seems peculiar that the inverse of a function should work better than the function itself, so it is only natural to seek conditions under which statement (b) of Theorem 5.12 can be made to jibe with parts (a), (c), and (d). This goal is pursued in Exercises 85 and 86.

Let us return to the definition of an indexed family. Let $\mathscr{A}$ be a nonempty family of sets. The statement that $\mathscr{A} = \{A_\alpha: \alpha \in \Lambda\}$ means that there is a nonempty set $\Lambda$, called the **index set**, and a function $f$ mapping $\Lambda$ onto $\mathscr{A}$, called the **indexing function**, such that for each $\alpha \in \Lambda$, $f(\alpha) = A_\alpha$. Thus, $f = \{(\alpha, A_\alpha): \alpha \in \Lambda\}$, so indexed sets are themselves functions.

**\MPLE 28** For each natural number $n$, let $A_n$ denote the closed interval $[n-1, n]$ and let $\mathscr{A} = \{A_n: n \in \mathbb{N}\}$. Then the indexing function is the function $f: \mathbb{N} \to \mathscr{A}$ defined by $f(n) = [n-1, n]$. □

**:orem 5.13** Let $\mathscr{A}$ be any nonempty family. Then there is an index set $\Lambda$ such that $\mathscr{A} = \{C_\alpha: \alpha \in \Lambda\}$.

**PROOF** For each $A \in \mathscr{A}$, let $C_A = A$, let $\Lambda = \mathscr{A}$, and define $f: \Lambda \to \mathscr{A}$ by $f(A) = C_A$. Since $f$ is the identity function, $f$ maps $\Lambda$ onto $\mathscr{A}$. Therefore, $f$ is an indexing function and hence $\mathscr{A} = \{C_A: A \in \Lambda\}$. □

**Theorem 5.14** It is impossible to index $\mathscr{P}(\mathbb{N})$ using $\mathbb{N}$ as the index set.

**PROOF** We prove that there does not exist a function that maps $\mathbb{N}$ onto $\mathscr{P}(\mathbb{N})$. The proof is by contradiction. Suppose that there is a function $f$ mapping $\mathbb{N}$ onto $\mathscr{P}(\mathbb{N})$. Let $A = \{n \in \mathbb{N}: n \notin f(n)\}$. Since $f$ maps $\mathbb{N}$ onto $\mathscr{P}(\mathbb{N})$, there is a $p \in \mathbb{N}$ such that $f(p) = A$. We consider the question: Is $p \in A$? Suppose that $p \in A$. Then $p \notin f(p)$. But $f(p) = A$, so $p \notin A$. We have reached a contradiction. Suppose that $p \notin A$. Then $p \in f(p)$, but $f(p) = A$. Thus, $p \in A$, and we have reached a contradiction. We have shown that it is impossible that $p \in A$ and also impossible that $p \notin A$. Therefore, there does not exist a function that maps $\mathbb{N}$ onto $\mathscr{P}(\mathbb{N})$. □

## EXERCISES 5.7

**82.** Prove Theorem 5.12(a).

**83.** Prove Theorem 5.12(b).

**84.** Prove Theorem 5.12(c).

**85.** Give an example of a function $f: X \to Y$ and a nonempty collection $\{A_\alpha: \alpha \in \Lambda\}$ of subsets of $X$ such that

$$f\left(\bigcap_{\alpha \in \Lambda} A_\alpha\right) \neq \bigcap_{\alpha \in \Lambda} f(A_\alpha)$$

**86.** Prove that if $f: X \to Y$ is a one-to-one function and $\{A_\alpha: \alpha \in \Lambda\}$ is any nonempty collection of subsets of $X$, then

$$f\left(\bigcap_{\alpha \in \Lambda} A_\alpha\right) = \bigcap_{\alpha \in \Lambda} f(A_\alpha)$$

## WRITING EXERCISE

Explain, as if to a puzzled classmate, what it means to say that a function is well defined. Illustrate your explanation with your own examples of functions that are, or are not, well defined. Is it fair to say that a function that is not well defined is just a relation masquerading as a function?

CHAPTER

# 6

# COMBINATORIAL PROOFS

Do not be fooled by the fancy title; this chapter is about counting. There are two basic ways to use counting as a method of proof. First, counting can reveal the existence of a solution to a problem by showing that when all nonsolutions are counted out, at least one object remains. Arguments of this type are said to use the Pigeonhole Principle. Second, some problems can be solved by counting a finite set in two different ways. The argument given in Chapter 3 that the sum of the first 100 natural numbers is 5050 is an example of this procedure. Obviously both mathematical induction and counting are fundamental techniques, and there are many results like the equation

$$\sum_{i=1}^{n} i = \frac{n(n+1)}{2}$$

that can be proved either by induction or by counting. But often counting reveals why a theorem is true when induction does not. Therefore, we ask that you put aside mathematical induction as a method of proof in the first three sections of this chapter, although we concede that formal proofs of underlying principles that we consider obvious would probably be proved by induction.

## 6.1 *The Sum and Product Rules*

Suppose that we have two boxes labeled $A_1$ and $A_2$. Box $A_1$ contains $m$ objects, box $A_2$ contains $n$ objects, and of course no object is contained in both boxes. There are $m$ **plus** $n$ ways to choose an object from box $A_1$ **or** an object from box $A_2$ and $m$ **times** $n$ ways to choose an object from box $A_1$ **and** an object from box $A_2$. We can extend and formalize these observations, as follows.

**The Sum and Product Rules**

Let $A_1, A_2, \ldots, A_n$ be pairwise disjoint sets and suppose that for $i = 1, 2, \ldots, n$, $A_i$ has $m_i$ elements. Then there are $\Sigma_{i=1}^{n} m_i$ ways in which to choose an element from any one of the sets $A_i$ and $m_1 m_2 \cdots m_n$ ways in which to choose an element from each of the sets $A_i$.

To illustrate the use of the Product Rule, we reconsider the problem of determining the number of subsets of an $n$-element set. We have already proved by mathematical induction that this number is $2^n$, but in this chapter, proofs by mathematical induction have been disallowed. We know from Chapter 2 that $\varnothing$ has $1 = 2^0$ subsets. Let $n$ be a natural number, and let $A = \{a_1, a_2, \ldots, a_n\}$ be an $n$-element set. Imagine $n$ boxes $A_1, A_2, \ldots, A_n$ and imagine that in each box $A_i$ there are two cards, one saying "yes, $a_i$ is an element" and the other saying "no, $a_i$ is not an element." Each subset of $A$ is determined by selecting one card from each box. For example, $\varnothing$ is determined by choosing the "no-card" from each box and the subset $A$ itself is determined by choosing the "yes-card" from each box. By the Product Rule, there are $2^n$ ways in which to choose a card from each of the $n$ boxes; thus $A$ has $2^n$ subsets.

Now we consider the number of arrangements of the $n$-element set $\{1, 2, \ldots, n\}$. It is easy to see that $\{1, 2, 3\}$ has 6 arrangements: $\langle 1, 2, 3\rangle$, $\langle 1, 3, 2\rangle$, $\langle 2, 1, 3\rangle$, $\langle 2, 3, 1\rangle$, $\langle 3, 1, 2\rangle$, and $\langle 3, 2, 1\rangle$. We can use the Product Rule to determine that for any natural number $n$, $\{1, 2, 3, \ldots, n\}$ has $n!$ arrangements. Imagine $n$ boxes $A_1, A_2, \ldots, A_n$. Each box contains $n$ cards and the $n$ cards in a typical box $A_i$ read, "1 is in the $i$th position," "2 is in the $i$th position," ..., "$n$ is in the $i$th position." To determine an ordering of $\{1, 2, \ldots, n\}$, we reach in the first box and select a card, which reads "$k$ is in the first position." Since we are not able to use $k$ again, we remove the card "$k$ is in the $i$th position" from all the remaining boxes. Then from the second box we select a card that reads "$j$ is in the second position." We remove the card "$j$ is in the $i$th position" from each of the remaining boxes, and select a card from the third box. When we have selected a card from each of the boxes in this way, we have determined an arrangement of $\{1, 2, \ldots, n\}$ and it is obvious that any arrangement can be so determined. By the time we select a card from the $i$th box, the box has $n - (i - 1)$ cards in it. In particular, when we select a card from the $n$th box, the box has $1 = (n - (n - 1))$ cards in it. By the Product Rule, the number of arrangements is

$$n[n-1][n-2]\cdots[n-(n-1)] = n!$$

The numbers we have been considering, the number of subsets of an $n$-element set, and the number of arrangements of $\{1, 2, \ldots, n\}$ can be generalized in a natural way.

**Notation** Let $n$ be a natural number, and let $r$ be an integer with $0 \leqslant r \leqslant n$. Then $C(n, r)$ denotes the number of $r$-element subsets of $\{1, 2, \ldots, n\}$. If $r \geqslant 1$, $P(n, r)$ denotes the number of $r$-element arrangements of $\{1, 2, \ldots, n\}$.

**Definition**

The number $P(n, r)$ is called an *r-permutation*, and an arrangement (that is, an $n$-permutation) is called simply a **permutation**.

The number $C(n, r)$, which is read "$n$-choose $r$," is often denoted by $\binom{n}{r}$. Given any $n$-element set $A$ and any integer $r$ with $0 \leqslant r \leqslant n$, $C(n, r)$ is the number of $r$-element subsets of $A$. Since each set has only one empty subset, $C(n, 0) = 1$. Similarly, if $A$ is any $n$-element set and $r$ is a natural number with $r \leqslant n$, then we can think of $P(n, r)$ as the number of $r$-element arrangements of $A$. For example, if $A = \{a, b, c\}$ is a three-element set and $r = 2$, then since $A$ has the two-element arrangements $ab$, $ac$, $ba$, $bc$, $ca$, and $cb$, $P(3, 2) = 6$. We adopt the following convention.

**Convention** For each natural number $n$, $P(n, 0) = C(n, 0) = 0! = 1$.

It is easy to adapt our use of the Product Rule to show that $P(n, n) = n!$ to the general problem of determining a formula for $P(n, r)$, where $0 \leqslant r \leqslant n$.

**Theorem 6.1**

Let $n$ be a natural number, and let $r$ be an integer with $0 \leqslant r \leqslant n$. Then $P(n, r) = n!/(n-r)!$.

**PROOF** When $r = 0$, $n!/(n-r)! = 1 = P(n, r)$. Let $r \geqslant 1$, and let $A$ be an $n$-element set. To determine an $r$-permutation of $A$, we choose any one of $n$ elements for the first position, any one of $n-1$ elements for the second position, and continue choosing until we choose $r$ elements. At our last choice, we are choosing the $r$th element and there are $n-r$ elements that will not be chosen. We must, as it were, be choosing from the $(n-r+1)$st box. By the Product Rule, there are $n(n-1)(n-2)\cdots(n-r+1)$ $r$-permutations of $A$. A comparison of $n(n-1)(n-2)\cdots(n-r+1)$ with

$$n! = n(n-1)(n-2)\cdots(n-r+1)(n-r)(n-r-1)\cdots(3)(2)(1)$$

yields

$$P(n, r) = \frac{n!}{(n-r)(n-r-1)\cdots(3)(2)(1)} = \frac{n!}{(n-r)!}. \qquad \square$$

It is not so easy to determine a formula for $C(n, r)$, but there is a trick, which the following example illustrates.

**EXAMPLE 1** We can construct the 3-permutations of the 4-element set $\{1, 2, 3, 4\}$ by selecting the arrangements of the 3-element subset of $\{1, 2, 3, 4\}$. There are four 3-element subsets of $\{1, 2, 3, 4\}$—namely, $\{1, 2, 3\}$, $\{1, 2, 4\}$, $\{1, 3, 4\}$, and $\{2, 3, 4\}$—and there are 3! arrangements associated with each of these sets. By the Product Rule, $P(4, 3) = (4)(3!) = 24$. ❑

Note that in Example 1 we are calculating $P(4, 3)$ differently from the way in which we calculated $P(4, 3)$ in the proof of Theorem 6.1.

**Theorem 6.2** Let $n$ be a natural number, and let $r$ be an integer with $0 \leqslant r \leqslant n$. Then

$$C(n, r) = \frac{P(n, r)}{r!} = \frac{n!}{r!(n-r)!}$$

**PROOF** We can easily see that the conventional values for 0! and $P(n, 0)$ have been chosen so that

$$C(n, 0) = \frac{P(n, 0)}{0!} = \frac{n!}{0!(n-0)!}$$

Therefore, we assume that $r \geqslant 1$. We calculate $P(n, r)$ as in Example 1. There are $C(n, r)$ $r$-element subsets of an $n$-element set $A$, and each of these subsets has $r!$ arrangements. By the Product Rule, $P(n, r) = C(n, r) \cdot r!$ Thus,

$$C(n, r) = \frac{P(n, r)}{r!}$$

and by Theorem 6.1,

$$\frac{P(n, r)}{r!} = \frac{n!}{(n-r)!r!}$$ □

Theorem 6.2 has the following corollary, which can also be established independent of the theorem (see Exercise 24).

**Corollary 6.3** Let $n$ be a natural number, and let $r$ be an integer such that $0 \leqslant r \leqslant n$. Then $C(n, r) = C(n, n-r)$.

**PROOF** By Theorem 6.2,

$$C(n, n-r) = \frac{n!}{[n-(n-r)]!(n-r)!} = \frac{n!}{r!(n-r)!} = C(n, r)$$ □

**EXAMPLE 2** The number of different ways in which 8 people can occupy a row of 8 seats is 8!. In how many different ways can 8 people be arranged in a circle?

The reason that the answer is not 8! is that two circular arrangements are considered to be the same if one can be obtained from the other by a rotation. We first calculate that there are 8! different arrangements, and then when we are told that two circular arrangements are considered the same if one can be obtained from the other by a rotation, we divide by the correct number to adjust for the change in the rules. Every time we are given an arrangement such as the one shown in Figure 6.1, there are really 8 arrangements that we consider the same. (The first has $P_1$ at the top, the second has $P_2$ at the top, and so forth.) Therefore, each really different arrangement has been counted 8 times, so we must divide 8! by 8. The result is 7!.

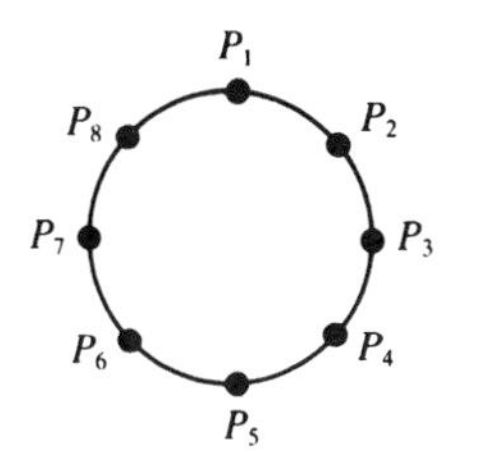


Figure 6.1 ❑

## EXERCISES 6.1

1. A standard deck of cards has 52 cards, which are made up of four suits (spades, hearts, diamonds, clubs), each having thirteen cards (Ace, King, Queen, Jack, 10, 9, 8, 7, 6, 5, 4, 3, 2). Using a standard deck of cards, given one draw,
   **a)** In how many ways can you draw a card that is either a 2 or a 3?
   **b)** In how many ways can you draw a card that is either a queen or a heart?
2. A die has the numbers 1 through 6, one on each of the 6 sides. Given 2 rolls, in how many ways can you roll a number greater than 4 and then a number less than 4?
3. A true–false test has 10 questions.
   **a)** In how many ways can a student answer the questions if every question is answered?
   **b)** In how many ways can a student answer the questions if there is a penalty for guessing and the student leaves some (or all) answers blank?
4. Two dice, one red and one green, are rolled.
   **a)** In how many ways can you get an even sum?
   **b)** In how many ways can you get an odd sum?
5. **a)** When you write the numbers from 1 to 100, how many times do you write the digit 5?
   **b)** When you write the numbers from 100 to 1000, how many times do you write the digit 5?
6. What is the least number of 3-digit area codes needed to guarantee that 40 million phones have distinct 10-digit numbers?

**7.** Let $S = \{a, b, c, d, e, f, g\}$ and assume that $a$, $b$, $c$, $d$, $e$, $f$, and $g$ are all different.

**a)** How many permutations are there of $S$?

**b)** How many permutations of $S$ end with $g$?

**c)** How many permutations of $S$ begin with $a$ and end with $g$?

**d)** How many 6-permutations are there of $S$?

**e)** How many 5-permutations are there of $S$?

**8.** Let $S = \{1, 2, 3, 4, 5\}$.

**a)** List all the 3-permutations of $S$.

**b)** List all the 4-permutations of $S$.

**9. a)** How many arrangements of the letters ABCDEF contain the word BAD? [*Hint*: The letters BAD must be kept together in this order.]

**b)** How many arrangements of the letters ABCDEF contain the letters BAD together in any order?

**10.** Find each of the following.

**a)** $P(7, 4)$ **b)** $P(10, 5)$ **c)** $P(12, 9)$

**11.** How many different numbers are there between 100 and 1000, each of whose digits is one of 1, 3, 5, 7, 8, 9 and in which no two digits are the same?

**12.** In how many ways can we seat 8 people in a straight line if some 2 of them refuse to sit next to each other?

**13.** In the game of Scrabble, a player tries to make words from certain letters. Suppose Susan has the letters A, I, H, U, C, S, and F. How many different 3-letter arrangements are there? Because Susan has an S, she will be able to play any word she can form, and there is a 50-point bonus for using all seven letters. How many 7-letter arrangements must Susan consider in order to be sure she has considered all possible 7-letter arrangements?

**14.** Four women and 3 men are to be seated in 7 chairs on the same side of the head table. How many seating arrangements are possible if no 2 women are to sit next to each other?

**15.** Four women and 3 men are to be seated in 7 chairs at a circular table. How many seating arrangements are possible if no 2 women are to sit next to each other?

**16.** If $P(n, 4) = 15{,}120$, find $C(n, 4)$.

**17.** If $C(n, 5) = 462$, find $P(n, 5)$.

**18.** Does there exist a natural number $n$ such that $P(n, 4) = 11{,}880$? If so, find $n$. If not, explain why not.

**19.** If a baseball league consists of 8 teams, how many games will be played during the course of a season if each team plays every other team exactly 22 times?

**20. a)** In how many ways can a student choose 10 questions from a 13-question exam?

**b)** In how many ways can a student choose 10 questions from a 13-question exam if 8 questions must be chosen from the first 10 and 2 from the last 3?

**c)** In how many ways can a student choose 10 questions from a 13-question exam if at least 5 questions must be chosen from the first 8 and at least 3 questions from the last 5?

**21.** How many different 2-element subsets of $\{n \in \mathbb{N}: n \leqslant 50\}$ are there such that the sum of the 2 elements is even?

**22.** Prove that if $n, r \in \mathbb{N}$ and $r \leqslant n$, then

$$C(n, r) = \binom{n}{r} C(n-1, r-1)$$

**23.** Prove that if $n \in \mathbb{N}$, then $C(2n, n)$ is even.

**24.** In 25 or fewer words, give a proof of Corollary 6.3 that does not use Theorem 6.2.

**25.** Explain why it is true that for each natural number $n$, $2^n = \Sigma_{i=0}^{n} C(n, i)$. (No fair using a theorem not proved in this section!)

**26.** Prove that if $C(n, r) = C(n, r+1)$, then $n$ is odd and $r = (n-1)/2$.

## Dirichlet's Pigeonhole Principle

**Dirichlet's Pigeonhole Principle**, named in honor of Peter Gustav Lejeune-Dirichlet (1805–1859), is a powerful tool for proving the existence of a required object. Intuitively this principle says that when a large collection of objects is divided into a small number of sets, one of the sets will contain a certain minimum number of objects. More precisely, it is stated as follows.

**Dirichlet's Pigeonhole Principle**

Let $k$ and $n$ be natural numbers. If $kn+1$ objects are distributed among $n$ sets, one of the sets will contain at least $k+1$ objects.

Suppose that 101 letters are distributed to 100 mailboxes. Then 1 box must contain at least 2 letters. If 25 letters are distributed among 6 mailboxes, then 1 box will contain at least 5 letters ($k = 4$, $n = 6$).

**PROOF** Suppose none of the sets contains at least $k+1$ objects. Then the number of objects in each set is at most $k$. Therefore, by the Sum Rule, the total number of objects is at most $kn$. Since we had $kn+1$ objects, this is a contradiction. □

As you can see, it is easy to prove the Pigeonhole Principle. The difficulty occurs in recognizing when and how to use it. Let's consider a simple example.

**Theorem 6.4**

Let $a_1, a_2, \ldots, a_{17}$ be a list of 17 positive integers. Then there are at least 4 integers in the list that have the same remainder when divided by 5.

**PROOF** We have 17 objects and only $n = 5$ possible remainders (namely, 0, 1, 2, 3, 4). For $k = 3$, we have $kn + 1$ (actually there are $kn + 2$) objects distributed among 5 sets. (These 5 sets are the set of numbers having remainder 0 when divided by 5, having remainder 1 when divided by 5, and so forth.) By the Pigeonhole Principle, some one of these 5 sets has $k + 1 = 4$ members from the list. □

**Theorem 6.5**

> Suppose that $P_1, P_2, \ldots, P_5$ are any five points in the interior of the unit square, and for each $i$ and $j$, let $d_{ij}$ denote the distance between $P_i$ and $P_j$. Then there exist $P_i$ and $P_j$ such that $d_{ij} < \sqrt{2}/2$.

**PROOF** Divide the unit square into four congruent squares, as shown in Figure 6.2. We have five objects and four squares. So, taking $n = 4$ and $k = 1$, and using the Pigeonhole Principle, we see that one of the squares must contain at least two of the points. The distance between these two points is less than $\sqrt{2}/2$ because the length of the diagonal of the square is $\sqrt{2}/2$.

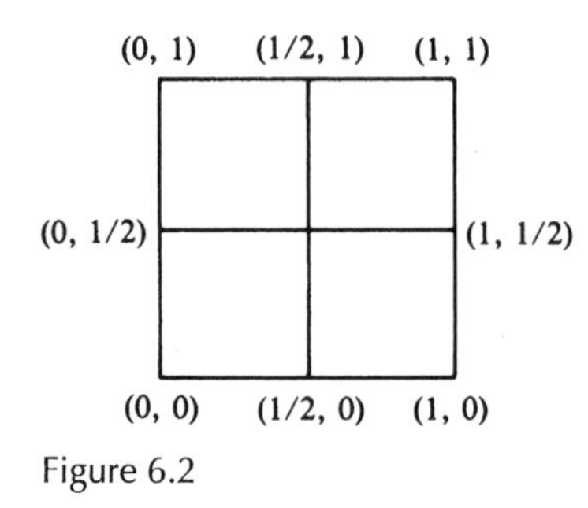


Figure 6.2 □

We now consider a more complicated example. Theorem 6.6. This example appeared in the 1976 U.S.A. Olympiad, and the proof is due to Loren Larson [*Problem-Solving Through Problems*, Springer-Verlag (1983), 80–81]. (The wording of this theorem is our own.)

**Theorem 6.6**

> If each square of a 3-by-7 chessboard is colored either black or white, then in any such coloring the board must contain a rectangle consisting of at least four squares whose corner squares are all the same color.

Before attempting to prove the theorem, let's draw a picture to make sure that we understand it (see Figure 6.3). In this example, the rectangle drawn with dotted lines satisfies the conclusion of the theorem.

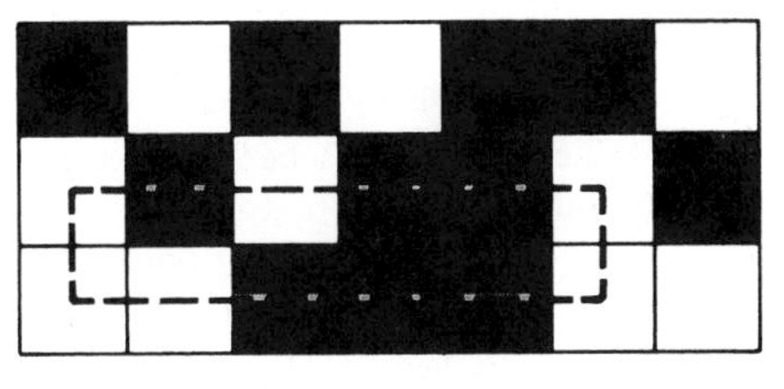

Figure 6.3

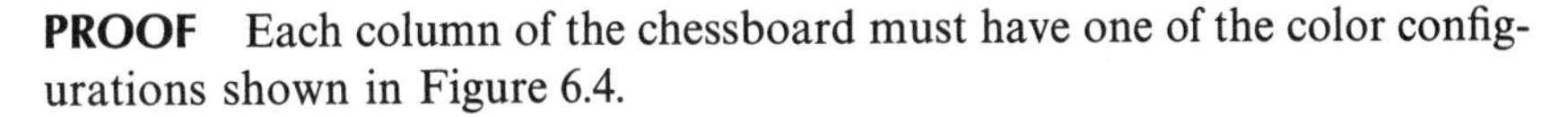

**PROOF** Each column of the chessboard must have one of the color configurations shown in Figure 6.4.

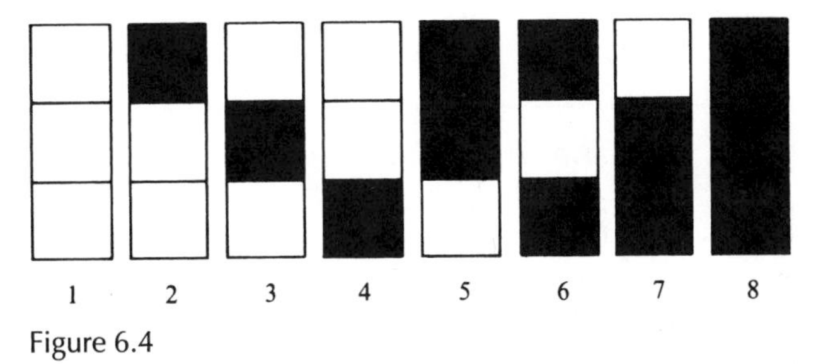


Figure 6.4

We consider three cases.

***Case I.*** Suppose that one of the columns of the chessboard is of type 1. If one of the remaining six columns is of type 1, 2, 3, or 4, then we can draw a rectangle whose corner squares are all white. So, suppose that all of the remaining six columns are of type 5, 6, 7, or 8. We have six columns (only five are needed) and only four types, so by the Pigeonhole Principle, two of the six columns must be of the same type, and we are finished.

***Case II.*** If one of the columns of the chessboard is of type 8, the proof is essentially the same as the proof for Case I.

***Case III.*** Suppose none of the six columns is of type 1 or type 8. Then we have seven columns and only six types, so by the Pigeonhole Principle, two columns have the same type. □

## EXERCISES 6.2

**27.** Use the division algorithm to prove that if $k$ pigeons nest in $n$ nests in the ceiling of a house, then some nest has $\lceil k/n \rceil$ pigeons. Deduce the Pigeonhole Principle from this result.

**28.** Argue that for any natural number $n$, either the decimal representation of $1/n$ terminates or the representation repeats in blocks of $n-1$ or fewer digits.

**29.** Let $n$ be a natural number greater than 2. Prove that if there are $n$ people at a party, then there are at least 2 people who know the same number of people at the party. *Assume* that if A knows B, then B knows A.

30. Let $n$ be a natural number greater than 1 and suppose that if $p$ is a prime and $p \leqslant \sqrt{n}$, then $p$ does not divide $n$. Prove that $n$ is a prime.
31. Let $P_1, P_2, \ldots, P_9$ be 9 distinct points in Euclidean 3-space whose coordinates are integers. Prove that there exist $i$ and $j$ such that the line segment joining $P_i$ and $P_j$ contains an interior point whose coordinates are integers.
32. Let $n$ be a positive integer, and let $P_1, P_2, \ldots, P_{n+1}$ be integers chosen from the integers between 1 and $2n$ (inclusive). Prove that there exist $i$ and $j$ such that $P_i$ divides $P_j$.
33. The final exam in a certain course consists of 3 true–false questions, and there are 7 students who take the exam. Prove that at least 2 students have the same answers to at least 2 questions.

## 6.3 *The Binomial Theorem*

The theorem that is the title of this section first appears in chapters such as this one, which is about counting finite sets. The Binomial Theorem plays a peripheral role in calculus when it is used to determine the derivative of the $n$th-root function and a central role when it is used in the study of power series. For these reasons, among others, the theorem does not fit the tiny compartment provided for it here. We begin by establishing Pascal's formula (Theorem 6.7), named in honor of Blaise Pascal (1623–1662).

**Theorem 6.7**

Let $r$ and $n$ be natural numbers with $r < n$. Then

$$C(n, r) = C(n-1, r) + C(n-1, r-1)$$

**PROOF** Let $S$ be a set with $n$ elements, and let $x \in S$. We divide the $r$-element subsets of $S$ into two collections $\mathscr{A}$ and $\mathscr{B}$, where $\mathscr{A}$ is the collection whose members do not contain $x$ and $\mathscr{B}$ is the collection whose members do contain $x$. It is easy to count the number of members of $\mathscr{A}$. It is just the number of $r$-element subsets of $S - \{x\}$, namely, $C(n-1, r)$. In order to count the members of $\mathscr{B}$, we introduce the collection $\mathscr{B}' = \{B - \{x\}: B \in \mathscr{B}\}$. Note that $\mathscr{B}$ and $\mathscr{B}'$ have the same number of members. Every set in $\mathscr{B}'$ is an $(r-1)$-element subset of $S - \{x\}$; what's more, every $(r-1)$-element subset of $S - \{x\}$ belongs to $\mathscr{B}'$. Therefore, both $\mathscr{B}'$ and $\mathscr{B}$ have $C(n-1, r-1)$ members. It follows that $C(n, r) = C(n-1, r) + C(n-1, r-1)$. □

Theorem 6.7 can also be proved by algebraic manipulation (see Exercise 51). We are ready to prove the Binomial Theorem. As we indicated in the introduction, we prefer an informal counting argument to a formal proof by induction, because the counting argument provides insight into why the theorem is true. In the single exception to the governing rule of the first three sections of this chapter, we ask you to supply a formal proof by induction in Exercise 49.

**Theorem 6.8**

**The Binomial Theorem**

Let $r$ and $n$ be natural numbers with $r \leqslant n$, and let $x$ and $y$ be real numbers. Then the following formula, called the Binomial Formula, holds:

$$(x+y)^n = \sum_{r=0}^{n} C(n,r)x^{n-r}y^r$$

**INFORMAL ARGUMENT** We first raise $x+y$ to the third power, and we use subscripts to keep track of the multiplications we are performing.

$$\begin{array}{l} x_1 + y_1 \\ x_2 + y_2 \\ \hline x_1x_2 + y_1x_2 + x_1y_2 + y_1y_2 \\ \qquad\qquad x_3 + y_3 \\ \hline x_1x_2x_3 + \underline{y_1x_2x_3} + \underline{x_1y_2x_3} + y_1y_2x_3 + \underline{x_1x_2y_3} + y_1x_2y_3 + x_1y_2y_3 + y_1y_2y_3 \end{array}$$

Of course, in calculating $(x+y)^3$, we know that all the $x$'s are really the same, as are all the $y$'s. Thus the three products that are underlined are all really $x^2y$. Why should there be exactly three $x^2y$'s? We have a 3-element set $\{x_1, x_2, x_3\}$ and each 2-element subset of this set gives rise to $x^2y$. So the coefficient of $x^2y$ in the expansion of $(x+y)^3$ is $C(3,2)$, where 3 is the power to which $x+y$ is raised and 2 is the power to which $x$ is raised.

We continue our imagined distinction with each multiplication and find the coefficient of $x^5y^2$ in the expansion of $(x+y)^7$. We have a 7-element set $\{x_1, x_2, x_3, x_4, x_5, x_6, x_7\}$ and each 5-element subset of this set gives rise to one of the $x^5y^2$'s. For example, the 5-element subset $\{x_1, x_2, x_4, x_6, x_7\}$ gives rise to $x_1x_2y_3x_4y_5x_6x_7$ in which we multiplied at $x_1$ and multiplied by $x_i$ in stages $i = 2$, 4, 6, and 7 and by $y_i$ in stages $i = 3$ and 5. If at any given stage we did not mutiply by $x_i$, then we must have multiplied by $y_i$. There are therefore as many $x^5y^2$'s as there are 5-element subsets of $\{x_1, x_2, x_3, x_4, x_5, x_6, x_7\}$. That is, there are $C(7,5)x^5y^2$'s in the expansion of $(x+y)^7$. For each natural number $n$ and each $r$ with $r = 0, 1, 2, \ldots, n$, we have an $x^{n-r}y^r$ in the expansion of $(x+y)^n$ and its coefficient is $C(n, n-r) = C(n,r)$. As a result

$$(x+y)^n = \sum_{r=0}^{n} C(n,r)x^{n-r}y^r \qquad \square$$

We would like to say now that numbers of the form $C(n,r)$, where $n \in \mathbb{N}$ and $r$ is an integer such that $0 \leqslant r \leqslant n$ are called **binomial coefficients**, because such numbers appear as coefficients in the right-hand side of the Binomial Formula. Unfortunately for any natural number $n$, $C(n, 1) = n$, so the definition we would like to make would lead to our using "binomial coefficient" as a rather awkward synonym for "natural number." The problem is that we would like to define a number not only by what it is but by how it is thought of or how we come upon it.

The binomial coefficients are frequently arranged in the form of a triangle, called **Pascal's triangle** (Figure 6.5). Pascal himself called the triangle the "arithmetic triangle" and discussed its properties in a treatise written in 1653. One could argue that "arithmetic triangle" is a better name for Pascal's triangle, because the triangle had been discovered by the Chinese at least 400 years before Pascal wrote his treatise.

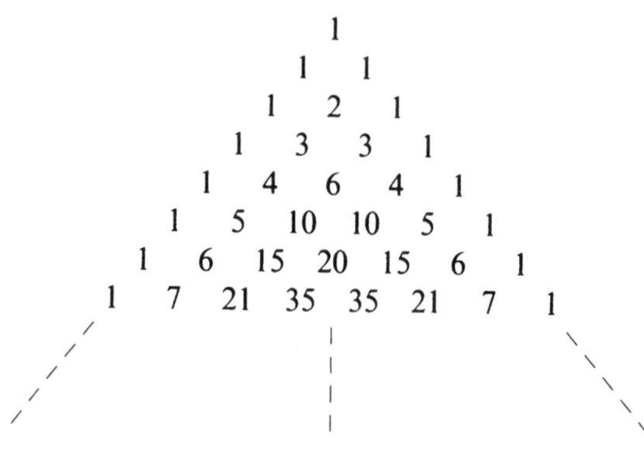

Figure 6.5

It is customary to call the top row of Pascal's triangle the 0th row, the next row the 1st row, and so on. With this understanding the $n$th row of Pascal's triangle consists of the terms of the expansion of $(1 + 1)^n$. Pascal's formula (Theorem 6.7) says that in order to obtain a particular entry (other than the ends) in this triangle, we simply add the two nearest binomial coefficients in the row immediately above it. Thus the 8th row of Pascal's triangle is

$$\begin{array}{ccccccccc} 1 & (1+7) & (7+21) & (21+35) & (35+35) & (35+21) & (21+7) & (7+1) & 1 \\ 1 & 8 & 28 & 56 & 70 & 56 & 28 & 8 & 1 \end{array}$$

Pascal's triangle motivates the following definition.

**Definition**

Let $A$ and $B$ be algebraic expressions, and let $n$ be a natural number. Let $r \in \{1, 2, 3, \ldots, n+1\}$. Then $C(n, r-1)A^{n-r+1}B^{r-1}$ is called the ***r*th term of the binomial expansion** of $(A + B)^n$. When $n = 2k$, the $(k + 1)$st term is called the **middle term**.

**EXAMPLE 3** Use the Binomial Formula to find each term of the binomial expansion of $(2x+3y)^5$. Evaluate each term.

$$\begin{aligned}(2x+3y)^5 &= C(5,0)(2x)^5 + C(5,1)(2x)^4(3y) + C(5,2)(2x)^3(3y)^2 \\ &\quad + C(5,3)(2x)^2(3y)^3 + C(5,4)(2x)(3y)^4 + C(5,5)(3y)^5 \\ &= (2x)^5 + 5(2x)^4(3y) + 10(2x)^3(3y)^2 + 10(2x)^2(3y)^3 \\ &\quad + 5(2x)(3y)^4 + (3y)^5 \\ &= 32x^5 + 240x^4y + 720x^3y^2 + 1080x^2y^3 + 810xy^4 + 243y^5\end{aligned}$$ ❑

**EXAMPLE 4** Write the 6th term of the binomial expansion of $(x^2+2y)^8$.

The 6th term contains $(2y)^5$, so it is

$$C(8,5)(x^2)^3(2y)^5 = \frac{8 \cdot 7 \cdot 6}{3 \cdot 2} x^6(32y^5) = 1792x^6y^5$$ ❑

Examples 3 and 4 illustrate a straightforward, if uninteresting, application of the Binomial Theorem. The interesting applications are more subtle.

**Theorem 6.9** Let $n$ be a natural number. Then the following are true:

**a)** $C(n,0)+C(n,1)+C(n,2)+\cdots+C(n,n)=2^n$

**b)** $C(n,0)-C(n,1)+C(n,2)-C(n,3)+\cdots \pm C(n,n)=0$

**c)** $C(n,0)+C(n,2)+C(n,4)+\cdots+C(n,2\lfloor n/2\rfloor)$
$= C(n,1)+C(n,3)+C(n,5)+\cdots+C(n,2\lfloor (n+1)/2\rfloor-1)$
$=2^{n-1}$

**PROOF** Part (a) is a restatement of Exercise 25 and can be proved without using the Binomial Theorem. In Exercises 40 and 41, however, you are asked to use the Binomial Theorem to prove both parts (a) and (b). The proof of part (c) does not use the Binomial Theorem, but it does use parts (a) and (b) (see Exercise 42). We outline a combinatorial proof of part (c) in Exercise 56. □

Note that Theorem 6.9(c) says that the sum of the coefficients of the odd terms equals the sum of the coefficients of the even terms and that this common sum is $2^{n-1}$.

Our final application of the Binomial Theorem is to use its consequence, Theorem 6.9(b), to prove an important principle of counting known as the Principle of Inclusion–Exclusion. This principle is also known by the acronym PIE. We illustrate PIE in the following simple example.

**EXAMPLE 5** Let $A=\{1,2,4,7\}$, $B=\{3,4,5,7\}$, and $C=\{2,4,6\}$. Then $|A\cup B\cup C|=7$, $|A|+|B|+|C|=11$, $-(|A\cap B|+|A\cap C|+|B\cap C|)=-5$, and $|A\cap B\cap C|=1$. Note that $7=11-5+1$. This equation is not a coincidence, for we could easily prove that for any three finite sets $A$, $B$, and $C$,

$$|A \cup B \cup C| = |A| + |B| + |C| - |A \cap B| - |A \cap C| - |B \cap C| + |A \cap B \cap C|$$

Indeed the latter equation is a special case of PIE. ❑

For the purpose of establishing PIE, we introduce the following notations. For any finite set $A$, we let $n(A)$ denote the number of members of $A$. Moreover, if we are given $n$ finite sets $A_1, A_2, \ldots, A_n$ and a natural number $r$ with $1 < r \leqslant n$, we let $m_r$ denote the sum of all $n(A)$ such that $A$ is an intersection of $r$ members of $\{A_1, A_2, \ldots, A_n\}$. For $r = 1$, we let $m_r = \Sigma_{i=1}^{n} n(A_i)$. Example 6 illustrates this notation.

**EXAMPLE 6** Let $A_1 = \{1, 2, 3, 6\}$, $A_2 = \{1, 2, 4, 5\}$, $A_3 = \{2, 4, 6\}$, $A_4 = \{1, 3\}$, and $A_5 = \{2, 4, 5, 7\}$. Then $n(A_1) = n(A_2) = n(A_5) = 4$, $n(A_3) = 3$, and $n(A_4) = 2$. Thus,

$$\begin{aligned} m_1 &= n(A_1) + n(A_2) + n(A_3) + n(A_4) + n(A_5) \\ &= 17 \end{aligned}$$

$$\begin{aligned} m_2 &= n(A_1 \cap A_2) + n(A_1 \cap A_3) + n(A_1 \cap A_4) + n(A_1 \cap A_5) + n(A_2 \cap A_3) \\ &\quad + n(A_2 \cap A_4) + n(A_2 \cap A_5) + n(A_3 \cap A_4) + n(A_3 \cap A_5) + n(A_4 \cap A_5) \\ &= (2 + 1 + 2 + 1) + (2 + 1 + 3 + 0 + 2 + 0) \\ &= 14 \end{aligned}$$

$$\begin{aligned} m_3 &= n(A_3 \cap A_4 \cap A_5) + n(A_2 \cap A_4 \cap A_5) + n(A_2 \cap A_3 \cap A_5) \\ &\quad + n(A_2 \cap A_3 \cap A_4) + n(A_1 \cap A_4 \cap A_5) + n(A_1 \cap A_3 \cap A_5) \\ &\quad + n(A_1 \cap A_3 \cap A_4) + n(A_1 \cap A_2 \cap A_5) + n(A_1 \cap A_2 \cap A_4) \\ &\quad + n(A_1 \cap A_2 \cap A_3) \\ &= (0 + 0 + 2) + (0 + 0 + 1) + (0 + 1 + 1 + 0) \\ &= 5 \end{aligned}$$

$$\begin{aligned} m_4 &= n(A_2 \cap A_3 \cap A_4 \cap A_5) + n(A_1 \cap A_3 \cap A_4 \cap A_5) \\ &\quad + n(A_1 \cap A_2 \cap A_4 \cap A_5) + n(A_1 \cap A_2 \cap A_3 \cap A_5) \\ &\quad + n(A_1 \cap A_2 \cap A_3 \cap A_4) \\ &= 0 + 0 + 0 + 1 + 0 \\ &= 1 \end{aligned}$$

$$\begin{aligned} m_5 &= n(A_1 \cap A_2 \cap A_3 \cap A_4 \cap A_5) \\ &= 0 \end{aligned}$$

It is noteworthy that since we have five sets, in order to calculate each $m_i$ we must consider $C(5, i)$ sets. What is certainly more curious is that

$$n\left(\bigcup_{i=1}^{5} A_i\right) = 7 = m_1 - m_2 + m_3 - m_4 + m_5$$ ❑

**Theorem 6.10**

**The Principle of Inclusion–Exclusion**

Let $A_1, A_2, \ldots, A_n$ be a list of finite sets. Then

$$n\left(\bigcup_{r=1}^{n} A_r\right) = m_1 - m_2 + m_3 - m_4 + \cdots + (-1)^{n+1} m_n$$

**PROOF** We want to prove that

$$n\left(\bigcup_{r=1}^{n} A_r\right) = \sum_{r=1}^{n} (-1)^{r+1} m_r$$

Let $x$ be an element of $\cup_{r=1}^{n} A_r$. Then $n(\cup_{r=1}^{n} A_r)$ counts this element $x$ once. We must make sure that $\Sigma_{r=1}^{n}(-1)^{r+1}m_r$ also counts $x$ once. Let us suppose that $x$ belongs to $k$ of the sets $A_1, A_2, \ldots, A_n$. Let $B = \{j \in \mathbb{N}$: $x$ belongs to $A_j\}$; that is, $B$ is the set whose members are those subscripts $j$ such that $x$ belongs to $A_j$. Since we are assuming that $x$ belongs to $k$ of the sets $A_1, A_2, \ldots, A_n$, $B$ has $k$ members, and the first term $m_1$ of $\Sigma_{r=1}^{n}(-1)^{r+1}m_r$ counts $x$ $k$ times. How many times does the second term $-m_2$ of $\Sigma_{r=1}^{n}(-1)^{r+1}m_r$ count $x$? The second term counts $x$ (negatively) just as many times as there are 2-element subsets of the set $B$, namely, $C(k, 2)$. The third term adds back $C(k, 3)$, but the fourth term subtracts $C(k, 4)$, and so forth. Summing and remembering that $k = C(k, 1)$, we find that the number of times $x$ is counted in $\Sigma_{r=1}^{n}(-1)^{r+1}m_r$ is $C(k, 1) - C(k, 2) + C(k, 3) - \cdots \pm C(k, k)$. It remains to show that this sum equals 1.

We have arrived at a sum that looks much like Theorem 6.9(b) when $n = k$. Indeed by Theorem 6.9(b),

$$0 = -C(k, 0) + C(k, 1) - C(k, 2) + C(k, 3) - \cdots \pm C(k, k)$$

and so

$$1 = C(k, 0) = C(k, 1) - C(k, 2) + C(k, 3) - \cdots \pm C(k, k)$$ □

**EXAMPLE 7** A bridge hand consists of 13 cards drawn from a standard 52-card deck. Thus there are $C(52, 13)$ possible bridge hands. How many of these are void in a suit (that is, contain no cards in a suit)?

Let $A$ denote the set of all bridge hands that are void in spades, let $B$ denote the set of all bridge hands that are void in hearts, let $C$ denote the set of all bridge hands that are void in clubs, and let $D$ denote the set of all bridge hands that are void in diamonds. Then by the preceding formula, $n(A \cup B \cup C \cup D) = m_1 - m_2 + m_3 - m_4$, and so we must determine $m_i$ for each $i = 1, 2, 3, 4$.

Since a hand that is void in spades consists of 13 cards chosen from the 39 cards that are not spades, $n(A) = C(39, 13)$. Of course, it is also true that $n(B) = n(C) = n(D) = C(39, 13)$. Therefore, $m_1 = 4 \times C(39, 13)$. Similarly, $n(A \cap B) = n(A \cap C) = n(A \cap D) = n(B \cap C) = n(B \cap D) = n(C \cap D) = C(26, 13)$, and so $m_2 = 6 \times C(26, 13)$. Moreover, $m_3 = 4$ and $m_4 = 0$ (see Exercise 53). By PIE the number we seek, $n(A \cup B \cup C \cup D)$, is $4 \times C(39, 13) - 6 \times C(26, 13) + 4 - 0$. (It turns out that this number is 32,427,298,180, and perhaps if you keep this in mind it will improve your bidding.) ❑

## EXERCISES 6.3

**34.** Which binomial coefficient is the middle term of the binomial expansion of $(1+1)^{100}$?

**35.** An unscrupulous stockbroker recommends the purchase or sale of 10 highly speculative stocks. He buys 1024 postcards, which he sends to 1024 prospective clients, and he gives a different recommendation on each card.

**a)** Assume that after six months, all ten stocks will change in value. Argue that 1024 is just the number of cards the stockbroker needs to be *sure* that some prospective client will have a recommendation that is right on all ten stocks.

**b)** How many prospective clients receive a card with at least eight out of the ten recommendations proving to be correct?

**36.** A jogger wishes to jog 14 blocks from point $A$ to point $B$.

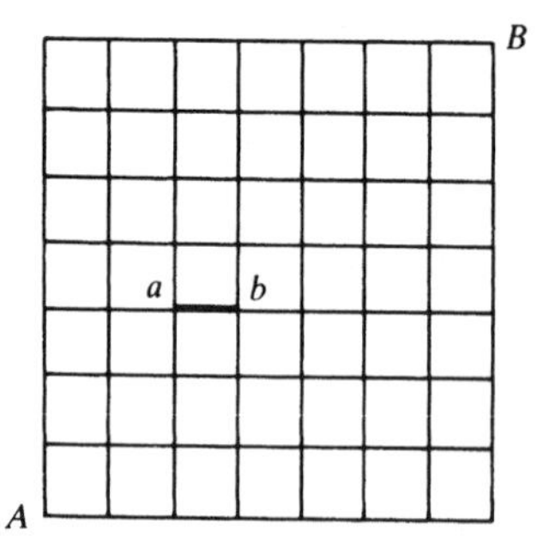


**a)** How many different shortest routes are possible?

**b)** Suppose that the street labeled $a$——$b$ is torn up and cannot be used. How many different shortest routes are possible?

**37. a)** Label the grid below with the number of shortest routes from point $A$ to each corner. (We have labeled a few corners to get you started.)

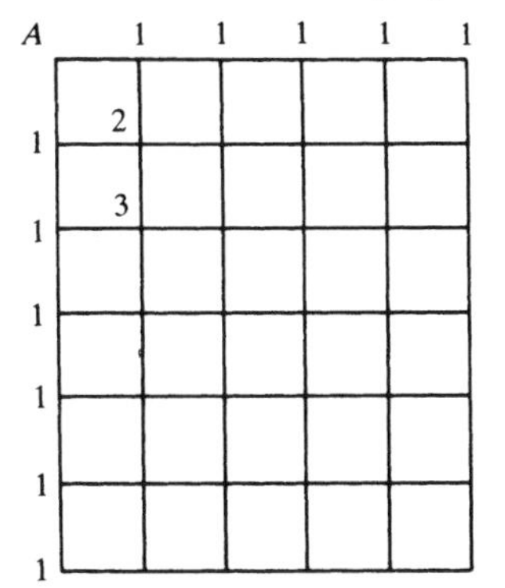


**b)** In light of your labeling of the grid in part (a), guess correctly the answer to this question: The answer to part (a) is a disguised presentation of ______.

**38. a)** A girl wishes to purchase three candy bars from a machine that sells only Hershey bars and Mars bars. How many different purchases of three candy bars can be made? List all possible purchases.

**b)** How many nonnegative integer-valued solutions does the equation $x_1 + x_2 = 3$ have? List them.

**39.** How many nonnegative integer-valued solutions does the equation $x_1 + x_2 = 17$ have?

**40.** Using the Binominal Theorem, prove that for each natural number $n$,

$$C(n, 0) + C(n, 1) + C(n, 2) + \cdots + C(n, n) = 2^n$$

**41.** Let $n \in \mathbb{N}$. Use the Binomial Theorem to prove that

$$C(n, 0) - C(n, 1) + C(n, 2) - C(n, 3) + \cdots \pm C(n, n) = 0$$

**42.** Prove Theorem 6.9(c).

**43.** Prove that for each natural number $n$,

$$\sum_{r=0}^{n} C(2n + 1, r) = 2^{2n}$$

**44.** Prove that for each natural number $n$, $C(2n, 2) = 2C(n, 2) + n^2$.

**45.** Prove that for each natural number $n$,

$$\sum_{r=0}^{n} (r + 1)C(n, r) = 2^n + n2^{n-1}$$

**46.** Let $S$ be a set with 16 elements.

**a)** How many subsets of $S$ have an even number of elements?

**b)** How many subsets of $S$ have an odd number of elements?

**47.** Let $S$ be a set with 17 elements.

**a)** How many subsets of $S$ have an even number of elements?

**b)** How many subsets of $S$ have an odd number of elements?

**48.** Let $S$ be the ordinary English alphabet.

**a)** How many 12-element subsets of $S$ do not contain the letter $Q$?

**b)** How many 12-element subsets of $S$ contain the letter $J$?

**49.** Prove the Binomial Theorem by induction.

**50.** Explain informally what would happen if in the proof of Theorem 6.8 you were to keep track of the $y$'s instead of the $x$'s in calculating the coefficient of $x^5y^2$ in the binomial expansion of $(x+y)^7$.

**51.** Prove Pascal's formula by simplifying

$$C(n-1,r)+C(n-1,r-1)=\frac{(n-1)!}{r!(n-1-r)!}+\frac{(n-1)!}{(r-1)!(n-r)!}$$

**52.** Let $A_1=\{2,7,11,12\}$, $A_2=\{2,3,8,11\}$, $A_3=\{2,8,13\}$, and $A_4=\{11,12,13\}$.

**a)** Find $m_1$ and $m_4$.

**b)** Find $m_2$.

**c)** Find $m_3$.

**53.** **a)** Argue that in Example 7, $n(A\cap B)=C(26,13)$.

**b)** Argue that in Example 7, $m_3=4$ and $m_4=0$.

**54.** How many natural numbers not exceeding 10,000 are divisible by 7, 17, or 23?

**55.** How many natural numbers not exceeding 10,000 are neither perfect squares nor perfect cubes?

**Warning** Some numbers such as 1 and 64 are both perfect squares and perfect cubes.

## 6.4 *Graphs*

In this section, we provide a brief introduction to graphs. Intuitively a **graph** is a finite set of points, called **vertices**, together with a finite set of lines, called **edges**, that join some or all of these points. We deal only with graphs that are sometimes called simple graphs.

**Definition**

A **graph** $C$ is an ordered pair $(V,E)$, where $V$ is a finite nonempty set whose members are called **vertices** and $E$ is a set of 2-element subsets of $V$ whose members are called edges.

**EXAMPLE 8**

Let $V=\{v_1,v_2,v_3,v_4,v_5\}$ and

$$E=\{\{v_1,v_2\},\{v_1,v_3\},\{v_2,v_4\},\{v_2,v_3\},\{v_2,v_5\},\{v_3,v_4\},\{v_4,v_5\}\}$$

Then $G=(V,E)$ is a graph. ❑

An appealing feature of graph theory is the geometric aspect of the subject. It is often useful to draw a picture in the plane of a graph, where each member of $V$ is represented by a point and each member of $E$ is represented by a line segment. We say that such a picture represents the graph $G$. Figure 6.6 represents the graph described in Example 8.

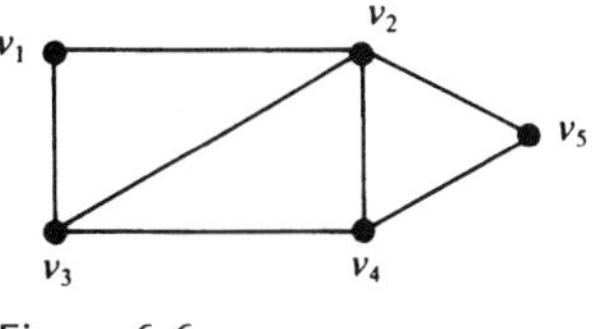

Figure 6.6

**Definition**

Let $G=(V, E)$ be a graph, let $e \in E$, and let $u \in V$. Then $e$ is **incident with** $u$ if $u \in e$. If $v \in V$, the **degree** of $v$, denoted by $\deg(v)$, is the number of edges that are incident with $v$.

In Example 8, $\deg(v_1)=\deg(v_5)=2$, $\deg(v_2)=4$, and $\deg(v_3)=\deg(v_4)=3$.

**Theorem 6.11**

If $G=(V, E)$ is a graph, where $V=\{v_1, v_2, \ldots, v_p\}$ and $E=\{e_1, e_2, \ldots, e_q\}$, then $\Sigma_{i=1}^{p} \deg(v_i)=2q$.

**PROOF** This result follows immediately from the fact that each edge has two vertices. □

**Theorem 6.12**

If $G=(V, E)$ is a graph, then the number of vertices of odd degree is even.

**PROOF** Let $G=(V, E)$ be a graph, let $v_1, v_2, \ldots, v_k$ be the vertices of odd degree, let $v_{k+1}, v_{k+2}, \ldots, v_p$ denote the vertices of even degree, and suppose that $q$ is the number of members of $E$. By Theorem 6.11,

$$2q=\sum_{i=1}^{p} \deg(v_i)=\sum_{i=1}^{k} \deg(v_i)+\sum_{i=k+1}^{p} \deg(v_i)$$

Since $\Sigma_{i=k+1}^{p} \deg(v_i)$ is even, $\Sigma_{i=1}^{k} \deg(v_i)$ is even, and therefore $k$ is even. □

**Definition**

Let $u$ and $v$ be vertices of a graph $G=(V, E)$. A **walk** of length $n$ from $u$ to $v$ is a finite list whose terms are alternately vertices and edges, $w=ue_1v_1e_2v_2\cdots v_{n-1}e_nv$, such that $u \in e_1$, $v \in e_n$, and for each $i=1, 2, \ldots, n-1$, $v_i \in e_i \cap e_{i+1}$. We say that $w$ is a **$(u, v)$-walk**.

**Definition**

A walk whose edges are distinct is called a **trail**. We use the notation $(u, v)$-trail to denote a trail from $u$ to $v$. A trail whose vertices are distinct is called a **path**. We use the notation $(u, v)$-path to denote a path from $u$ to $v$. If $u$ and $v$ are vertices of a graph $G$ and there is a $(u, v)$-path, we say that $u$ and $v$ are **joined by** a path.

**Theorem 6.13**

If $u$ and $v$ are distinct vertices of a graph $G=(V, E)$ and there is a $(u, v)$-walk in $G$, then there is a $(u, v)$-path in $G$.

**PROOF** Let $u$ and $v$ be distinct vertices of a graph $G=(V, E)$, and let $T=\{n \in \mathbb{N}: \text{there is a } (u, v)\text{-walk in } G \text{ of length } n\}$. By the Least-Natural-Number Principle, $T$ has a smallest member $p$. Let $w=ue_1v_1e_2v_2\cdots v_{p-1}e_pv$ be a $(u, v)$-walk in $G$ of length $p$. We claim that $w$ is a $(u, v)$-path in $G$. To see this, suppose it is not. Then $v_i=v_j$ for some $i$ and $j$, where $i<j$. Then $w=ue_1v_1e_2v_2\cdots v_{i-1}e_iv_je_{j+1}\cdots v_{p-1}e_pv$ is a $(u, v)$-walk in $G$ of length $q$, where $q<p$, and we have a contradiction. □

In Exercise 70, you are asked to prove a stronger version of Theorem 6.13.

**Definition**

Vertices $u$ and $v$ of a graph $G$ are **connected** if $u=v$ or if there is a $(u, v)$-path in $G$, and the graph $G$ is **connected** if each pair of vertices of $G$ is connected.

**Theorem 6.14**

Let $G=(V, E)$ be a graph. The relation $R=\{(u, v) \in V \times V: u \text{ and } v \text{ are connected}\}$ is an equivalence relation on $V$.

**PROOF** Let $v \in V$. By definition, $v$ and $v$ are connected. Therefore, $(v, v) \in R$, and hence $R$ is reflexive on $V$.

Let $(u, v) \in R$. If $u=v$, then $(v, u) \in R$. Suppose that $u \neq v$. Then there is a $(u, v)$-path, $ue_1v_1e_2v_2\cdots v_{n-1}e_nv$, in $G$. Since $ve_nv_{n-1}\cdots v_2e_2v_1e_1u$ is a $(v, u)$-path in $G$, $(v, u) \in R$. Therefore, $R$ is symmetric.

Suppose that $(u, v), (v, w) \in R$. If $u=w$, then $(u, w) \in R$. Suppose that $u \neq w$. If $u=v$, then the $(v, w)$-path in $G$ is a $(u, w)$-path. Also, if $v=w$, then the $(u, v)$-path in $G$ is a $(u, w)$-path. Suppose that $u \neq v$ and $v \neq w$. Then there is a $(u, v)$-path, $ue_1v_1e_2v_2\cdots v_{n-1}e_nv$, and a $(v, w)$-path, $vf_1w_1f_2w_2\cdots w_{n-1}f_nw$, in $G$. Now $ue_1v_1e_2v_2\cdots v_{n-1}e_nvf_1w_1f_2w_2\cdots w_{n-1}f_nw$ is a $(u, w)$-walk in $G$. By

Theorem 6.13, there is a $(u, w)$-path in $G$. Thus, $(u, w) \in R$, and hence $R$ is transitive. □

**EXAMPLE 9** Let $V = \{v_1, v_2, v_3, v_4, v_5, v_6, v_7, v_8, v_9\}$ and

$$E = \{\{v_1, v_2\}, \{v_1, v_3\}, \{v_1, v_6\}, \{v_2, v_3\}, \{v_2, v_7\}, \{v_4, v_5\}, \{v_4, v_8\}\}$$

With respect to the relation on $V$ defined in Theorem 6.14, there are three equivalence classes: $[v_1] = \{v_1, v_2, v_3, v_6, v_7\}$, $[v_4] = \{v_4, v_5, v_8\}$, and $[v_9] = \{v_9\}$. ❑

**Definition** If $G = (V, E)$ is a graph and $n$ is the number of members of $V$, then we say that $G$ is a **graph on *n* vertices.**

**Definition** A graph $H = (W, F)$ is a **subgraph** of a graph $G = (V, E)$ provided $W \subseteq V$ and $F \subseteq E$.

**Theorem 6.15** If $G$ is a connected graph on $n$ vertices, then $G$ has at least $n - 1$ edges.

**PROOF** The proof is by induction on the number of vertices of $G$. Let $S = \{n \in \mathbb{N}$: if $G$ is a connected graph on $n$ vertices, then $G$ has at least $n - 1$ edges$\}$. Clearly, $1 \in S$. Let $n \in \mathbb{N}$, and assume that $n \in S$. Suppose that $n + 1 \notin S$. Then there is a connected graph $G$ on $n + 1$ vertices with fewer than $n$ edges. By Theorem 6.11, the sum of the degrees of the vertices of $G$ is less than $2n$, which is less than $2(n + 1)$. Therefore, the degree of some vertex $v$ is less than 2. Since $G$ is connected, $v$ must have degree 1. Let $G'$ be the graph obtained from $G$ by deleting $v$ and the edge that is incident with $v$. Then $G'$ is a connected graph on $n$ vertices with fewer than $n - 1$ edges. Since $n \in S$, this is a contradiction. Therefore, $n + 1 \in S$, and by the Principle of Mathematical Induction, $S = \mathbb{N}$. □

Let $G = (V, E)$ be a graph. By Theorem 4.4, the set of equivalence classes $\{V_1, V_2, \ldots, V_n\}$ of $V$ with respect to the equivalence relation on $V$ defined in Theorem 6.14 is a partition of $V$. For each $i = 1, 2, \ldots, n$, let $E_i = \{e \in E$: the vertices of $e$ are members of $V_i\}$. Then for each $i = 1, 2, \ldots, n$, $G_i = (V_i, E_i)$ is a subgraph of $G$, and it is called a **component** of $G$.

Note that each component of a graph is a connected graph, and if $G$ has one component, then $G$ is connected. If $G$ is a graph, we let $\omega(G)$ denote the number of components of $G$. Also, if $e$ is an edge of a graph $G$, we let $G - \{e\}$ denote the subgraph of $G$ that is obtained by deleting $e$. We give a combinatorial proof of the following theorem.

**Theorem 6.16**

If $G=(V, E)$ is a graph and $e \in E$, then $\omega(G) \leqslant \omega(G-\{e\}) \leqslant \omega(G)+1$.

**PROOF** Let $G=(V, E)$ be a graph. Since adding an edge to a graph cannot increase the number of components, $\omega(G) \leqslant \omega(G-\{e\})$ for each $e \in E$. Let $H$ be any subgraph of $G$, and suppose that we add an edge $e$ to $H$. Now $e = \{u, v\}$, where $u$ and $v$ are vertices of $H$. If there is a $(u, v)$-path in $G$, then adding $e$ does not change the number of components of $H$. If there is no $(u, v)$-path in $H$, then $u$ is a member of one component $H_1$ of $H$ and $v$ is a member of a different component $H_2$ of $H$. So $H_1 \cup H_2 \cup \{e\}$ is now one component of $H$, and hence $H$ has one less component. Therefore, $\omega(G-\{e\}) \leqslant \omega(G)+1$. □

**Definition**

A walk $v_0 e_1 v_1 e_2 v_2 \cdots v_{n-1} e_n v_n$ in a graph $G$ is **closed** if $v_0 = v_n$. A closed walk $v_0 e_1 v_1 e_2 v_2 \cdots v_{n-1} e_n v_0$ is a **cycle** if its $n$ vertices are distinct.

**Theorem 6.17**

Let $G$ be a connected graph, let $e$ be an edge of $G$ that is not part of any cycle in $G$, and let $G'$ be the graph obtained from $G$ by deleting $e$. Then $G'$ is not connected.

**PROOF** Suppose that $G'$ is connected, and let $u$ and $v$ be the vertices of $e$. Then there is a path $u e_1 v_1 e_2 v_2 \cdots v_{n-1} e_n v$ in $G'$. But this path together with $e$ forms a cycle in $G$. This is a contradiction, so $G'$ is not connected. □

As we mentioned briefly in Chapter 3, Sir William Rowen Hamilton invented a puzzle, which he called the "Around the World Game." From the perspective of graph theory, Hamilton's puzzle was to find a cycle that contained every vertex of the dodecahedron graph. This graph is the graph of the edges of the dodecahedron (the unique twelve-sided regular three-dimensional polyhedron on which calendars are sometimes printed). A *Hamilton cycle* in a graph is a cycle that contains every vertex of the graph, and a *Hamiltonian graph* is a graph that contains a Hamiltonian cycle.

*Personal Note*

## Sir William Rowan Hamilton (1805–1865)

Hamilton was orphaned at the age of three and brought up by his uncle, who was a linguist. At age five, Hamilton could read and translate Latin, Greek, and Hebrew; by the age of ten he had added Italian, French, Arabic, and Sanskrit. He liked writing poetry and was friends with both Coleridge and Wordsworth. Wordsworth maintained that Hamilton and Coleridge were the two most wonderful men he had ever known. Wordsworth also had kind words for Hamilton's poetry: "Again I do venture to submit to your consideration, whether the poetical parts of your nature would not find a field more favorable to their exercise in the regions of prose...." Although Hamilton never managed to write a memorable poem, he did write the world's most famous graffiti: "[As] I was walking ... an undercurrent of thought was going on in my mind .... An electric circuit seemed to close and a spark flashed forth .... Nor could I resist the impulse ... to cut with a knife on a stone of Brougham Bridge ... the fundamental formula [of quaterions] ... $i^2 = j^2 = k^2 = ijk = -1$."

We have already proved (Theorem 6.15), by induction, that a connected graph on $n$ vertices has at least $n-1$ edges. In Exercise 69, you are asked to give a combinatorial proof of this theorem.

The following definitions and examples show that partitions, even two-element partitions, can be used to point out a commonality among graphs that appear at first to be quite unrelated.

**Definition**

Let $G = (V, E)$ be a graph. Two vertices $u$ and $v$ of $V$ are **adjacent** provided that $\{u, v\} \in E$.

**Definition**

A graph $G = (V, E)$ is said to be a **complete bipartite graph** if

**a)** there are nonempty disjoint subsets $V_1$ and $V_2$ such that $V = V_1 \cup V_2$, and

**b)** $u$ and $v$ are adjacent vertices only if one of them belongs to $V_1$ and the other to $V_2$.

The set $\{V_1, V_2\}$ is called a **bipartition** of $G$. A complete bipartite graph with bipartition $\{V_1, V_2\}$, where $m$ is the number of members of $V_1$ and $n$ is the number of members of $V_2$, is denoted by $K_{m,n}$.

The following drawings represent $K_{1,1}$, $K_{1,2}$, $K_{2,2}$, $K_{2,3}$, and $K_{3,3}$.

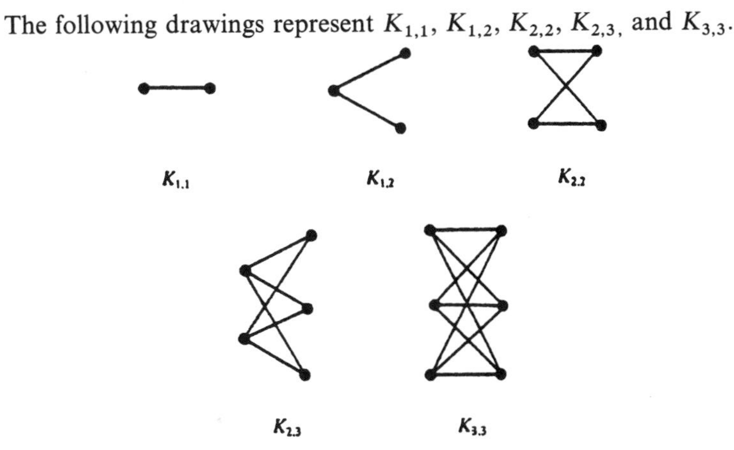

## EXERCISES 6.4

**56.** Let $V$ be the set of all odd prime numbers less than 20. Let $E_1 = \{(a, b) \in V \times V$: $(a, b)$ is a twin-prime pair) and let $E_2 = \{(a, b) \in V \times V: |b - a| \text{ divides } b - 1\}$.

**a)** Find the component of 7 in $(V, E_1)$.

**b)** Find the component of 7 in $(V, E_2)$.

**c)** Find the component of 19 in $(V, E_1)$.

**d)** Find the component of 19 in $(V, E_2)$.

**57.** Show that for each natural number $n \geqslant 2$, the number of graphs with $n$ vertices is $2^{C(n,2)}$.

**58.** Let $G$ be a graph of order $n$. Prove that the length of a path in $G$ is at most $n - 1$.

**59.** Let $G$ be a graph of order $n \geqslant 2$. Prove that there are two vertices that have the same degree.

**60.** Suppose there are six people—Tom, Sue, Maxine, David, Fred, and Beth—who pass rumors among themselves. Each day Tom talks with Sue and Beth; Sue talks with Tom, Maxine, and David; Maxine talks with Sue, David, and Fred; David talks with Sue, Maxine, Fred, and Beth; Fred talks with Maxine, David, and Beth; and Beth talks with Tom, David, and Fred. Whatever people hear one day, they pass on to others the next day.

**a)** Construct a graph that models this rumor-passing situation.

**b)** How many days does it take for a rumor to pass from Tom to David? Who will tell it to David?

**c)** Is there any way that if two people stopped talking to each other, it would take three days to pass a rumor from one person to each of the others?

**61.** Show that the sum over the set of people at a party, of the number of people a person has shaken hands with, is even.

**62.** Let $G$ be a graph, with $p$ vertices and $q$ edges, let $M$ be the maximum degree of the vertices of $G$, and let $m$ be the minimum degree of the vertices of $G$. Show that $2q/p \geqslant m$ and that $2q/p \leqslant M$.

**63.** A football league has two divisions and each division has 13 teams. Show that it is not possible for each team to play 11 games with teams in their own division and 3 games with teams outside their division.

**64.** Suppose that a graph has $p$ vertices and the degree of each vertex is greater than or equal to 1. What is the smallest number of edges that the graph can have?

**65.** Suppose that a graph has $p$ vertices and the degree of each vertex is greater than or equal to 2. What is the smallest number of edges that the graph can have?

**66.** A graph has $q$ edges, where $q \geqslant 2$. What is the smallest number of vertices that the graph can have?

**67.** Let $n \in \mathbb{N}$ $(n \geqslant 4)$, let $V = \{1, 2, \ldots, n\}$, and let $E = \{\{p, q\}: p, q \in V$ and $p$ divides $q$ or $q$ divides $p\}$. Which vertices of the graph $(V, E)$ have degree 1?

**68.** Are the graphs represented by each of the following drawings bipartite? Prove your answer.

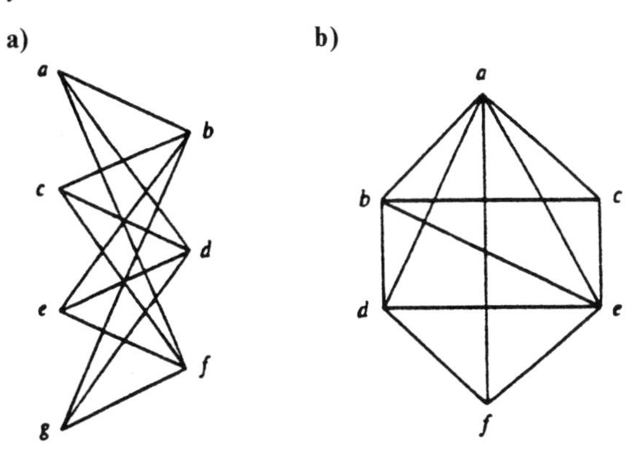


**69.** Give a combinatorial proof that a connected graph on $n$ vertices has at least $n-1$ edges.

**70.** There are four married couples, including Mr. and Mrs. Smith, at a bridge party. No spouses shook hands, no two people shook hands more than once, and no one shook hands with herself or himself. When the other people (including Mr. Smith) told Mrs. Smith how many hands he or she shook, the answers were all different. How many hands did Mr. and Mrs. Smith each shake?

**71.** Prove or give a counterexample to the following proposition: If a graph has exactly two vertices $u$ and $v$ of odd degree, then there is a path from $u$ to $v$.

**72.** Let $u$ and $v$ be distinct vertices of a graph $G$, and let $W$ be a $(u, v)$-walk in $G$. Prove that $W$ contains a $(u\text{-}v)$-path.

**73.** **a)** Show that the dodecahedron graph shown below is a Hamiltonian graph.

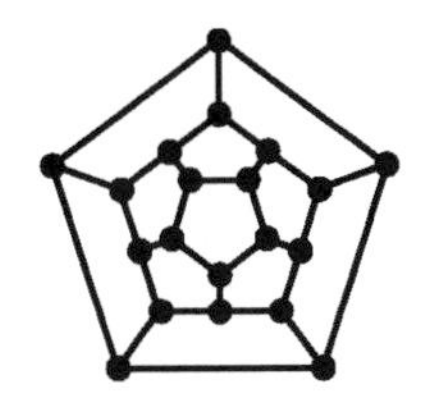

**b)** Draw a connected graph that is not a Hamiltonian graph.

---

## Writing Exercise

Let $S$ be a nonempty set and let $x \in S$. Note that if $A$ is a subset of $S$ with an even number of elements and $x \in A$, then $A - \{x\}$ has an odd number of elements, whereas if $x \notin A$ then $A \cup \{x\}$ has an odd number of elements. Use this observation to argue that any nonempty set has the same number of even subsets as odd subsets. First give a formal argument, in which you define a function, and then give an informal argument that is designed to persuade an intelligent reader who has never heard of one-to-one functions or bijections.

# CHAPTER 7

# COUNTABLE AND UNCOUNTABLE SETS

The reunification of mathematics under the empire of set theory proved short-lived, and there are ever as many mathematical fiefdoms today as there were before Zermelo and Fraenkel codified the theory of infinite sets, which had been developed by Georg Cantor during the late nineteenth century. As this chapter concerns the downfall of the set-theoretic empire, we include a brief discussion of its history. The source of the theory of infinite sets is a conversation among three seventeenth-century Italians, Filippo Salviati, Giovanni Sagredo, and Simplicio. The conversation is an imaginary one, with the parts of all three participants written by Galileo Galilei (1564–1642), but Salviati and Sagredo are real names, which Galileo used in the hopes of immortalizing two of his close friends. Salviati expresses Galileo's present view, Sagredo raises objections that Galileo thought reasonable or at least plausible, and, as the name suggests, Simplicio raises objections that Galileo thought to be without merit. There is a translation of the complete conversation in Galileo Galilei's *Two New Sciences* (University of Wisconsin Press, 1974). We are content here to paraphrase a piece of Galileo's imaginary discussion.

SIMP. I don't see how it can make any sense to say that one infinite set is greater than another.

SALV. It's clear to me that we should talk about greater and smaller sets only when the sets are finite. I have thought of an argument that even you, Simplicio, should be able to follow. I assume you know which positive integers are squares and will agree that there are just as many squares as there are positive integers, because every square has exactly one positive root and every root has exactly one square.

SAGR. So what?

SALV. The inescapable conclusion is that any two infinite sets are equally numerous, neither one being more or less than the other. So the notions of equal, greater and smaller only make sense for finite sets. Simplicio, you probably believe that there are more points on a longer line than on a shorter line. But I tell you that there are neither more nor less; we can say only that each line has infinitely many points. In fact, Simplicio, I can say that there are exactly as many points on one of the two lines as there are positive integers that are squares and exactly as many points on the other line as there are positive integers that are cubes. But still it does not make sense to say that one of the lines has more points than the other. All we can say is that both have infinitely many points.

With twentieth-century hindsight, it is not difficult to scoff at Galileo's imaginary conversation. Salviati has observed correctly that there is a one-to-one function from the set of all natural numbers onto the set of all squares and so the natural numbers and their squares are equally numerous. In the last paragraph, Salviati leaps from this one example to the conclusion that any two infinite sets are equally numerous; in particular, he suggests that the points in a line [segment] are "as many as the square numbers." We will see shortly that the points of a line segment and the squares of natural numbers are not equally numerous. So Galileo, who held a chair in mathematics at Padua, was flat wrong.

But this scoffing at Galileo is grossly unfair. In the first place, Galileo is careful to point out that his conversations represent his considered opinions rather than mathematically demonstrated truth. We also need to consider the circumstances under which Galileo worked. *Two New Sciences*, from which our conversation is excerpted, was written when Galileo was 70 years old, blind (from having viewed the sun with his newly improved telescope), and living under house arrest (after having confessed to having held the heretical view that the earth is not the center of the universe). Furthermore, in his confession to the Inquisition, Galileo had written: "I have been judged to be suspected of heresy.... I swear that in the future I will never again say or assert verbally or in writing anything that might cause a similar suspicion against me." Under these circumstances, it is amazing that Galileo did not content himself with puttering in his garden raising petunias, but rather spent three years writing a classic text in mathematics and mathematical physics. Moreover, Galileo had obviously put considerable thought into the question of deciding what it should mean for two infinite sets to be equally numerous, and Salviati expresses the definition that is universally accepted nowadays. His phrase "nor is there any square that has more than just one root" demands that $f(n) = \sqrt{n}$ be a *function* with its domain the set of all perfect squares, and his phrase "or any root that has more than just one square" demands that this function be one-to-one.

## 7.1 Finite and Infinite Sets

**Definition**

(Galileo) Let $S$ and $T$ be sets. We say that $S$ and $T$ **have the same cardinality**, and write $S \sim T$, provided that there is a **one-to-one** function mapping $S$ **onto** $T$.

Our notation $S \sim T$ certainly suggests that we are considering an equivalence relation. Alas, we would like the domain of $\sim$ to be the "set" of all sets, and we recall from Chapter 2 that we cannot allow this. The following proposition is worded cagily to duck the problem.

**Proposition 7.1**

**a)** For any set $A$, $A \sim A$.

**b)** If $A$ and $B$ are sets and $A \sim B$, then $B \sim A$.

**c)** If $A$, $B$, and $C$ are sets, $A \sim B$, and $B \sim C$, then $A \sim C$.

**PROOF** See Exercise 1. □

Since we will not need to consider an equivalence relation in this chapter, whenever $A$ and $B$ are sets and we write $A \sim B$ we will mean that $A$ and $B$ have the same cardinality. We have, of course, already applied this convention in stating the preceding proposition.

One problem Galileo did not consider. Just what is an infinite set? A set that is not finite. Well, okay, but what is a finite set? To our knowledge, the first sensible answers to this question were given by the American mathematician C. S. Peirce in 1881 and by R. Dedekind in 1888; but we will use neither Dedekind's nor Peirce's definition.

**Definition**

A set $S$ is **finite** if it is empty or if there is a natural number $n$ such that $\{1, 2, \ldots, n\}$ and $S$ have the same cardinality. A set that is not finite is said to be **infinite**. A set $S$ is **countably infinite** provided that $S$ and $\mathbb{N}$ have the same cardinality.

**Proposition 7.2**

(As the terminology suggests), every countably infinite set is infinite.

**PROOF** Let $A$ be a countably infinite set. Then there is a one-to-one function $f$ mapping $A$ onto $\mathbb{N}$. Suppose that $A$ is finite. Since $A \neq \varnothing$, there is a natural number $m$ such that $A \sim \{1, 2, \ldots, m\}$. Let $g$ be a one-to-one map from $\{1, 2, \ldots, m\}$ onto $A$. Then, by Theorem 5.7, $f \circ g$ is a one-to-one function from $\{1, 2, \ldots, m\}$ onto $\mathbb{N}$. Let

$$k = 17 + \sum_{i=1}^{m} f \circ g(i)$$

Then for each $i \in \{1, 2, \ldots, m\}$, $f \circ g(i) < k$, and so $f \circ g$ does not map onto $\mathbb{N}$, a contradiction. □

In the exercises that follow, you are asked to expand your knowledge of finite and infinite sets, but before considering these exercises, let us jot down two theorems, one of which you already know. The proof of the first is due to Galileo Galilei; the proof of the second is due to Georg Cantor.

*Personal Note*

## Georg Cantor (1845–1918)

In his biography of Georg Cantor, Joseph Dauben writes: "George Cantor ... is one of the most ... controversial figures in the history of mathematics.... Leopold Kronecker considered Cantor a scientific charlatan, a renegade, a 'corrupter of youth'.... Henri Poincaré thought ... Cantor's transfinite numbers represented a grave mathematical malady, a perverse pathological illness that would one day be cured. In his own time and in the years since, Cantor's name has signified both controversy and schism."

There is one area of agreement: Cantor suffered terribly from a mental illness, probably what we now call manic depression. He was hospitalized for this illness for at least part of the years 1884, 1899, 1902, 1903, 1904, 1905, 1907, 1908, 1911, and 1912. He was admitted to the Halle Nervenklinik for the last time in 1917 and died there in 1918. Cantor felt that he was being persecuted for his mathematical beliefs, which were for him nearly religious beliefs. Like many manic depressives (c.f., van Gogh), Cantor had few lasting friendships. Friendships with Kronecker, Richard Dedekind, M. A. Swartz, and Gustav Mittag-Leffler were short-lived. In 1874 Cantor published a paper whose title announced the countability of the algebraic numbers. There is no mention in Cantor's paper that the result touted in its title was based on collaboration with Dedekind. Cantor's friendship with Mittag-Leffler, who had published Cantor's mathematics over the objections of Kronecker and others, ended abruptly in 1905 when Mittag-Leffler, as editor of *Acta Mathematica*, urged Cantor to withdraw two letters to the editor. In a famous remark, David Hilbert assured the mathematical community, "No one shall expel us from the paradise Cantor has created for us." It is all the more amazing that Cantor created this paradise while living in a hell of suspicion, recrimination and unforgiveness.

**Proposition 7.3**

The set of all perfect squares is countably infinite.

**PROOF** Let $S$ denote the set of all perfect squares, and let $f\colon S \to \mathbb{N}$ be defined by $f(n) = \sqrt{n}$. Then $f$ is a one-to-one function from $S$ onto $\mathbb{N}$. □

**Proposition 7.4**

The sets $\mathbb{N}$ and $\mathscr{P}(\mathbb{N})$ do not have the same cardinality.

Proposition 7.4 is a special case of the following proposition.

**Proposition 7.5**

Let $S$ be a set. Then $S$ and $\mathscr{P}(S)$ do not have the same cardinality.

**PROOF** If $S = \varnothing$, then $\mathscr{P}(S) = \{\varnothing\}$ and so $S$ and $\mathscr{P}(S)$ do not have the same cardinality. Suppose that $S \neq \varnothing$ and that there is a one-to-one function $f$ mapping $S$ onto $\mathscr{P}(S)$. Let $A = \{s \in S: s \notin f(s)\}$. Since $f$ maps $S$ onto $\mathscr{P}(S)$, there is an $x \in S$ such that $f(x) = A$. We reach a contradiction by arguing that it is impossible that $x \in A$ and also impossible that $x \notin A$. Suppose that $x \in A$. Then $x \in \{s \in S: s \notin f(s)\}$. Therefore, $x \notin f(x) = A$. Suppose that $x \notin A$. Then $x \notin f(x)$ and so $x \in \{s \in S: s \notin f(s)\} = A$. □

The contradiction arrived at in Cantor's proof of Proposition 7.5 is that if $x \in A$, then $x \notin A$ and if $x \notin A$, then $x \in A$. This strange form of a contradiction bewildered the mathematical community when it first appeared.

**Corollary 7.6**

$\mathscr{P}(\mathbb{N})$ is an infinite set that is not countably infinite.

**PROOF** By Proposition 7.4, $\mathscr{P}(\mathbb{N})$ is not countably infinite. Suppose that $\mathscr{P}(\mathbb{N})$ is finite. Then, as $\mathscr{P}(\mathbb{N}) \neq \varnothing$, there is a natural number $m$ and a one-to-one function $f$ mapping $\{1, 2, \ldots, m\}$ onto $\mathscr{P}(\mathbb{N})$. Let $k = 17 \sum_{j=1}^{m} a_j$, where $a_j = 1$ if $f(j)$ is not a singleton set and $a_j = n_j$ if $f(j) = \{n_j\}$. Since $f$ maps onto $\mathscr{P}(\mathbb{N})$, there is a natural number $p \leq m$ such that $f(p) = \{k\}$. By definition, $a_p = k$, which contradicts the equation $k = 17 \sum_{j=1}^{m} a_j$. This contradiction establishes that $\mathscr{P}(\mathbb{N})$ is infinite. □

**Definition**

A set that is either finite or countably infinite is said to be **countable**. A set that is not countable is said to be **uncountable**.

With our new terminology, Corollary 7.6 can be stated succinctly: $\mathscr{P}(\mathbb{N})$ is an uncountable set.

## EXERCISES 7.1

1. Prove Proposition 7.1.
2. Let $A$, $A'$, $B$, and $B'$ be sets. Suppose that $A \sim A'$ and $B \sim B'$ but that $A$ and $B$ do not have the same cardinality. Prove that $A'$ and $B'$ do not have the same cardinality.
3. Prove that the set of all perfect squares and the set of all perfect cubes have the same cardinality.
4. Prove that $\mathbb{N}$ and $\mathbb{Z}$ (the set of all integers) have the same cardinality. (In other words, prove that $\mathbb{Z}$ is countably infinite.)
5. Prove that if $(a, b)$ and $(c, d)$ are any two intervals of real numbers with the same length (that is, $b - a = d - c$), then these intervals have the same cardinality.

**6.** Prove that the intervals $(0, 2)$ and $(0, 11)$ have the same cardinality.

**7.** Use the result of Exercise 5 to prove that any two open intervals $(a, b)$ and $(c, d)$ have the same cardinality.

**8.** Prove that if $[a, b]$ and $[c, d]$ are any two closed intervals, then $[a, b] \sim [c, d]$.

**9.** Prove in one line that the open interval $(-1, 1)$ and $\mathbb{R}$ have the same cardinality.

**10.** Prove in one line that $(-\pi/2, \pi/2)$ and $\mathbb{R}$ have the same cardinality.

**11.** Prove that $\mathbb{R}$ is an infinite set.

**12.** Use Exercise 9 to prove that $(0, 1)$ and $\mathbb{R}$ have the same cardinality.

## 7.2 The Schroeder–Bernstein Theorem

One reason that the study of infinite sets is difficult is that a good many intuitively obvious statements are, in fact, quite tricky to prove. Here is a list of such intuitively obvious propositions:

1. Every infinite set contains a countably infinite subset.
2. If $a$ and $b$ are real numbers and $a < b$, then $(a, b) \sim [a, b]$.
3. $\mathbb{R}$ is uncountable.
4. If $\Lambda$ is countable and for each $\alpha \in \Lambda$, $A_\alpha$ is countable, then $\cup\{A_\alpha : \alpha \in \Lambda\}$ is countable.

It is known that two of these innocuous-sounding propositions cannot be proved without using a special case of Rule 10 of Chapter 2, known as the Countable Axiom of Choice. For this reason, we will have to interrupt our program in the next section in order to discuss Rule 10 in more detail. There are, however, several important results, some listed above, and some not, that we can get without worrying about the (countable) Axiom of Choice. The purpose of this section is to dispose of these results, and the major tool we use is the theorem indicated in the title of this section. Although Dedekind certainly knew a version of this theorem, it is usually attributed to some combination of G. Cantor, Felix Bernstein (1878–1956), and E. Schroeder.

What the Schroeder–Bernstein Theorem says is that two sets $S$ and $T$ have the same cardinality provided that there is a one-to-one function $f_1$ mapping $S$ *into* $T$ and a one-to-one function $f_2$ mapping $T$ *into* $S$. In other words, it allows us to swap one hard problem, finding a one-to-one function from $S$ *onto* $T$, for two easier problems. Let us apply this theorem to problem 2 given at the beginning of this section. Clearly, $i_{(a,b)}$ is a one-to-one function mapping $(a, b)$ *into* $[a, b]$. Let $c$ and $d$ be real numbers such that $a < c < d < b$. We have already seen that $[c, d] \sim [a, b]$ (Exercise 8), and so there is a one-to-one function $f$ mapping $[a, b]$ onto $[c, d]$. Clearly, $f$ maps *into* $(a, b)$, and so by the Schroeder–Bernstein Theorem, $(a, b)$ and $[a, b]$ have the same cardinality.

You will gain some appreciation for the value of the Schroeder–Bernstein Theorem by establishing that $(a, b)$ and $[a, b]$ have the same cardinality without using the Schroeder–Bernstein Theorem (see Exercise 21).

**Lemma 7.7**

Let $X$ be a set, and let $F: \mathscr{P}(X) \to \mathscr{P}(X)$ be a function with the property that $F(B) \subseteq F(D)$ whenever $B \subseteq D \subseteq X$. Then there is a subset $A$ of $X$ such that $A = F(A)$.

**PROOF** We show that $A = \cup\{B \subseteq X: B \subseteq F(B)\}$ is a subset of $X$ such that $A = F(A)$. Let $x \in A$. Then there is a subset $B$ of $X$ such that $B \subseteq F(B)$ and $x \in B$. Since $B \subseteq A$, $x \in B \subseteq F(B) \subseteq F(A)$, and so $A \subseteq F(A)$.

Now that we know that $A \subseteq F(A)$, we also have that $F(A) \subseteq F(F(A))$, and so $F(A)$ is one of the $B$'s whose union forms $A$. Thus, $F(A) \subseteq A$. □

**Theorem 7.8**

**The Schroeder–Bernstein Theorem**

Let $X_1$ and $X_2$ be sets and suppose that

**a)** There is a one-to-one function $f_1: X_1 \to X_2$.

**b)** There is a one-to-one function $f_2: X_2 \to X_1$.

Then $X_1 \sim X_2$.

**PROOF** [The proof we give is due to Ignace I. Kolodner, *Amer. Math. Monthly* **74** (1967), 995–996.] We need a one-to-one function $f$ mapping $X_1$ *onto* $X_2$. Suppose your fairy godfather were to grant you that there are subsets $A_1$ of $X_1$ and $A_2$ of $X_2$ such that (see Figure 7.1) $f_1(A_1) = X_2 - A_2$ and $f_2(A_2) = X_1 - A_1$. Presumably you would be happy, for in this case you would define $f: X_1 \to X_2$ as follows:

$$f(x) = \begin{cases} f_1(x) & \text{if } x \in A_1 \\ f_2^{-1}(x) & \text{if } x \in X_1 - A_1 \end{cases}$$

Then $f$ is the desired one-to-one function mapping $X_1$ onto $X_2$ (see Exercise 80 in Section 5.6).

Moreover, if we let $C_1: \mathscr{P}(X_1) \to \mathscr{P}(X_1)$ be the complement function for $X_1$ [defined by $C_1(S) = X_1 - S$ for each $S \in \mathscr{P}(X_1)$] and let $C_2: \mathscr{P}(X_2) \to \mathscr{P}(X_2)$ be the complement function for $X_2$, we would have

$$F_1(A_1) = C_2(A_2) \quad \text{and} \quad F_2(A_2) = C_1(A_1)$$

(where $F_1: \mathscr{P}(X_1) \to \mathscr{P}(X_2)$ and $F_2: \mathscr{P}(X_2) \to \mathscr{P}(X_1)$ are defined from $f_1$ and $f_2$ as in Section 5.6).

But there are sets $A_1$ of $X_1$ and $A_2$ of $X_2$ such that $F_1(A_1) = C_2(A_2)$ and $F_2(A_2) = C_1(A_1)$ if and only if there is a subset $A_1$ of $X_1$ such that

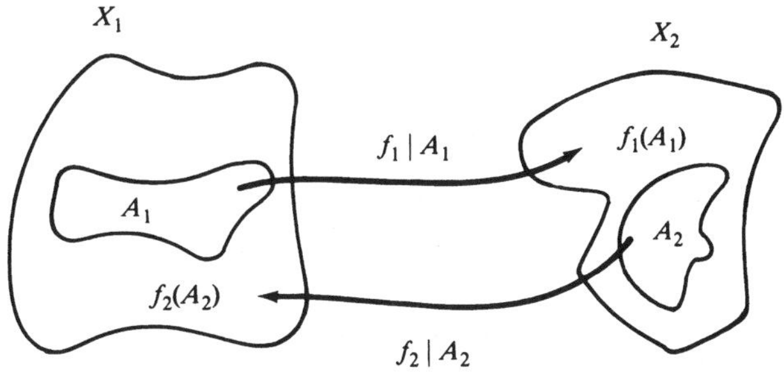


Figure 7.1

$A_1 = C_1 \circ F_2 \circ C_2 \circ F_1(A_1)$ (Exercise 14). Thus we have reduced the problem somewhat.

"Oh fairy godfather, I see it was presumptuous and greedy of me to ask for sets $A_1$ and $A_2$ such that $F_1(A_1) = C_2(A_2)$ and $F_2(A_2) = C_1(A_1)$. How about your giving me just one set $A_1$ such that $A_1 = C_1 \circ F_2 \circ C_2 \circ F_1(A_1)$? If you do this one little favor for me, I'll find a subset $A_2$ of $X_2$ such that $F_2(A_2) = C_1(A_1)$ and $F_1(A_1) = C_2(A_2)$, and then I'll define a one-to-one function $f$ mapping $X_1$ onto $X_2$ by

$$f(x) = \begin{cases} f_1(x) & \text{if } x \in A_1 \\ f_2^{-1}(x) & \text{if } x \in X_1 - A_1 \end{cases}$$

So I'm not asking for a coach-and-four from a pumpkin. Just one little set $A_1$ is all I want."

"Okay, I'll tell you what. First you prove that if $B$ and $D$ are subsets of $X_1$ and $B \subseteq D$, then

$$C_1 \circ F_2 \circ C_2 \circ F_1(B) \subseteq C_1 \circ F_2 \circ C_2 \circ F_1(D)$$

(Exercise 15). Then by Lemma 7.7, you will have the set you need:

$$A_1 = \cup\{B \subseteq X_1 \colon B \subseteq C_1 \circ F_2 \circ C_2 \circ F_1(B)\}\text{"}$$ □

## EXERCISES 7.2

**13.** **a)** Let $S = \{(x, y) \in \mathbb{R} \times \mathbb{R} \colon |x| \leqslant 1$ and $|y| \leqslant 1\}$, and let $D = \{(x, y) \in \mathbb{R} \times \mathbb{R} \colon x^2 \times y^2 \leqslant 1\}$. (Thus $S$ is a square with center at the origin and $D$ is a disk with center at the origin.) Prove that $S$ and $D$ have the same cardinality.

**b)** Let $A = \{(x, y) \in S \colon |x| = 1$ or $|y| = 1\}$, and let $B = \{(x, y) \in \mathbb{R} \times \mathbb{R} \colon x^2 \times y^2 = 1\}$. Draw a picture that illustrates how to define a one-to-one function mapping $A$ onto $B$.

**14.** Prove that there are subsets $A_1$ and $A_2$ of $X_1$ and $X_2$ such that $F_1(A_1) = C_2(A_2)$ and $F_2(A_2) = C_1(A_1)$ if and only if there is a subset $A_1$ of $X_1$ such that

$$C_1 \circ F_2 \circ C_2 \circ F_1(A_1) = A_1$$

**15.** Prove that $C_1 \circ F_2 \circ C_2 \circ F_1$ is order preserving in the sense that if $B \subseteq D \subseteq X_1$, then

$$C_1 \circ F_2 \circ C_2 \circ F_1(B) \subseteq C_1 \circ F_2 \circ C_2 \circ F_1(D)$$

**16.** Let $X = \{1, 2, 3\}$. Find a function $F: \mathscr{P}(X) \to \mathscr{P}(X)$ satisfying the conditions of Lemma 7.7 such that $F(\varnothing) \neq \varnothing$ and $F(X) \neq X$. List all subsets $A$ of $X$ such that for your function $F$, $A = F(A)$.

**17.** Let $\mathbb{F}$ be the set of all fractions.

**a)** Find a one-to-one function $f: \mathbb{N} \to \mathbb{F}$ that does not map $\mathbb{N}$ onto $\mathbb{F}$.

**b)** Find a one-to-one function $g: \mathbb{F} \to \mathbb{N}$ that does not map $\mathbb{F}$ onto $\mathbb{N}$.

**c)** Is the set $\mathbb{F}$ countably infinite? Explain briefly.

**18.** What's wrong with the following argument, which purports to show that $\mathbb{N} \sim \mathscr{P}(\mathbb{N})$? Clearly, the map $f: \mathbb{N} \to \mathscr{P}(\mathbb{N})$ defined by $f(n) = \{n\}$ is one-to-one. Define $g: \mathscr{P}(\mathbb{N}) \to \mathbb{N}$ as follows: Given $A \in \mathscr{P}(\mathbb{N})$, set $g(A) = 2^{a_1} 3^{a_2} 5^{a_3} \cdots P_n^{a_n} \cdots$, where $a_n$ is the $n$th smallest member of $A$ and $P_n$ is the $n$th prime number. By the Fundamental Theorem of Arithmetic, $g$ is a one-to-one function and so the Schroeder–Bernstein Theorem applies.

**19.** Prove that the set of all finite subsets of $\mathbb{N}$ is countably infinite.

**20.** In this exercise, you are asked to give an informal argument (based on the Schroeder–Bernstein Theorem) that $\mathscr{P}(\mathbb{N}) \sim \mathbb{R}$. It is not required that you define the following functions $f_1$ and $f_2$ in a formal manner; but be prepared to argue that the functions about which you are thinking are one-to-one.

**a)** Indicate a way of matching each member of $\mathscr{P}(\mathbb{N})$ with a real number in such a way that no two members of $\mathscr{P}(\mathbb{N})$ are matched to the same real number ($f_1: \mathscr{P}(\mathbb{N}) \to \mathbb{R}$).

**b)** Indicate a way of matching each real number with a subset of $\mathbb{N}$ in such a way that no two real numbers are matched to the same subset of $\mathbb{N}$.

***21.** In the text, we used the Schroeder–Bernstein Theorem to show that the open interval $(a, b)$ and the closed interval $[a, b]$ have the same cardinality. Prove this result without using the Schroeder–Bernstein Theorem.

## 7.3 The Well-ordering Principle and the Axiom of Choice

Recall that a partial order $\leqslant$ on a set $S$ is a well order provided that each nonempty subset of $S$ has a least member. Suppose that we are given some finite set $S$. Does there necessarily exist a well order $\leqslant$ on $S$? It certainly seems so. If $S$ has $n$ elements, we ought to be able to choose one member of $S$ and label it the first member, then choose another member of $S$ and label it the second member, and so on. We can even outline how a finite set can be well ordered. If we assume that any $k$-element set can be well ordered and we wish to well order a $(k+1)$-element set $S$, we can choose some member $x$ of $S$ to be the largest member of $S$ and then well order the $k$-element set $S - \{x\}$.

Here is a similar problem. Suppose that we are given a finite collection $\{A_1, A_2, A_3, \ldots, A_n\}$ of nonempty sets. Does there exist a function $f\colon \{1, 2, 3, \ldots, n\} \to \cup_{i=1}^{n} A_i$ such that for each $i \in \{1, 2, 3, \ldots, n\}$, $f(i) \in A_i$? It certainly seems so. Since $A_1 \neq \varnothing$, choose $a_1 \in A_1$ and throw $(1, a_1)$ into the hopper. Since $A_2 \neq \varnothing$, choose $a_2 \in A_2$ and throw $(2, a_2)$ into the hopper. After you have thrown $(n, a_n)$ into the hopper, you will have a function $f = \{(1, a_1), (2, a_2), \ldots, (n, a_n)\}$ with the required properties. We can even outline how a formal proof by induction would proceed (see Exercise 22). As we mentioned at the end of Chapter 4, Zermelo, in the theorem that bears his name, answered a more difficult question that the two questions we have been considering.

**Zermelo's Theorem** Let $S$ be a nonempty set. Then there is a well order $\leqslant$ on $S$.

Zermelo's Theorem, which was proved in 1904, took the mathematical community by surprise. No one knew how to well order $\mathbb{R}$, let alone how to well order *any* nonempty set. In the proof of his theorem, Zermelo used the Axiom of Choice, and it is now known that Zermelo's Principle, the principle that every set can be well ordered, and the Axiom of Choice are equivalent axioms.

With regard to this situation, the mathematical community is divided into roughly three camps: the liberals, the moderates, and the conservatives. The liberal view runs something like this: "What a mess. This reminds me of the scene from Gilbert & Sullivan's *The Mikado*, when the Mikado tells poor Pooh-bah that there ought to be a law but there just isn't. I was born knowing that the following axiom is true: Given any nonempty collection of nonempty sets $\mathscr{C} = \{C_\alpha : \alpha \in \Lambda\}$, there is a function $f\colon \Lambda \to \cup\{C_\alpha : \alpha \in \Lambda\}$ such that for each $\alpha \in \Lambda$, $f(\alpha) \in C_\alpha$. The function $f$ 'chooses' a member of $C_\alpha$ for me for each $\alpha \in \Lambda$, and so I refer to such a function as a choice function, and call my axiom the Axiom of Choice. Since I believe the Axiom of Choice, I also believe Zermelo's Theorem."

The conservative would reply: "I've learned from bitter experience that endowing infinite sets with properties that hold for finite sets can be disastrous, and some of the consequences of your Axiom of Choice, including Zermelo's Theorem, strike me as bizarre indeed. Further, while I cannot get results that are consequences of the Axiom of Choice, I have found that I can do all the mathematics that is useful and interesting to me without having to rely on your unwarranted intrusion. Therefore, I do not accept the Axiom of Choice nor any 'theorem' that relies on this axiom for its proof."

Here is an expression of the moderate point of view: "It would be nice if we could derive the Axiom of Choice from some other axioms that the conservative would accept. But the axiom is so different from our other rules that we are not likely to assuage the conservative in this way. Kurt Gödel

(1906–1978) has shown, however, that the Axiom of Choice is consistent with Zermelo–Fraenkel Set Theory in that we cannot use the rules of Zermelo–Fraenkel Set Theory to disprove the Axiom of Choice (unless by some bizarre twist of fate there already exists a contradiction within Zermelo–Fraenkel Set Theory anyway). Thus we cannot inflict any devastating harm to mathematics by assuming this axiom, and so the choice between this axiom and some other consistent axiom boils down to aesthetics. It is obvious to me that the benefits of the Axiom of Choice are many. We get such important results as the Tychonoff Product Theorem in topology and the Hahn–Banach Theorem in analysis, and so I accept the Axiom of Choice until such time as someone can produce some other axiom that requires the negation of the Axiom of Choice and leads to prettier mathematics."

We presume that you have had enough of imaginary conversations in this chapter and are eager to get on with it. Thus we will not indulge in a Galilean dialogue in which we try to decide who is "right." But it is only fair to give a few warnings. (1) The vast majority of present-day mathematicians are either liberals or moderates and so work in ZFC (Zermelo–Fraenkel Set Theory with the Axiom of Choice). (2) Often people think that they are using the Axiom of Choice when they are not. If $\mathscr{C} = \{C_\alpha : \alpha \in \Lambda\}$ is a nonempty collection of nonempty sets of natural numbers, then for each $\alpha \in \Lambda$ we can let $f(\alpha)$ be the least member of $C_\alpha$ and so a choice function already exists. Even if $\mathscr{C} = \{C_\alpha : \alpha \in \Lambda\}$ is a nonempty collection of intervals in $\mathbb{R}$ we can, for each $\alpha \in \Lambda$, let $f(\alpha)$ be the midpoint of the interval $C_\alpha$ and thereby define a choice function.

Although it is natural for mathematicians to write "Choose $x$ in $X$" as an alternative way of saying "Let $x$ be an element of $X$," this choice of $x$ from $X$ can be made without appealing to the Axiom of Choice. It's only when we have infinitely many nonempty sets in which there is no way to "distinguish" a particular member, from each set, that we need to appeal to the Axiom of Choice.

In Section 7.4, we will consider some problems concerning countable sets. There we will need a much weaker version of the Axiom of Choice, which is almost universally accepted as a reasonable axiom. Even this weaker axiom cannot be proved in Zermelo–Fraenkel Set Theory (if Zermelo–Fraenkel Set Theory is consistent).

**The Countable Axiom of Choice**

Let $\mathscr{C} = \{C_n ; n \in \mathbb{N}\}$ be a countably infinite collection of nonempty sets. Then there is a choice function $f: \mathbb{N} \to \cup\mathscr{C}$ (that is, $f(n) \in C_n$ for each $n \in \mathbb{N}$).

**EXERCISES 7.3**

**22.** Let $\mathscr{C} = \{C_1, C_2, \ldots, C_n\}$ be a finite collection of nonempty sets. Prove by induction that there exists a choice function $f\colon \{1, 2, \ldots, n\} \to \cup\mathscr{C}$ such that $f(i) \in C_i$ for $i = 1, 2, \ldots, n$.

**23.** Without using the (Countable) Axiom of Choice, define a choice function for each of the following infinite collections of nonempty sets.

**a)** $\mathscr{C}$ is the collection of all intervals of real numbers of the form $(-\infty, a]$.

**b)** $\mathscr{C}$ is the collection of all intervals of real numbers of the form $(-\infty, a)$.

**c)** $\mathscr{C}$ is the collection of all nonempty subsets of $\mathbb{N}$.

**d)** $\mathscr{C}$ is the collection of all lines in the plane with slope 1.

## 7.4 Countable Sets

In this section, we prove that every infinite subset of $\mathbb{N}$ is countably infinite. Consequently, every subset of $\mathbb{N}$ is either finite or countably infinite.

**Notation** For the remainder of this chapter, $I_n$ denotes the subset of $\mathbb{N}$ consisting of the first $n$ natural numbers; that is, $I_n = \{1, 2, \ldots, n\}$.

**Lemma 7.9**

Every subset of $\mathbb{N}$ is countable.

**PROOF** Let $X$ be a subset of $\mathbb{N}$. If $X$ is finite, there is nothing to prove. Suppose, therefore, that $X$ is infinite. Inductively we can define a one-to-one function $f$ mapping $\mathbb{N}$ onto $X$, where $f(1)$ is the least member of $X$ and, for each $n \in \mathbb{N}$, $f(n+1)$ is the least member of $X - \{f(1), f(2), \ldots, f(n)\}$. Since $f$ is one-to-one, if we show that $X = f(\mathbb{N})$, then by definition, $X$ will be countably infinite and hence countable.

To see that $X = f(\mathbb{N})$, we first observe that for all $n \in \mathbb{N}$, $f(n) \geqslant n$ (see Exercise 24).

Now suppose that $X \neq f(\mathbb{N})$ and let $z$ be the least member of $X - f(\mathbb{N})$. Then $z < f(z)$ and $z \neq 1$. (Why?) Since $z \in X$ and $z$ is less than the least member of $X - f(I_{z-1})$, $z \in f(I_{z-1}) \subseteq f(\mathbb{N})$. We have contradicted the assumption that $X \neq f(\mathbb{N})$, so we know that $f$ is a one-to-one function mapping $\mathbb{N}$ onto $X$. □

**Proposition 7.10**

A set $S$ is countable if and only if there is a one-to-one function $f$ mapping $S$ into $\mathbb{N}$.

**PROOF** Suppose that $S$ is a set and $h: S \to \mathbb{N}$ is a one-to-one function. Since we are trying to prove that $S$ is countable, we assume that $S$ is an infinite set. Then $h(S)$ is an infinite subset of $\mathbb{N}$, and by the previous lemma there is a one-to-one function $f$ mapping $\mathbb{N}$ onto $h(S)$. It follows from Theorem 5.7 that $h^{-1} \circ f$ is a one-to-one function from $\mathbb{N}$ onto $S$. Thus $S$ is countably infinite and so countable.

Now suppose that $T$ is a countable set. Since we are trying to find a one-to-one function mapping $T$ into $\mathbb{N}$, we assume that $T$ is finite. By definition, there is an $n \in \mathbb{N}$ such that $T$ and $I_n$ have the same cardinality. It follows that there is a one-to-one function mapping $T$ onto $I_n$. Clearly, this function maps $T$ into $\mathbb{N}$. □

In the proof of Lemma 7.9, we showed that every infinite subset of $\mathbb{N}$ is countably infinite. Note that our proof used the assumption that the usual order on $\mathbb{N}$ is a well order. (Every nonempty subset of $\mathbb{N}$ has a least member.) Suppose that we are given an infinite set $S$. Recall that by Zermelo's Well-Ordering Principle, which is known to be equivalent to the Axiom of Choice, there is a well order on $S$. With the Well-Ordering Principle at hand, we are able to prove the following proposition in essentially the same way that we established Lemma 7.9.

**Proposition 7.11**

Every infinite set contains a countably infinite subset.

**PROOF** Let $S$ be an infinite set, and let $\leqslant$ be a well order on $S$. Inductively we define a one-to-one function $f: \mathbb{N} \to S$ as follows. Let $f(1)$ be the least member of $S$, and for each $n \in \mathbb{N}$, let $f(n+1)$ be the least member of $S - f(I_n)$. To see that $f(\mathbb{N})$ is a countably infinite subset of $S$, it suffices to show that $f$ is a one-to-one function (see Exercise 27). □

Stated informally, our next proposition says that the countable union of countable sets is countable. In order to establish this result, we use the Countable Axiom of Choice, which sneaks in when we choose certain functions. Do not be misled by the casual way in which the Countable Axiom of Choice enters our proof. It is known that the proposition cannot be proved using only the axioms of Zermelo–Fraenkel set theory, unless these axioms are inconsistent. Note that the proposition we are able to prove can be thought of as an infinite version of the Pigeonhole Principle: If uncountably many pigeons roost in countably many pigeonholes, one of these pigeonholes has uncountably many pigeons. This way of thinking about Proposition 7.12 is sometimes very useful.

**Proposition 7.12**

Let $B = \cup\{A_\alpha: \alpha \in \Lambda\}$. If $\Lambda$ is countable and for each $\alpha \in \Lambda$, $A_\alpha$ is countable, then $B$ is countable.

**PROOF** What's fun about this proof is that it uses all sorts of results we have established previously.

By Proposition 7.10, it suffices to find a one-to-one function $g$ mapping $B$ into $\mathbb{N}$. Since $\Lambda$ is countable, by Proposition 7.10 there is a one-to-one function $h$ mapping $\Lambda$ into $\mathbb{N}$. It is helpful to think of $B$ as a hotel in which each set $A_\alpha$ represents a floor. Let $x \in B$. The element $x$ may belong to many $A_\alpha$'s; $x$ may even have rented rooms on infinitely many floors. Anyway, let $\Lambda_x = \{\alpha \in \Lambda: x \in A_\alpha\}$ and let $n(x)$ be the least natural number belonging to $h(\Lambda_x)$. Then $x$ has rented a room on floor $A_{\beta_x}$, where $\beta_x$ is *the* member of $\Lambda$ such that $h(\beta_x) = n(x)$. [Suppose that $h(\gamma) = n(x) = h(\beta_x)$. Since $h$ is one-to-one, $\gamma = \beta_x$.] Since $A_{\beta_x}$ is countable, by Proposition 7.10, the collection of one-to-one functions mapping $A_{\beta_x}$ into $\mathbb{N}$ is nonempty. *Using the Countable Axiom of Choice*, choose one one-to-one function $m_{\beta_x}: A_{\beta_x} \to \mathbb{N}$ for each $\beta_x$. (By Exercise 29, we are choosing only countably many functions.)

Define $g: B \to \mathbb{N}$ by $g(x) = 2^{n(x)}(3)^{m_{\beta_x}(x)}$. By Proposition 7.10, it suffices to show that $g$ is a one-to-one function. Suppose that $x$ and $y$ belong to $B$ and $g(x) = g(y)$. Then $2^{n(x)}(3)^{m_{\beta_x}(x)} = 2^{n(y)}(3)^{m_{\beta_y}(y)}$, and by the Fundamental Theorem of Arithmetic, $n(x) = n(y)$ and $m_{\beta_x}(x) = m_{\beta_y}(y)$. Since $h$ is one-to-one and $h(\beta_x) = n(x) = n(y) = h(\beta_y)$, $\beta_x = \beta_y$. By substitution, $m_{\beta_x} = m_{\beta_y}$ and so $m_{\beta_x}(x) = m_{\beta_x}(y)$. Finally, since $m_{\beta_x}$ is one-to-one, $x = y$. □

We turn from considering subsets of infinite sets to considering products.

**Proposition 7.13**

The product of two countable sets is countable.

**PROOF** Let $A$ and $B$ be countable sets. For each $a \in A$, let $B_a = \{(a, b): b \in B\}$. Since $B$ is countable, it follows from Proposition 7.10 that each $B_a$ is countable (see Exercise 33). Moreover, by Proposition 7.12, $A \times B = \cup\{B_a: a \in A\}$ is countable. □

**Proposition 7.14**

Let $X$ be a countable set. Then $X \times \mathbb{R}$ and $\mathbb{R}$ have the same cardinality.

**PROOF** We use the Schroeder–Bernstein Theorem. Let $x \in X$. Then $j: \mathbb{R} \to X \times \mathbb{R}$ defined by $j(r) = (x, r)$ is one-to-one. Since $X$ is countable, by Proposition 7.10 there is a one-to-one function $g: X \to \mathbb{N}$ and by Exercise 12 there is a one-to-one function $h: \mathbb{R} \to (0, 1)$. Define $f: X \times \mathbb{R} \to \mathbb{R}$ by $f((x, r)) = g(x) + h(r)$. Then, as you are asked to show in Exercise 34, $f$ is a one-to-one function, so by the Schroeder–Bernstein Theorem $X \times \mathbb{R}$ and $\mathbb{R}$ have the same cardinality. □

**Notation** Let $A$ and $B$ be sets. Then $B^A$ denotes the set of all functions mapping $A$ into $B$.

Recall that $2 = \{0, 1\}$. Thus for any set $X$, $2^X$ denotes the set of all functions mapping $X$ into $\{0, 1\}$. According to Exercise 35, for any set $X$, $2^X$ and $\mathscr{P}(X)$ have the same cardinality. For this reason, in purely set-theoretic arguments, $2^X$ is sometimes used in place of $\mathscr{P}(X)$. There is a similar blurring of the distinction between $X^2$ and $X \times X$. These two sets have the same cardinality (see Exercise 35) and even when $X = \mathbb{R}$, it is sometimes useful to think of $X \times X$ as though it were $X^2$.

## EXERCISES 7.4

**24.** Prove by induction that the one-to-one function $f: \mathbb{N} \to X$ of Lemma 7.9 has the property that for all $n \in \mathbb{N}$, $f(n) \geqslant n$.

**25.** Argue that the set of all terminating decimals between 0 and 1 is countable.

**26.** Argue that the set of all terminating decimals is countable.

**27.** Prove that the function $f$ defined inductively in Proposition 7.11 is a one-to-one function.

**28.** Let $A$ be an infinite set and suppose that there is a one-to-one function $f: A \to \mathbb{N}$. Prove that $A$ is countably infinite.

**29.** Prove that every subset of a countable set is countable.

**30.** Prove that if $A$ is a countable set and $x \in A$, then $A - \{x\}$ is countable.

***31.** **a)** Prove that the set of all rational numbers is the countable union of countable sets.

**b)** Let $\mathscr{A}$ be a pairwise disjoint collection of intervals of real numbers. Prove that $\mathscr{A}$ is countable. Assume that between any two real numbers there is a rational number.

**32.** Let $S$ be a countable set and consider the following two statements:

**i)** For each $n \in \mathbb{N}$ the collection of all $n$-element subsets of $S$ is countable.

**ii)** The collection of all finite subsets of $S$ is countable.

**a)** Argue that (i) holds if, and only if, (ii) holds, and then prove one of the two statements.

**b)** Is the collection of all countable subsets of a countably infinite set countable? Justify your answer.

**33.** Let $B$ be a countable set, and let $a$ be an element of another set $A$. Prove that $\{(a, b): b \in B\}$ is countable.

**34.** Show that the function $f: X \times \mathbb{R} \to \mathbb{R}$ defined in the proof of Proposition 7.14 is one-to-one.

**35.** Let $X$ be a nonempty set. Show that $\mathscr{P}(X)$ and $2^X$ have the same cardinality. [*Remark*: $\mathscr{P}(\varnothing)$ has one member, namely $\varnothing$, and $2^\varnothing$ has one member, namely $\varnothing$. Here we break our pledge given on page 148 not to consider the empty function.]

**36.** Let $X$ be a nonempty set. Prove that $X^2$ and $X \times X$ have the same cardinality.

***37.** Let $B$ be a countable set.

**a)** Let $A = \{a_1, a_2\}$ be a two-element set. Show that $B^A$ is countable.

**b)** Prove that if $F$ is a finite set, then $B^F$ is countable.

## 7.5 Uncountable Sets

Galileo's mistaken assumption that any two infinite sets have the same cardinality went unchallenged for nearly two and one-half centuries. It was challenged first by Bernhard Bolzano (1781–1848) and subsequently by Cantor. In 1873, Cantor wrote to Dedekind:

> Take the collection of all positive whole numbers and denote it by $(n)$, then think of the collection of all real numbers and denote it by $(x)$. The question is simply whether $(n)$ and $(x)$ may be corresponded so that each individual of one collection corresponds to one and only one of the other?

In 1874, Cantor published his proof that the answer to the question he had asked Dedekind is no. His proof, given in Proposition 7.15, which is known as the Cantor diagonal argument, is illustrated in Figure 7.2.

$f(1) = .b_{11}\ b_{12}\ b_{13}\ b_{14} \cdots$ $\qquad$ $a_1$ is 7 or 5 but is not $b_{11}$.

$f(2) = .b_{12}\ b_{22}\ b_{23}\ b_{24} \cdots$ $\qquad$ $a_2$ is 7 or 5 but is not $b_{22}$.

$f(3) = .b_{31}\ b_{32}\ b_{33}\ b_{34} \cdots$ $\qquad$ $a_3$ is 7 or 5 but is not $b_{33}$.

$f(4) = .b_{41}\ b_{42}\ b_{43}\ b_{44} \cdots$ $\qquad$ $a_4$ is 7 or 5 but is not $b_{44}$.

$\vdots$ $\qquad$ $\vdots$

$f(n) = .b_{n1}\ b_{n2}\ b_{n3}\ b_{n4} \cdots$ $\qquad$ $a_n$ is 7 or 5 but is not $b_{nn}$.

Figure 7.2

Personal Note

### Bernhard Bolzano (1781–1848)

If ever there was a mathematician who received scant recognition for major contributions to mathematics, it was the Czechoslovakian priest, Bernhard Bolzano. Bolzano, around 1840, was the first to understand the difference between countable and uncountable sets; he was the first (1834) to find a function continuous on an interval $[a, b]$ that fails to be differentiable anywhere; he was the first to state Cauchy's criterion for convergence of sequences. Moreover, he proved the Bolzano–Weierstrass theorem and the intermediate-value theorem long before these theorems were rediscovered by Weierstrass and Cauchy. A convenient explanation for the scant credit given Bolzano during his lifetime is that his work was inaccessible. The case against this explanation is given by Gurt Schubring ["Bernhard Bolzano—Not so unknown to his contemporaries as is commonly believed," *Historia Math.* **20** (1993), 45–53]. Another plausible explanation is that Bolzano's political and religious views, like his mathematics, were too radical for his times. We know that Bolzano lost his Professorship of Theology at the University of Prague because of his religious views and that after 1817 his work was censored by the authorities in Vienna. It may have been expedient to view Bolzano's mathematics as interesting, but far out. Even Abel's statement "Bolzano is a clever man" seems deliberately ambiguous and noncommital.

**Proposition 7.15**

The interval $(0, 1)$ and $\mathbb{N}$ do not have the same cardinality.

**PROOF** Suppose to the contrary that $f\colon \mathbb{N} \to (0, 1)$ is a one-to-one map of $\mathbb{N}$ onto $(0, 1)$. Consider the real number

$$\sum_{i=1}^{\infty} a_i/10^i$$

where $a_i = 7$ so long as the $i$th digit of the decimal $f(i)$ is not the digit 7 and $a_i = 5$ if, perchance, the $i$th digit of the decimal $f(i)$ is a seven. (The choice of 5's and 7's is arbitrary.) The real number under consideration is itself a decimal between 0 and 1. Moreover, it is not the decimal $f(1)$ because it differs from $f(1)$ in its first digit. In general, it is not the decimal $f(n)$, because it differs from $f(n)$ in the $n$th digit. Thus $f$ does not map onto $(0, 1)$ because we have exhibited a member of $(0, 1)$ that does not belong to the range of $f$. We have reached a contradiction. □

**Corollary 7.16**

The set of all real numbers is an uncountable set.

**PROOF** See Exercise 39. □

**Corollary 7.17**

The set of all irrational numbers is an uncountable set.

**PROOF** See Exercise 40. □

**Corollary 7.18**

The collection of all infinite sets of natural numbers is an uncountable collection.

**PROOF** See Exercise 41. □

Suppose that $C$ is an infinite subset of $\mathbb{R}$. By Proposition 7.11, $C$ contains a countably infinite subset. Must it be true that $C$ is either countably infinite or that $C$ has the cardinality of $\mathbb{R}$? In 1884, Cantor announced a positive solution to this problem in the prestigious journal, *Mathematische Annalen.* Alas, Cantor's proof fell through, and he was to spend the rest of his life trying to establish the result he had announced. In those days the real line was commonly referred to as the continuum. Consequently, the conjecture that Cantor was trying to prove became known as the *Continuum Hypothesis.* (Also for this reason, any set having the same cardinality as $\mathbb{R}$ was said to *have cardinality* $c$. We know, for example, that $\mathscr{P}(\mathbb{N})$ has cardinality $c$, as does any interval of real numbers.) Cantor's announcement of a solution to the Continuum Hypothesis is full of irony. Of all the contributions Cantor made to twentieth-century mathematics—and we have already seen these were many—his insistence on trying to resolve the Continuum Hypothesis led to the most profound change in the philosophy of mathematics. Moreover, the ultimate import of the Continuum Hypothesis lay in Cantor's having no real hope of redeeming his pledge by establishing this result, which he had announced in 1884.

In 1900, David Hilbert addressed the Second International Congress of Mathematicians and listed 23 problems that he deemed to be the most important problems facing the twentieth-century mathematician. The problem he chose to mention first was Cantor's Continuum Hypothesis. Here are a few bits and pieces of what Hilbert had to say:

> Who of us would not be glad to lift the veil behind which the future lies hidden; to cast a glance at the next advances of our science and at the secrets of its development during future centuries?...
>
> If we would obtain an idea of the probable development of mathematical knowledge in the immediate future, we must let the unsettled questions pass before our minds and look over the problems which the science of to-day sets and whose solution we expect from the future. To such a review of problems the present day, lying at the meeting of the centuries, seems to me well adapted. For the close of a great epoch not only invites us to look back into the past but also directs our thoughts to the unknown future....
>
> Take any definite unsolved problem, such as the question as to the irrationality of the Euler–Mascheroni constant $C$, or the existence of an infinite number of prime numbers of the form $2^n + 1$. However unapproachable these problems may seem to us and however helpless we stand before them, we have, nevertheless, the firm conviction that their solution must follow by a finite number of purely logical processes....
>
> This conviction of the solvability of every mathematical problem is a powerful incentive to the worker. We hear within us the perpetual call: There is the problem. Seek its solution. You can find it by pure reason, for in mathematics there is no *ignorabimus*....
>
> Let us look at the principles of analysis and geometry. The most suggestive and notable achievements of the last century in this field are, as it seems to

> me, the arithmetical formulation of the concept of the continuum in the works of Cauchy, Bolzano, and Cantor, and the discovery of non-euclidean geometry by Gauss, Bolyai, and Lobachevsky. I therefore first direct your attention to some problems belonging to these fields.
>
> 1. Cantor's Problem of the Cardinal Number of the Continuum
>
> Two systems, i.e., two assemblages of the ordinary real numbers or points, are said to be (according to Cantor) equivalent or of equal *cardinal number*, if they can be brought into a relation to one another such that to every number of the one assemblage corresponds one and only one definite number of the other. The investigations of Cantor on such assemblages of points suggest a very plausible theorem, which nevertheless, in spite of the most strenuous efforts, no one has succeeded in proving. This is the theorem:
>
> Every system of infinitely many real numbers, i.e., every assemblage of numbers (or points), is either equivalent to the assemblage of natural integers, 1, 2, 3,... or to the assemblage of all real numbers and therefore to the continuum, that is, to the points of a line; *as regards equivalence there are, therefore, only two assemblages of numbers, the countable assemblage and the continuum.*
>
> From this theorem it would follow at once that the continuum has the next cardinal number beyond that of the countable assemblage; the proof of this theorem would, therefore, form a new bridge between the countable assemblage and the continuum.

Positioned as we are at the other end of the twentieth century, it is not hard to lift the veil, as Hilbert would say, and see what happened. The first sign of trouble came in 1931, when Gödel published his Incompleteness Theorem. This theorem says roughly that any consistent mathematical system containing the rules of ordinary arithmetic also contains a proposition that can be neither proved nor refuted by working within the given system. In short, Hilbert's firm conviction that any well-posed problem can be solved was misplaced. Gödel's proof only showed the existence of a proposition that could be neither proved nor refuted. Yes, there was a tiger out there, but there was no reason to suppose that the tiger was in our backyard. For twenty or thirty years the mathematical community went blithely on presuming that whatever propositions were unprovable, they were not likely to be propositions of central importance. Then, in 1963, Paul Cohen proved that the Continuum Hypothesis was just such an unprovable proposition. ZFC, the system of axioms of Zermelo–Fraenkel set theory together with the Axiom of Choice, may already be an inconsistent system, but Cohen proved that if ZFC is consistent so are the axiom systems ZFC with the Continuum Hypothesis and ZFC with the negation of the Continuum Hypothesis. Where does this leave set theory? Well, what set theory do you mean? Do you accept the Continuum Hypothesis? Perhaps you accept Martin's axiom and the negation of the Continuum Hypothesis. Set theory, which was to unify mathematics, has left our discipline divided. A natural reaction to this situation is to try to work in areas of mathematics that seem far removed from the chaos and turmoil of set theory. Stay close to shore and never sail out to sea. It remains to be seen how mathematics will reward such timidity.

## EXERCISES 7.5

**38.** Prove that if $S$ is an uncountable set and $A$ is a countable set, then $S - A$ is an uncountable set.

**39.** Prove that $\mathbb{R}$ is an uncountable set using Proposition 7.15 (rather than Exercise 20).

**40.** Prove that the set of all irrational numbers is an uncountable set.

**41.** Prove that the collection of all infinite sets of natural numbers is an uncountable set.

**42.** Let $I$ denote the closed interval $[0, 1]$, and let $S = \{x \in I$: the digit 7 does not appear in at least one decimal expansion of $x\}$. (Note that $.7 \in S$). Let $T = I - S$. Without defining the required functions formally, use the Schroeder–Bernstein Theorem to argue that $S$ and $T$ have the same cardinality. Are the sets $S$ and $T$ uncountable? Explain.

**43.** Give an example to show that two uncountable sets do not necessarily have the same cardinality.

***44.** Prove that if $A$ is any countable subset of the plane $\Pi = \mathbb{R} \times \mathbb{R}$ and $x$ and $y$ are any two points of $\Pi - A$, then there is a path from $x$ to $y$ missing $A$ that uses at most two line segments.

***45.** Let $\Phi$ be the collection of all polynomial functions with integer coefficients. (A typical member of $\Phi$ is $f(x) = a_n x^n + a_{n-1} x^{n-1} + \cdots + a_1 x + a_0$, where each $a_i$ is an integer.) Prove that there are uncountably many real numbers that are not roots of any polynomial function belonging to $\Phi$.

## 7.6 Any Infinite Set Contains a Countably Infinite Subset

Using the Well-Ordering Principle, we have already established the result that any infinite set contains a countably infinite subset (see Proposition 7.11). Since we have not established the equivalence of the Well-Ordering Principle and the Axiom of Choice in this text, we thought you might be interested in the following proof, which uses the (Countable) Axiom of Choice, rather than the Well-Ordering Principle, to establish that each infinite set contains a countably infinite set. The proof highlights the relationship between this result and the equally important result that the countable union of countable sets is countable.

**Proposition 7.19**

> Every infinite set contains a countably infinite set.

**PROOF** Let $T$ be an infinite set. For each $n \in \mathbb{N}$, let $\mathscr{A}_n$ be the collection of all subsets $A$ of $T$ such that $A \sim I_n$ (where, as usual, $I_n$ denotes $\{1, 2, \ldots, n\}$).

Let $S = \{n \in \mathbb{N}: \mathscr{A}_n \neq \varnothing\}$. Evidently $1 \in S$. Suppose that $n \in S$. Then there exists $A_n \in \mathscr{A}_n$. Because $T$ is infinite, $T - A_n \neq \varnothing$. Let $x \in T - A_n$. Then $A_n \cup \{x\} \in \mathscr{A}_{n+1}$. Therefore, by induction, each $\mathscr{A}_n$ is a nonempty family of sets. Using the (Countable) Axiom of Choice, for each $n \in \mathbb{N}$, choose $A_n \in \mathscr{A}_n$ and set $T' = \cup\{A_n: n \in \mathbb{N}\}$. Clearly, $T' \subseteq T$. We show that $T'$ is infinite. Suppose, to the contrary, that $j \in \mathbb{N}$ and $T' \sim I_j$. Then $A_{j+1}$ is a subset of $T'$ with $j+1$ members whereas $T'$ itself has only $j$ members. This contradiction establishes that $T'$ is an infinite subset of $T$.

Since $T'$ is the countable union of finite sets, by Proposition 7.12 $T'$ is an infinite set that is countable. Such a set is countably infinite, and we are finished. □

The alert reader will perhaps look back at the last line of the proof of Proposition 7.19 and say, "We are not quite finished, because we have never proved that a set with only $j$ members cannot have $j+1$ members." This and other "obvious" facts about finite sets are more difficult to prove than our intuition would lead us to expect. We believe you are willing to connive at this small gap in the proof of Proposition 7.19. For the curious and carping reader, the following two lemmas plug the hole.

**Lemma 7.20**

Let $m$ be a natural number, let $S$ be a nonempty set, and let $x$ be a member of $S$. Then $S$ has exactly $m+1$ members if and only if $S - \{x\}$ has exactly $m$ members.

**PROOF** Suppose that $S - \{x\} \sim I_m$, and let $f: S - \{x\} \rightarrow I_m$ be a one-to-one function mapping onto $I_m$. Then $f \cup \{(x, m+1)\}$ is a one-to-one function mapping $S$ onto $I_{m+1}$ (compare Proposition 5.10). Thus, $S \sim I_{m+1}$ and, by definition, $S$ has exactly $m+1$ members.

Now suppose that $S$ has exactly $m+1$ members, and let $g$ be a one-to-one function mapping $S$ onto $I_{m+1}$. We consider two cases.

***Case I.*** Suppose that $g(x) = m+1$. Then $g|(S - \{x\})$ is a one-to-one function mapping $S - \{x\}$ onto $I_m$, and so $S - \{x\}$ has exactly $m$ members.

***Case II.*** Suppose that $g(x) \neq m+1$. There is a natural number $k$ with $1 \leqslant k \leqslant m$ such that $g(x) = k$, and there exists $y \neq x$ such that $g(y) = m+1$. We define a one-to-one function $f$ mapping $S - \{x\}$ onto $I_m$ as follows: $f(s) = g(s)$ if $s \neq y$. We define $f(y) = k$. Since $I_m$ and $S - \{x\}$ have the same cardinality, $S$ has exactly $m$ members. □

**Lemma 7.21**

If $A$ is a set, and $A$ has exactly $n$ members ($n \in \mathbb{N} \cup \{0\}$), then $A$ does not have $n+1$ members.

**PROOF** It is obvious that any set $A$ that has exactly 0 members does not have $0+1$ members, because $A=\emptyset$ is the only such set. The remainder of the proof is by induction. Let $S$ be the set of natural numbers to which $n$ belongs provided that if $A$ is any set with exactly $n$ members, then $A$ does not have $n+1$ members.

We first show that $1 \in S$. Suppose that $A$ has exactly one member but that $A$ also has two members. There is a subset $B$ of $A$ that has exactly two members, and since $A$ itself has only one member, there is a one-to-one function mapping $\{1\}$ onto $A$. It follows that $A=\{f(1)\}$. Since $A$ is the only nonempty subset of $A$, $A=B$. By the preceding lemma, $\emptyset = B-\{f(1)\}$ has exactly one member—a contradiction. Thus, $1 \in S$.

Suppose that $n \in S$. To see that $n+1 \in S$, let $A$ be a set with exactly $n+1$ members, and suppose that $B$ is a subset of $A$ that has exactly $n+2$ members. Let $x \in B$. Then $B-\{x\} \subseteq A-\{x\}$. By the lemma, $B-\{x\}$ has exactly $n+1$ members, and $A-\{x\}$ has exactly $n$ members. Thus, $A-\{x\}$ is a set with exactly $n$ members that has $n+1$ members. But this last statement contradicts the assumption that $n \in S$. Therefore, $n+1 \in S$, and so $S=\mathbb{N}$, as we needed to prove. □

## 7.7 Cardinal Arithmetic

The study of complex analysis was a well-developed subject long before anyone knew what a complex number was. Indeed, when the definition of a complex number finally arrived it must have seemed vaguely irrelevant, for by that time everyone knew how the field of complex numbers behaved. In this section, we discuss the arithmetic of cardinal numbers, and even though we never define a cardinal number, we hope to provide some understanding of how the "system" (we can't even say "set") of cardinal numbers behaves. We begin with the obvious nondefinition, and from the beginning we assume the Axiom of Choice.

**Nondefinition**

We presume that there is associated with each set $A$ something called the **cardinal number** of $A$, denoted by $|A|$, and we presume that for any two sets $A$ and $B$, $|A|=|B|$ if and only if $A \sim B$.

Given two cardinal numbers $\alpha$ and $\beta$, how should we add them? Well, because $\alpha$ and $\beta$ are cardinal numbers, there are sets $A$ and $B$ such that $\alpha = |A|$ and $\beta=|B|$. We are not promised that $A$ and $B$ are disjoint, but that is a quibble, because it is at least intuitively obvious that we can swap in $A$ and $B$ for sets of the same cardinality that are disjoint. (We believe worrying about how to effect this swap properly belongs in a course in set theory but the idea is to take a rabbit  , which is known not to be the first term of any ordered pair that happens to belong to $A$, and then swap $B$ for $\{(\ , b): b \in B\}$. Then $A$ and the "rabbitted" set $B$ are disjoint.) So without loss of generality, we assume that $A \cap B = \emptyset$. Then $\alpha + \beta$ is defined to be $|A \cup B|$. This definition is wonderfully clever, if it makes sense. There is a danger, however, that it is not well defined. Perhaps there are disjoint sets $A^*$ and $B^*$ such that $A^* \sim A$ (that is, $|A^*| = \alpha$) and $B^* \sim B$ (that is, $|B^*| = \beta$) and yet $|A^* \cup B^*| \neq |A \cup B|$ (that is, $A^* \cup B^*$ is not equivalent to $A \cup B$). If this were to happen, our definition of $\alpha + \beta$ would make no sense at all. But it doesn't happen (see Exercise 46).

**Definition**

Let $\alpha$ and $\beta$ be cardinal numbers, and let $A$ and $B$ be disjoint sets such that $\alpha = |A|$ and $\beta = |B|$. Then $\alpha + \beta = |A \cup B|$.

**Definition**

The cardinality of $\emptyset$ is the cardinal number 0. The cardinality of $\{\emptyset\}$ is the cardinal number 1. The cardinality of $I_n$ is $n$.

The definitions of the cardinal numbers 0 and 1 say that the cardinality of a set with no elements is 0 and the cardinality of a set with exactly one element is 1. Since for each natural number $n$, $I_n = \{1, 2, \ldots, n\}$, the cardinality of any $n$-element set is $n$. We leave the proof of the following proposition to Exercises 47 and 48.

**Proposition 7.22**

Let $\alpha$, $\beta$, and $\gamma$ be cardinal numbers. Then

**a)** $\alpha + \beta = \beta + \alpha$

**b)** $\alpha + (\beta + \gamma) = (\alpha + \beta) + \gamma$

**c)** $\alpha + 0 = \alpha$

**Definition**

A cardinal number $\alpha$ is a **finite cardinal** number provided that there is a finite set $A$ such that $\alpha = |A|$.

The cardinal number $|\mathbb{N}|$ is denoted by $\aleph_0$. The cardinal number $|\mathbb{R}|$ is denoted by $\boldsymbol{c}$. We have already explained the choice of $\boldsymbol{c}$ to denote $|\mathbb{R}|$. Cantor chose $\aleph_0$ because $\aleph$ is the first letter of the Hebrew alphabet and $\aleph_0$ turns out to be the first infinite cardinal number. (A cardinal number that is not a finite cardinal number is said to be an infinite cardinal number.)

**Proposition 7.23**

Let $n$ be a finite cardinal number. Then:

**a)** $\aleph_0 + n = \aleph_0$

**b)** $\aleph_0 + \aleph_0 = \aleph_0$

**c)** $\boldsymbol{c} + \boldsymbol{c} = \boldsymbol{c}$

**PROOF** We prove part (c) and leave parts (a) and (b) as Exercises 50 and 51.

Let $A = [0, 1]$ and $B = [2, 3]$. Then $\boldsymbol{c} = |A| = |B|$ and so $\boldsymbol{c} + \boldsymbol{c} = |A \cup B| = \boldsymbol{c}$. □

We are now faced with two natural ways to define what it means to say that for cardinal numbers $\alpha$ and $\beta$, $\alpha \leqslant \beta$.

*The arithmetic approach*: $\alpha \leqslant \beta$ provided that there is a cardinal number $\gamma$ such that $\alpha + \gamma = \beta$.

*The set-theoretic approach*: $\alpha \leqslant \beta$ provided that whenever $A$ and $B$ are sets such that $\alpha = |A|$ and $\beta = |B|$, there is a subset $C$ of $B$ such that $A \sim C$.

Since $B = C \cup (B - C)$ and $C \cap (B - C) = \varnothing$, you should be able to check out that the two approaches lead to the same definition of $\alpha \leqslant \beta$.

**Definition**

Let $\alpha$ and $\beta$ be cardinal numbers. Then $\alpha \leqslant \beta$ provided there is a cardinal number $\gamma$ such that $\alpha + \gamma = \beta$.

Part (c) of the following theorem is the Schroeder–Bernstein Theorem in disguise. Parts (a) and (b) are consequences of Proposition 7.22 (see Exercise 49).

**Theorem 7.24**

Let $\alpha$ and $\beta$ be cardinal numbers. Then

**a)** $\alpha \leqslant \alpha$.

**b)** If $\alpha \leqslant \beta$ and $\beta \leqslant \gamma$, then $\alpha \leqslant \gamma$.

**c)** If $\alpha \leqslant \beta$ and $\beta \leqslant \alpha$, then $\alpha = \beta$ (the Schroeder–Bernstein Theorem in disguise).

Note that Theorem 7.24 says that "$\leqslant$" has all the properties of a partial order except that, since there are too many cardinal numbers, $\leqslant$ is not an order because it's not a set. It's natural to ask if "$\leqslant$" behaves like a linear order or a well order.

We state the following theorem, which depends on the Axiom of Choice, without proof.

**Theorem 7.25**

Let $S$ be a set of cardinal numbers. Then $\{(\alpha, \beta) \in S \times S: \alpha \leqslant \beta\}$ is a well order on $S$.

**Corollary 7.26**

Let $\alpha$ and $\beta$ be cardinal numbers. Then $\alpha \leqslant \beta$ or $\beta \leqslant \alpha$.

**PROOF** See Exercise 106 in Section 4.7. □

In Exercise 56, you are asked to prove that there is no largest cardinal number. This important result is a consequence of Proposition 7.5. Since there is no largest cardinal number, it follows from Theorem 7.25 that given any cardinal number $\alpha$, we can consider the least cardinal number greater than $\alpha$. In particular, we let $\aleph_1$ denote the least cardinal number greater than $\aleph_0$. Then Cantor's Continuum Hypothesis is the conjecture, undecidable in the usual set theory with the Axiom of Choice, that $\aleph_1 = c$.

As with any arithmetic, once we have for cardinal numbers $\alpha$ and $\beta$ the notion of $\alpha + \beta$ and $\alpha \leqslant \beta$, it is natural to seek a notion of the product $\alpha\beta$. In Exercise 52, we ask you to show that the definition of multiplication of cardinal numbers $\alpha$ and $\beta$ given next is well defined.

**Definition**

Let $\alpha$ and $\beta$ be cardinal numbers, and let $A$ and $B$ be sets such that $|A| = \alpha$ and $|B| = \beta$. Then $\alpha\beta = |A \times B|$ (where $A \times B$ denotes the usual Cartesian product of sets).

Suppose for the nonce that $\alpha$, $\beta$, and $\gamma$ denote **natural numbers**. Here is a list of facts about their multiplication.

**Theorem 7.27**

**a)** $\alpha\beta = \beta\alpha$ Multiplication is commutative.
**b)** $\alpha(\beta\gamma) = (\alpha\beta)\gamma$ Multiplication is associative.
**c)** $\alpha(\beta + \gamma) = \alpha\beta + \alpha\gamma$ Multiplication distributes over addition.
**d)** If $\alpha\beta = 0$, then $\alpha = 0$ or $\beta = 0$.
**e)** If $\alpha\beta = \alpha\gamma$ and $\alpha \neq 0$, then $\beta = \gamma$. Cancellation law
**f)** If $\beta \leqslant \gamma$, then $\alpha\beta \leqslant \alpha\gamma$.
**g)** $\alpha \cdot 0 = 0$ and $\alpha \cdot 1 = \alpha$

All but one of the statements of Theorem 7.27 hold for multiplication of cardinal numbers $\alpha$, $\beta$, and $\gamma$, and the behavior of $\aleph_0$ and $\boldsymbol{c}$ under multiplication discussed below should enable you to spot the impostor (see Exercise 53).

**Proposition 7.28**

Let $n$ be a finite cardinal number. Then

**a)** $n\aleph_0 = \aleph_0$
**b)** $\aleph_0\aleph_0 = \aleph_0$
**c)** $n\boldsymbol{c} = \boldsymbol{c}$
**d)** $\boldsymbol{c}\boldsymbol{c} = \boldsymbol{c}$
**e)** $\aleph_0\boldsymbol{c} = \boldsymbol{c}$

**PROOF** The first two equalities are consequences of the theorem that the countable union of countable sets is countable (see Exercises 54 and 55). We omit the proof of part (c) because it is similar to the proof of part (e).

**d)** Let $A$ be the open interval $(0, 1)$. By Exercise 12, $A \sim \mathbb{R}$, $\boldsymbol{c} = |A|$, and $\boldsymbol{cc} = |A \times A|$. Thus, it suffices to show that $A \times A \sim A$. Clearly, $f: A \to A \times A$ defined by $f(x) = (1/2, x)$ is one-to-one. There is an ingenious trick of Cantor in defining a one-to-one function mapping $A \times A$ into $A$. Given $(x, y) \in A \times A$—say, $x = \Sigma_{i=1}^{\infty} x_i/10^i$ and $y = \Sigma_{i=1}^{\infty} y_i/10^i$—define $f(x, y) = z$, where $z = x_1y_1x_2y_2x_3y_3x_4y_4\ldots$ (formally $z = \Sigma_{i=1}^{\infty} z_i$, where $z_i = x_{(i+1)/2}$ when $i$ is odd and $z_i = y_i/2$ when $i$ is even). By the Schroeder–Bernstein Theorem, $A \times A \sim A$ and so $\boldsymbol{cc} = \boldsymbol{c}$.

**e)** Again, let $A = (0, 1)$. Since $\aleph_0 \cdot \boldsymbol{c} = |\mathbb{N} \times A|$, it suffices to show that $\mathbb{N} \times A \sim \mathbb{R}$. But $\mathbb{N} \times A \sim \cup_{n=1}^{\infty} (n, n+1)$ and so $|\mathbb{N} \times A| \leqslant \boldsymbol{c}$. Moreover, $\boldsymbol{c} \leqslant |A| \leqslant |\mathbb{N} \times A|$, and so by Theorem 7.24(b), $\boldsymbol{c} \leqslant |\mathbb{N} \times A|$. The result then follows from Theorem 7.24(c). □

It has been suggested to us that it would be a sin, presumably a cardinal sin, not to confess to you that for those who accept the Axiom of Choice, cardinal arithmetic is greatly simplified by the following proposition, whose proof can be found in C. Pinter's *Set Theory* [24], page 161.

**Proposition 7.29**

Let $\alpha$ and $\beta$ be cardinal numbers, at least one of which is infinite. Then $\alpha + \beta = \alpha\beta = \max\{\alpha, \beta\}$.

## EXERCISES 7.7

**46.** Let $A$ and $B$ be disjoint sets, and let $A^*$ and $B^*$ be disjoint sets such that $A \sim A^*$ and $B \sim B^*$. Prove that $A \cup B \sim A^* \cup B^*$.

**47. a)** Let $\alpha$, $\beta$, and $\gamma$ be cardinal numbers. Prove that $\alpha + \beta = \beta + \alpha$ and that $(\alpha + \beta) + \gamma = \alpha + (\beta + \gamma)$.

**b)** Why in your opinion didn't this question read: Prove that addition of cardinal numbers is a commutative and associative operation?

**48.** Prove that for any cardinal number $\alpha$, $\alpha + 0 = \alpha$.

**49.** Use Proposition 7.22 to prove that for any cardinal numbers $\alpha$, $\beta$, and $\gamma$,

**a)** $\alpha \leqslant \alpha$.

**b)** If $\alpha \leqslant \beta$ and $\beta \leqslant \gamma$, then $\alpha \leqslant \gamma$.

**50.** Let $n$ be a finite cardinal number. Prove that $\aleph_0 + n = \aleph_0$.

**51.** Prove that $\aleph_0 + \aleph_0 = \aleph_0$.

**52.** Let $\alpha$ and $\beta$ be cardinal numbers. Prove that $\alpha\beta$ is well defined.

**53.** Let $\alpha$, $\beta$, and $\gamma$ be cardinal numbers. Prove or give a counterexample to each of the following statements.

**a)** $\alpha\beta = \beta\alpha$

**b)** $\alpha(\beta\gamma) = (\alpha\beta)\gamma$

**c)** $\alpha(\beta + \gamma) = \alpha\beta + \alpha\gamma$

**d)** If $\alpha\beta = 0$, then $\alpha = 0$ or $\beta = 0$.

**e)** If $\alpha\beta = \alpha\gamma$ and $\alpha \neq 0$, then $\beta = \gamma$.

**f)** If $\beta \leqslant \gamma$, then $\alpha\beta \leqslant \alpha\gamma$.

**g)** $\alpha \cdot 0 = 0$ and $\alpha \cdot 1 = \alpha$

**54.** Let $n$ be a nonzero finite cardinal number. Assuming that the finite union of countably infinite sets is countably infinite, prove that $n\aleph_0 = \aleph_0$.

**55.** Assuming that the countable union of countably infinite sets is countably infinite, prove that $\aleph_0\aleph_0 = \aleph_0$.

**56.** Prove that if $\alpha$ is a cardinal number, there is a cardinal number $\beta$ such that $\alpha < \beta$.

## Writing Exercise

The following proof that the set of all rational numbers is countable appeared in "Countability of Sets" by Stephen L. Campbell [*Amer. Math. Monthly*, Vol. 93 (1986), 480–481]. "Proof: Clearly the integers can be mapped into the rationals. Now observe that each rational number $a/b$ is a distinct integer written in base 11 with / as the extra symbol for 10." Explain Campbell's argument in detail and in particular explain which natural numbers (in base 10) are associated with 2/3, 2/11, and 2/1.

CHAPTER

# 8

# INTRODUCTION TO GROUPS

In this chapter, we give an introduction to modern algebra. In particular, we study an important algebraic structure, the group. The concept of a group emerged from the independent work of Evariste Galois (1811–1832) and Niels Abel (1802–1829). Galois was a brilliant Frenchman who was killed in a duel at age 21; Abel, a brilliant Norwegian, died of tuberculosis at age 26. Groups were introduced by Galois and Abel as a result of their studies of algebraic equations. More recently, many applications of group theory outside of mathematics have been found. Groups are useful, for example, in the study of symmetry. Crystallographers use group theory to study the symmetries of crystals and molecular structures, and physicists use group theory to study the symmetries of particles and fields of force. Symmetry is also found in art, and group theory can be used to study art forms as diverse as Arabic mosaics and twelve-tone music. Thus, while modern algebra is abstract, it has important applications in both art and science.

## 8.1 *Operations*

If $m$ and $n$ are natural numbers, then $m + n$ and $m \times n$ are natural numbers. Similarly, if $A$ and $B$ are $n$-by-$n$ matrices, then $A + B$ and $A \times B$ are $n$-by-$n$ matrices. These (addition and multiplication of natural numbers and matrices) are examples of operations.

*Personal Note*

### Evariste Galois (1811–1832)

Had Theodore Dreiser decided to write "An Algebraic Tragedy," he could hardly have concocted a more romantic and tragic tale than the traditional telling of the life of Evariste Galois. The legend goes something like this. As a student at Louis-le-Grand, Galois's mathematical genius is unrecognized. After failing the entrance exam to l'Ecole Polytechnique, Galois writes a paper, initiating Galois theory, but the referee, Augustin Cauchy, is too dimwitted to understand the work and tosses it in the wastebasket. Galois takes the entrance exam to l'Ecole Polytechnique a second time, and as it turns out the exam comes just days after Galois's father has commited suicide. Again Galois fails the exam, not because he is distracted by family tragedy, but entirely because of the stupidity of his examiners. Although Cauchy has given Galois no encouragement, Galois submits his work for the Grand Prize in mathematics. The committee—Fourier, Lacroix, Legendre, Poinsot, and Poisson—cannot understand Galois's work, and perhaps one of them, presumably Fourier, also throws Galois's paper away. Embittered by the lack of recognition of his mathematics, which Galois sees as the suppression of truth by the authorities, Galois turns to radical politics. He is arrested for proposing a toast to the death of Louis Phillipe, but a sympathetic jury accepts the argument that Galois has merely stood with a dagger in his hand and said, "To Louis Phillipe, if he betrays," the clause, "if he betrays" having been drowned out by the crowd. Meanwhile, Poisson admits that Galois's paper has been lost and asks Galois to submit a revision. Galois is arrested a second time, this time for parading in Paris on Bastille Day, armed to the teeth, and in the uniform of the outlawed Artillery Guard. By now the secret police view Galois as such a threat that they decide to have him assassinated. When an assassination attempt fails, the police hire an unknown woman to seduce Galois and involve him in a duel. Galois falls madly in love with this woman and makes some untoward remarks about her after she suddenly drops him.

In the final chapter of our historical romance, Poisson's rejection of Galois's paper arrives, and Galois, although denying that he has dishonored the woman he loves, accepts a challenge to a duel. In the night before the duel, he writes up all of Galois theory in a letter to his friend Chevalier. The next morning he is shot and later taken to a hospital, where he dies in the arms of his brother.

The account above is rife with inconsistencies and inaccuracies. Rothman [29] sorts facts from fiction. He exonerates the duelist [Pescheux d'Herbinville] and the mysterious woman [Stéphanie Dumotel] as well as Cauchy, who read and admired Galois's first paper, and Fourier, whose only misstep in the affair was to die at the wrong time. Curiously, the one reputation Rothman cannot rescue is that of Galois himself.

**Definition**

An **operation on a set** $X$ is a function from $X \times X$ into $X$.

Some authors use the term *binary operation* in place of *operation*.

So the operation of addition of natural numbers is the function $S\colon \mathbb{N} \times \mathbb{N} \to \mathbb{N}$ defined by $S(m, n) = m + n$. Note that the function $T\colon \mathbb{Z} \times \mathbb{Z} \to \mathbb{Z}$ defined by $T(m, n) = m - n$ is an operation on $\mathbb{Z}$; that is, subtraction is an operation on the set $\mathbb{Z}$ of integers. However, the restriction of $T$ to the positive integers is not an operation on that set; that is, subtraction is not an operation on the positive integers because if $m$ and $n$ are positive integers, $m - n$ is not necessarily a positive integer.

**Definition**

Let $f$ be an operation on a set $X$. A subset $A$ of $X$ is **closed** with respect to $f$ if $a, b \in A$ implies that $f(a, b) \in A$.

It should be observed that, under our definition, if $f$ is an operation on a set $X$, then $X$ is closed with respect to $f$. However, we have introduced the term *closed* for convenience, as is illustrated in the next sentence. Note that the integers are closed with respect to the operation of subtraction on $\mathbb{Z}$, whereas the positive integers are not closed with respect to this operation.

The function $D$ defined by $D(a, b) = a/b$ is not an operation on $\mathbb{R}$ because if $b = 0$, $a/b \notin \mathbb{R}$. However, the function $D: (\mathbb{R} - \{0\}) \times (\mathbb{R} - \{0\}) \to \mathbb{R} - \{0\}$ defined by $D(a, b) = a/b$ is an operation on $\mathbb{R} - \{0\}$. Why did we consider the function $D$ whose domain is $(\mathbb{R} - \{0\}) \times (\mathbb{R} - \{0\})$ rather than the function whose domain is $\mathbb{R} \times (\mathbb{R} - \{0\})$?

**EXAMPLE 1**

If $X$ is a finite set, we can specify an operation on $X$ using a table. For example, suppose that $X = \{a, b, c\}$ and that we wish to define an operation $*$ on $X$. (In this and other examples, we tacitly assume that $a$, $b$, and $c$ are three different entities.) We list the elements of $X$ down the left side and across the top of the table. Then, if $x$ and $y$ are members of $X$, we place the element $x * y$ in the row headed by $x$ and the column headed by $y$.

| $*$ | $a$ | $b$ | $c$ |
|---|---|---|---|
| $a$ | $a$ | $b$ | $c$ |
| $b$ | $b$ | $c$ | $a$ |
| $c$ | $c$ | $a$ | $b$ |

This table tells us: $a*a = a$, $a*b = b$, $a*c = c$, $b*a = b$, $b*b = c$, $b*c = a$, $c*a = c$, $c*b = a$, and $c*c = b$. ❑

**EXAMPLE 2**

We give another example of an operation on a set with three elements. Again, let $X = \{a, b, c\}$ and define an operation $*$ on $X$ using the following table.

| $*$ | $a$ | $b$ | $c$ |
|---|---|---|---|
| $a$ | $c$ | $b$ | $a$ |
| $b$ | $a$ | $c$ | $b$ |
| $c$ | $b$ | $c$ | $b$ |

This table tells us: $a*a = c$, $a*b = b$, $a*c = a$, $b*a = a$, $b*b = c$, $b*c = b$, $c*a = b$, $c*b = c$, and $c*c = b$. ❑

Tables like those in Examples 1 and 2 that define an operation are called **Cayley tables**, in honor of the British mathematician Arthur Cayley (1821–1895).

**EXAMPLE 3**

Let $X$ be the set of all functions from $\mathbb{R}$ into $\mathbb{R}$, and define $*$ on $X$ to be the usual sum of two functions; that is, $f*g = h$, where $h(x) = f(x) + g(x)$. Then $*$ is an operation on $X$. ❑

**EXAMPLE 4** Let $X$ be a nonempty set, and let $Y$ be the set of all functions from $X$ onto $X$. If $\circ$ denotes the usual composition of functions, then $\circ$ is an operation on $Y$. ❑

**EXAMPLE 5** If $X$ is the set of congruence classes mod 9, then $\oplus$ (defined in Section 4.5) is an operation on $X$. ❑

**Definition** An operation $*$ on a set $X$ is **associative** provided that $a*(b*c)=(a*b)*c$ for all $a, b, c \in X$.

The operations defined in Examples 1, 3, 4 (see Theorem 4.21), and 5 (see Exercise 1) are associative. However, the operation defined in Example 2 is not associative because $(a*a)*a=c*a=b$, whereas $a*(a*a)=a*c=a$. Is ordinary subtraction on $\mathbb{Z}$ associative? (See Exercise 7.)

**EXAMPLE 6** Let $*$ be the operation on $\mathbb{N}$ defined by $m*n=m^n$. Then $(m*n)*p=m^n*p=(m^n)^p$, whereas $m*(n*p)=m*(n^p)=m^{(n^p)}$. So this operation is not associative, since $(2^3)^2=8^2=64$ and $2^{(3^2)}=2^9=512$. ❑

**Definition** If $*$ is an operation on a set $X$, then an element $e$ of $X$ is called an **identity with respect to** $*$ provided that $e*x=x*e=x$ for all $x \in X$.

In Example 1, $a$ is the identity element with respect to $*$. The set $X$ in Example 2 has no identity element with respect to the operation $*$ (see Exercise 2). In Example 3, the identity element with respect to $*$ is the constant function $c$ defined by $c(x)=0$ for all $x \in \mathbb{R}$. In Example 4, the identity element with respect to composition is the identity function $i$ defined by $i(x)=x$ for all $x \in X$.

**Definition** Let $X$ be a set, let $*$ be an operation on $X$, let $e$ be an identity with respect to $*$, and let $a \in X$. An element $b$ of $X$ is an **inverse of a relative to** $*$ and $e$ provided that $a*b=b*a=e$.

In Example 1, $c$ is an inverse of $b$ relative to $*$, and $b$ is an inverse of $c$. In Example 3, if $f \in X$, then the function $g$ defined by $g(x)=-f(x)$ is an inverse of $f$.

**Definition** An operation $*$ on a set $X$ is **commutative** provided that $a*b=b*a$ for all $a, b \in X$.

The operations defined in Examples 1, 3, and 5 are commutative, but the operation defined in Example 2 is not commutative because $b*c=b$, whereas $c*b=c$. Matrix multiplication is an operation on the set of all 2-by-2 matrices with real entries that is not commutative. For example,

$$\begin{bmatrix} 2 & -1 \\ 3 & -4 \end{bmatrix}\begin{bmatrix} 2 & 1 \\ 0 & 4 \end{bmatrix} = \begin{bmatrix} 4 & -2 \\ 6 & -13 \end{bmatrix}$$

$$\text{whereas} \quad \begin{bmatrix} 2 & 1 \\ 0 & 4 \end{bmatrix}\begin{bmatrix} 2 & -1 \\ 3 & -4 \end{bmatrix} = \begin{bmatrix} 7 & -6 \\ 12 & -16 \end{bmatrix}$$

The operation defined in Example 6 is not commutative because $2^3 \neq 3^2$.

## EXERCISES 8.1

1. Prove that the operation defined in Example 5 is associative.
2. Prove that the operation defined in Example 2 has no identity element.
3. a) What is the identity element with respect to the operation defined in Example 5?
   b) Is there an identity element with respect to the operation defined in Example 6? If so, what is it? If not, prove this fact.
4. Let $X$ be a set, and let $*$ be an operation on $X$. Prove that if $e$ and $e'$ are identities with respect to $*$, then $e=e'$.
5. In Example 4, does there exist a member of $Y$ that does not have an inverse with respect to composition? Prove your answer.
6. Prove that each member of the set of congruence classes mod 9 has an inverse with respect to the operation $\oplus$ defined in Example 5.
7. Is ordinary subtraction on $\mathbb{Z}$ associative? Prove your answer.
8. Is the operation defined in Example 4 commutative? Prove your answer.
9. Complete the following table in such a way that $b$ is an identity element and each element has an inverse with respect to $*$.

| $*$ | $a$ | $b$ | $c$ | $d$ |
|---|---|---|---|---|
| $a$ | | | | |
| $b$ | | | | |
| $c$ | $d$ | | | $a$ |
| $d$ | | | $a$ | |

10. Complete the following table in such a way that $*$ is commutative.

| $*$ | $a$ | $b$ | $c$ | $d$ |
|---|---|---|---|---|
| $a$ | $d$ | $c$ | $a$ | $b$ |
| $b$ | | $d$ | | |
| $c$ | | $b$ | $c$ | $d$ |
| $d$ | | $a$ | | $c$ |

**11.** Complete the following table in such a way that $*$ is commutative and has an identity element, and each element has an inverse with respect to $*$.

| $*$ | $a$ | $b$ | $c$ | $d$ |
|---|---|---|---|---|
| $a$ | $d$ | | $b$ | |
| $b$ | $c$ | $a$ | | |
| $c$ | | | $a$ | |
| $d$ | | | | |

**12.** Give an example of an operation on $\mathbb{N}$ that is commutative but not associative.

## 8.2 *Integers Modulo n*

In this section, we will make use of concepts that were introduced in Chapter 4. Recall that if $n \in \mathbb{N}$ and $a, b \in \mathbb{Z}$, then $a$ is congruent to $b$ modulo $n$, written $a \equiv b \pmod{n}$, provided that $a - b$ is divisible by $n$. Theorem 4.8 establishes that congruence modulo $n$ is an equivalence relation on the set of integers. Also recall that the equivalence classes for this equivalence relation are called congruence classes modulo $n$, and each integer is congruent, modulo $n$, to precisely one of the integers $0, 1, 2, \ldots, n-1$. Let $\mathbb{Z}_n$ denote the set $\{[0], [1], [2], \ldots, [n-1]\}$ of congruence classes modulo $n$.

**Theorem 8.1**

Let $n \in \mathbb{N}$, and let $a, b, c \in \mathbb{Z}$.

**a)** If $a + c \equiv b + c \pmod{n}$, then $a \equiv b \pmod{n}$.

**b)** If $n$ and $c$ are relatively prime and $a \cdot c \equiv b \cdot c \pmod{n}$, then $a \equiv b \pmod{n}$.

**c)** If $\gcd(n, c) = d$, and $a \cdot c \equiv b \cdot c \pmod{n}$, then $a \equiv b \pmod{n/d}$.

**PROOF** See Exercise 13. □

Recall from Section 4.5 that if $[a], [b] \in \mathbb{Z}_n$, then $[a] \oplus [b]$ and $[a] \odot [b]$ are defined by

$$[a] \oplus [b] = [a + b] \quad \text{and} \quad [a] \odot [b] = [a \cdot b]$$

By Theorem 4.10, these operations are well defined.

**Theorem 8.2**

If $n \in \mathbb{N}$, then the operations $\oplus$ and $\odot$ on $\mathbb{Z}_n$ are commutative and associative.

**PROOF** Let $[a], [b], [c] \in \mathbb{Z}_n$. We must prove four things:

**a)** $[a] \oplus [b] = [b] \oplus [a]$
**b)** $[a] \odot [b] = [b] \odot [a]$
**c)** $[a] \oplus ([b] \oplus [c]) = ([a] \oplus [b]) \oplus [c]$
**d)** $[a] \odot ([b] \odot [c]) = ([a] \odot [b]) \odot [c]$

We prove part (a) and leave the other three parts as exercises (see Exercise 14).

$$\begin{aligned} [a] \oplus [b] &= [a+b] && \text{Definition of } \oplus \\ &= [b+a] && \text{Commutativity of } + \\ &= [b] \oplus [a] && \text{Definition of } \oplus \end{aligned}$$ □

**Theorem 8.3**

If $n \in N$, then the element $[0]$ of $\mathbb{Z}_n$ is an identity with respect to $\oplus$ and the element $[1]$ of $\mathbb{Z}_n$ is an identity with respect to $\odot$.

**PROOF** See Exercise 15. □

**Theorem 8.4**

If $n \in \mathbb{N}$ and $[a] \in \mathbb{Z}_n$, then the inverse of $[a]$ relative to $\oplus$ and $[0]$ is $[-a]$.

**PROOF** See Exercise 16. □

**Theorem 8.5**

If $n \in \mathbb{N}$ and $[a], [b], [c] \in \mathbb{Z}_n$, then

$$[a] \odot ([b] \oplus [c]) = [a] \odot [b] \oplus [a] \odot [c]$$

**PROOF** See Exercise 17. □

## EXERCISES 8.2

**13.** Prove Theorem 8.1.

**a)** Part (a) **b)** Part (b) **c)** Part (c)

**14.** Let $n \in \mathbb{N}$.

**a)** Prove that the operation $\odot$ on $\mathbb{Z}_n$ is commutative.

**b)** Prove that the operation $\oplus$ on $\mathbb{Z}_n$ is associative.

**c)** Prove that the operation $\odot$ on $\mathbb{Z}_n$ is associative.

**15.** Prove Theorem 8.3.

***16.** Prove Theorem 8.4.

17. Prove Theorem 8.5.
18. Find a natural number $n$ and members $[a]$ and $[b]$ of $\mathbb{Z}_n$ such that $[a] \odot [b] = [0]$ but $[a] \neq [0]$ and $[b] \neq [0]$.
19. Let $n \in \mathbb{N}$ such that $n > 1$. Prove that $n$ is prime if and only if for all $[a], [b] \in \mathbb{Z}_n$, $[a] \odot [b] = [0]$ implies either $[a] = [0]$ or $[b] = [0]$.
20. If $[a] \in \mathbb{Z}_n$ and $[a] \neq [0]$, then $[a]$ is a *divisor of zero* in $\mathbb{Z}_n$ if there is a nonzero element $[b]$ of $\mathbb{Z}_n$ such that $[a] \odot [b] = [0]$. Characterize those integers $a$ such that the element $[a]$ of $\mathbb{Z}_n$ is a divisor of zero in $\mathbb{Z}_n$.
21. Prove that the nonzero element $[a]$ of $\mathbb{Z}_n$ has a multiplicative inverse in $\mathbb{Z}_n$ if and only if $n$ and $a$ are relatively prime.

## 8.3 *Groups*

In Section 8.1, we gave a number of examples of sets and operations on these sets, and we observed that some of these operations do not possess such properties as associativity or the existence of an identity element. We define a group to be a set together with an operation such that certain specified properties hold. These properties have been chosen because they occur naturally in many important places.

**Definition**

A **group** $(G, \circ)$ is a nonempty set $G$ with an operation $\circ$ on $G$ such that the following conditions hold:

**a)** $(a \circ b) \circ c = a \circ (b \circ c)$ for all $a, b, c \in G$.

**b)** There exists an element $e$ of $G$ such that $a \circ e = e \circ a = a$ for every $a \in G$.

**c)** For each $a \in G$, there exists $x \in G$ such that $a \circ x = x \circ a = e$.

As we will see (Theorem 8.6), there is a unique element $e$ of a group $(G, \circ)$ such that $a \circ e = e \circ a = a$ for every $a \in G$. This element $e$ is called the **identity element** of $G$. Also by Theorem 8.6, for each element $a$ in a group $(G, \circ)$, there is a unique member $x$ of $G$ such that $a \circ x = x \circ a = e$. The element $x$ is called the **inverse** of $a$ and is denoted by $a^{-1}$.

We are already familiar with several examples of groups.

**EXAMPLE 7**

The set $\mathbb{Z}$ of all integers with the usual operation of addition is a group. The identity of the group is the integer 0, and the inverse of an integer $a$ is the integer $-a$. ❑

**EXAMPLE 8**

The set $\mathbb{Z}_n$ defined in Section 8.2 with the operation $\oplus$ defined in that section is a group. The necessary properties are proved in Theorems 8.2–8.4. ❑

**EXAMPLE 9** The set $X$ of all nonzero rational numbers with the usual operation of multiplication is a group. Since the product of two nonzero rational numbers is also a nonzero rational number, ordinary multiplication is an operation on $X$. We know that multiplication of real numbers is associative, and $X$ is a subset of the real numbers. The identity of the group is the rational number 1, and the inverse of a member $a$ of $X$ is the nonzero rational number $1/a$. ❑

**EXAMPLE 10** The set of all nonzero integers with the usual operation of multiplication is not a group because if $a$ is an integer such that $|a| > 1$, then $a$ does not have an inverse in the set. ❑

**EXAMPLE 11** The set whose only member is the integer 0 with the usual operation of addition is a group. ❑

**EXAMPLE 12** The set whose only member is the integer 1 with the usual operation of multiplication is a group. ❑

**EXAMPLE 13** Let $X$ be the set $\{a, b, c, d\}$, and let $*$ be the operation on $X$ defined using the following table:

| $*$ | $a$ | $b$ | $c$ | $d$ |
|---|---|---|---|---|
| $a$ | $a$ | $b$ | $c$ | $d$ |
| $b$ | $b$ | $a$ | $d$ | $c$ |
| $c$ | $c$ | $d$ | $a$ | $b$ |
| $d$ | $d$ | $c$ | $b$ | $a$ |

The element $a$ is the identity, and each element is its own inverse. It can be verified that the associative law holds. Therefore, $(X, *)$ is a group. ❑

**EXAMPLE 14** If $S$ is a nonempty set, then, by Proposition 5.8, Sym $(S)$ with the usual operation of composition of functions is a group. ❑

**Definition** Let $(G, \circ)$ be a group. If $a \circ b = b \circ a$ for all $a, b \in G$, then $G$ is said to be a **commutative** or **abelian group**.

The term *abelian group* is the one more often used. The name is derived from Niels Abel.

With the exception of Example 14, all the examples we have given thus far are examples of abelian groups. The group in Example 14 is not necessarily abelian, as we now show.

**EXAMPLE 15** Let $X = \{x \in \mathbb{R} : x \geqslant 0\}$. If $f: X \to X$ is defined by $f(x) = 2x$ and $g: X \to X$ is defined by $g(x) = x^2$, then $f$ and $g$ are members of Sym $(X)$. However, $(f \circ g)(x) = 2x^2$ and $(g \circ f)(x) = 4x^2$, so $f \circ g \neq g \circ f$. ❑

We establish the following properties of a group.

**Theorem 8.6**

Let $(G, \circ)$ be a group.

**a)** The identity is unique.
**b)** If $a, b, c \in G$ and $a \circ b = a \circ c$, then $b = c$.
**c)** If $a, b, c \in G$ and $b \circ a = c \circ a$, then $b = c$.
**d)** Each element $a$ of $G$ has a unique inverse, denoted by $a^{-1}$.
**e)** If $a, b \in G$, there exists a unique element $x$ of $G$ such that $a \circ x = b$ and a unique element $y$ of $G$ such that $y \circ a = b$.
**f)** If $a, b \in G$, then $(a \circ b)^{-1} = b^{-1} \circ a^{-1}$.

**PROOF**

**a)** See Exercise 4.

**b)** Suppose that $a \circ b = a \circ c$. There exists an element $x$ of $G$ such that $a \circ x = x \circ a = e$. So $x \circ (a \circ b) = x \circ (a \circ c)$, $x \circ (a \circ b) = (x \circ a) \circ b = e \circ b = b$, and $x \circ (a \circ c) = (x \circ a) \circ c = e \circ c = c$. Therefore, $b = c$.

**c)** Suppose that $b \circ a = c \circ a$. There exists an element $x$ of $G$ such that $a \circ x = x \circ a = e$. So $(b \circ a) \circ x = (c \circ a) \circ x$, $(b \circ a) \circ x = b \circ (a \circ x) = b \circ e = b$, and $(c \circ a) \circ x = c \circ (a \circ x) = c \circ e = c$. Therefore, $b = c$.

**d)** Let $a \in G$. Suppose that $x$ and $x'$ are members of $G$ such that $a \circ x = x \circ a = e$ and $a \circ x' = x' \circ a = e$. Then $x = x \circ e = x \circ (a \circ x') = (x \circ a) \circ x' = e \circ x' = x'$.

**e)** Since $a^{-1} \in G$, $a^{-1} \circ b \in G$. Also, $a \circ (a^{-1} \circ b) = (a \circ a^{-1}) \circ b = e \circ b = b$. So $x = a^{-1} \circ b$ is a member of $G$ such that $a \circ x = b$. Suppose that $x'$ is a member of $G$ such that $a \circ x' = b$. Then $a \circ x = a \circ x'$, and hence by part (b), $x = x'$. Now $b \circ a^{-1} \in G$ and $(b \circ a^{-1}) \circ a = b \circ (a^{-1} \circ a) = b \circ e = b$. Therefore, $y = b \circ a^{-1}$ is a member of $G$ such that $y \circ a = b$. Suppose that $y'$ is a member of $G$ such that $y' \circ a = b$. Then $y \circ a = y' \circ a$, and hence by part (c), $y = y'$.

**f)** $(a \circ b)^{-1}$ is an inverse of $a \circ b$. By part (d), in order to show that $(a \circ b)^{-1} = b^{-1} \circ a^{-1}$, it is sufficient to show that $b^{-1} \circ a^{-1}$ is an inverse of $a \circ b$. Now

$$(b^{-1} \circ a^{-1}) \circ (a \circ b) = [(b^{-1} \circ a^{-1}) \circ a] \circ b = [b^{-1} \circ (a^{-1} \circ a)] \circ b$$
$$= (b^{-1} \circ e) \circ b = b^{-1} \circ b = e,$$

and

$$(a \circ b) \circ (b^{-1} \circ a^{-1}) = a \circ [b \circ (b^{-1} \circ a^{-1})] = a \circ [(b \circ b^{-1}) \circ a^{-1}]$$
$$= a \circ (e \circ a^{-1}) = a \circ a^{-1} = e.$$

Therefore, $b^{-1} \circ a^{-1}$ is an inverse of $a \circ b$. □

It is an immediate consequence of Theorem 8.6(e) that each member of a finite group appears exactly once in each row and in each column of the table.

**Definition**

If $(G, \circ)$ is a group and $G$ is a finite set, then the number of elements in $G$ is called the **order** of the group. If $G$ is infinite, then $(G, \circ)$ is said to have infinite order. The order of $(G, \circ)$ is denoted by $|G|$. A group is said to be **finite** or **infinite** depending on whether its order is finite or infinite.

The groups in Examples 7 and 9 are infinite. The groups in Examples 11 and 12 have order 1. The group $\mathbb{Z}_n$ in Example 8 has order $n$. The group in Example 13 has order 4.

Recall that a permutation on a nonempty set $S$ is a one-to-one function mapping $S$ onto $S$ and that Sym $(S)$ denotes the set of all permutations on $S$. We have seen, in Example 14, that Sym $(S)$ with the usual operation of composition is a group. If $R$ is a relation on a set $A$, then for each $n \in \mathbb{N}$, $R^n$ is defined as in Exercise 86 of Section 4.6. Thus, if $S$ is a nonempty set and $f \in \text{Sym}(S)$, then for each $n \in \mathbb{N}$, $f^n$ is defined inductively as follows: $f^1 = f$ and $f^{n+1} = f^n \circ f$. We also define $f^0$ to be $i_S$ and $f^{-n}$ to be $(f^{-1})^n$.

**EXAMPLE 16**

Let $S = \mathbb{R} \times \mathbb{R}$ and define $f, g \in \text{Sym}(S)$ by $f(x, y) = (-x, y)$ and $g(x, y) = (-y, x)$. (Note that $f$ is the reflection about the $y$-axis and $g$ is the rotation through 90° in a counterclockwise direction about the origin.) Let $G = \{f^i \circ g^j : i = 0, 1 \text{ and } j = 0, 1, 2, 3\}$. Then $G \subseteq \text{Sym}(S)$. We leave as Exercise 32 the proof that $(G, \circ)$ is a nonabelian group of order 8. This group is called the **dihedral group** of order 8. ❑

**Theorem 8.7**

Let $(G, \circ)$ and $(H, *)$ be groups. If $\#$ is the operation on $G \times H$ defined by

$$(a_1, b_1) \# (a_2, b_2) = (a_1 \circ a_2, b_1 * b_2)$$

for all $a_1, a_2 \in G$ and $b_1, b_2 \in H$, then $(G \times H, \#)$ is a group.

**PROOF** See Exercise 28. □

**Definition**

The group $(G \times H, \#)$ is called the **direct product** of $(G, \circ)$ and $(H, *)$.

**EXAMPLE 17**

The members of $\mathbb{Z}_2 \times \mathbb{Z}_3$ are $a = ([0], [0])$, $b = ([0], [1])$, $c = ([0], [2])$, $d = ([1], [0])$, $e = ([1], [1])$, and $f = ([1], [2])$, and the operation $\#$ on $\mathbb{Z}_2 \times \mathbb{Z}_3$ is the operation in the following table.

| # | a | b | c | d | e | f |
|---|---|---|---|---|---|---|
| a | a | b | c | d | e | f |
| b | b | c | a | e | f | d |
| c | c | a | b | f | d | e |
| d | d | e | f | a | b | c |
| e | e | f | d | b | c | a |
| f | f | d | e | c | a | b |

❑

## EXERCISES 8.3

**22.** Let $X$ be a set, and let $*$ be an operation on $X$. Suppose that $e$ is the identity on $X$ with respect to $*$. That is, suppose that $a*e = e*a = a$ for each $a \in G$. Let $a, b, c \in X$. Prove that $a*(b*c) = (a*b)*c$ if any one of $a$, $b$, or $c$ is equal to $e$.

**23.** For each of the following examples, decide whether the given set with respect to the given operation forms a group. If it does, give the identity element and the inverse of each element. If it does not, give the reasons.

**a)** All positive real numbers, multiplication

**b)** All even integers, addition

**c)** $\{-1, 1\}$, multiplication

**d)** $\{a, b, c\}$, operation $*$ defined by the table

| * | a | b | c |
|---|---|---|---|
| a | b | c | a |
| b | a | b | c |
| c | c | a | b |

**e)** $\{a, b\}$, operation $*$ defined by the table

| * | a | b |
|---|---|---|
| a | b | a |
| b | a | b |

**f)** All integers, subtraction

**g)** All nonzero real numbers, division

**h)** All 2-by-2 matrices with integer entries, matrix addition

**i)** All 2-by-2 matrices except the matrix whose entries are all zeros, matrix multiplication

**24.** Complete the following table so that $\{e, a, b\}$ with the operation $*$ is a group.

| * | e | a | b |
|---|---|---|---|
| e | e | a | b |
| a | a | | |
| b | b | | |

**25.** Is there more than one correct answer to Exercise 24? Prove your answer.

**26.** Complete the following table so that $\{e, a, b, c\}$ with the operation $*$ is a group.

| $*$ | $e$ | $a$ | $b$ | $c$ |
|---|---|---|---|---|
| $e$ | $e$ | $a$ | $b$ | $c$ |
| $a$ | $a$ | | | |
| $b$ | $b$ | | | |
| $c$ | $c$ | | | |

**27.** Is there more than one correct answer to Exercise 26? Prove your answer.

**28.** Prove Theorem 8.7.

**29.** If $(G, \circ)$ is a group in which $(a \circ b)^{-1} = a^{-1} \circ b^{-1}$ for all $a, b \in G$, show that $(G, \circ)$ is abelian.

**30.** If $(G, \circ)$ is a group of even order, show that at least one element of $G$ different from the identity is its own inverse.

**31.** Prove that every group of order 4 is abelian.

**32.** For each two real numbers $a$ and $b$ with $a \neq 0$, define $f_{a,b} \in \text{Sym}(\mathbb{R})$ by $f_{a,b}(x) = ax + b$. Let $G = \{f_{a,b}: a, b \in \mathbb{R} \text{ and } a \neq 0\}$. Prove that $(G, \circ)$ is a nonabelian group.

**33.** Let $(G, \circ)$ be the group given in Exercise 32, and let $H = \{f_{a,b} \in G: a \text{ is rational}\}$. Prove that $(H, \circ)$ is a nonabelian group.

**34.** Let $(G, \circ)$ be the group given in Exercise 32, and let $K = \{f_{a,b} \in G: a = 1\}$. Prove that $(K, \circ)$ is an abelian group.

**35.** Let $n$ be a prime. Prove that $(\mathbb{Z}_n - \{[0]\}, \odot)$ is a group.

**36.** Let $G$ be a nonempty set, and let $\circ$ be an associative operation on $G$ such that (1) there exists $e \in G$ such that $a \circ e = a$ for each $a \in G$, and (2) for each $a \in G$, there exists $a^* \in G$ such that $a \circ a^* = e$. Prove that $(G, \circ)$ is a group.

**37.** Let $G$ be a nonempty finite set, and let $\circ$ be an associative operation on $S$ such that (1) if $a, b, c \in G$ and $a \circ b = a \circ c$, then $b = c$, and (2) if $a, b, c \in G$ and $b \circ a = c \circ a$, then $b = c$. Prove that $(G, \circ)$ is a group.

## 8.4 Permutation Groups

In the preceding section, we gave examples of groups that are subsets of Sym $(S)$, where $S$ is a nonempty set. In this section, we study Sym $(S)$, where $S$ is a finite nonempty set.

**Definition**

A **permutation group** is any group whose elements are permutations and whose operation is composition. If $n \in \mathbb{N}$, then the group $S_n$ of all permutations of the set consisting of the first $n$ natural numbers is called the **symmetric group** on $n$ symbols.

Now $S_1$ is clearly abelian, and it is easy to prove that $S_2$ is also abelian. However, we have a different result for $n > 2$.

**Theorem 8.8**

> If $n > 2$, then $S_n$ is not abelian.

**PROOF** Let $\alpha$ and $\beta$ be defined as follows:

$$\alpha(1) = 1,\ \alpha(2) = 3,\ \alpha(3) = 2, \text{ and } \alpha(i) = i \text{ for all } i \geqslant 4.$$
$$\beta(1) = 3,\ \beta(2) = 2,\ \beta(3) = 1, \text{ and } \beta(i) = i \text{ for all } i \geqslant 4.$$

Then $(\beta \circ \alpha)(1) = 3$, but $(\alpha \circ \beta)(1) = 2$, so $\beta \circ \alpha \neq \alpha \circ \beta$. □

The symmetric group on three symbols, $S_3$, is given by the following table. As is customary, we suppress writing single cycles. For example, (12) is short for (12)(3) and (1) is short for (1)(2)(3).

| ∘ | (1) | (12) | (13) | (23) | (123) | (132) |
|---|---|---|---|---|---|---|
| (1) | (1) | (12) | (13) | (23) | (123) | (132) |
| (12) | (12) | (1) | (132) | (123) | (23) | (13) |
| (13) | (13) | (123) | (1) | (132) | (12) | (23) |
| (23) | (23) | (132) | (123) | (1) | (13) | (12) |
| (123) | (123) | (13) | (23) | (12) | (132) | (1) |
| (132) | (132) | (23) | (12) | (13) | (1) | (123) |

We can construct permutation groups using the properties of symmetry of certain geometric figures. For example, consider a square and all rigid motions of it into itself. Number the four corners of the square as shown in Figure 8.1.

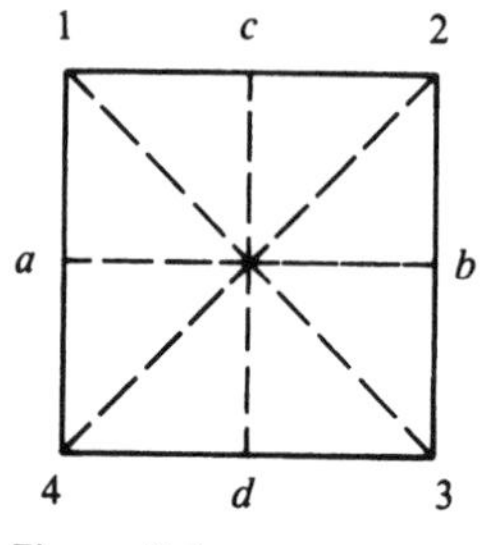

Figure 8.1

If we rotate the square 90° in a clockwise direction about its center, we can interpret this rotation as mapping 1 into 2, 2 into 3, 3 into 4, and 4 into 1. That is, this rotation is the permutation $\alpha = (1234)$.

If we rotate the square 180° in a clockwise direction about its center, we can interpret this rotation as mapping 1 into 3, 2 into 4, 3 into 1, and 4 into 2. This rotation is $\alpha \circ \alpha$, which we denote by $\alpha^2$.

If we rotate the square 270° in a clockwise direction about its center, we can interpret this rotation as mapping 1 into 4, 2 into 1, 3 into 2, and 4 into 3. This rotation is $\alpha \circ \alpha^2$, which we denote by $\alpha^3$.

If we rotate the square 360° in a clockwise direction about its center, we have the identity permutation $e$.

We also have rigid motions of the square that consist of rotations in space about a line of symmetry of the square. If we rotate the square 180° about the horizontal line $ab$, we have the permutation $\beta = (14)(23)$.

If we rotate the square 180° about the vertical line $cd$, we have the permutation $\gamma = (12)(34)$.

We also have two permutations that result from rotations of 180° about the diagonals of the square. These are $\delta = (13)(2)(4)$ and $\sigma = (1)(24)(3)$.

Let $D_4$ denote the set $\{e, \alpha, \alpha^2, \alpha^3, \beta, \gamma, \delta, \sigma\}$ with the operation of composition. The table for $D_4$ is given as follows:

| $\circ$ | $e$ | $\alpha$ | $\alpha^2$ | $\alpha^3$ | $\beta$ | $\gamma$ | $\delta$ | $\sigma$ |
|---|---|---|---|---|---|---|---|---|
| $e$ | $e$ | $\alpha$ | $\alpha^2$ | $\alpha^3$ | $\beta$ | $\gamma$ | $\delta$ | $\sigma$ |
| $\alpha$ | $\alpha$ | $\alpha^2$ | $\alpha^3$ | $e$ | $\sigma$ | $\delta$ | $\beta$ | $\gamma$ |
| $\alpha^2$ | $\alpha^2$ | $\alpha^3$ | $e$ | $\alpha$ | $\gamma$ | $\beta$ | $\sigma$ | $\delta$ |
| $\alpha^3$ | $\alpha^3$ | $e$ | $\alpha$ | $\alpha^2$ | $\delta$ | $\sigma$ | $\gamma$ | $\beta$ |
| $\beta$ | $\beta$ | $\delta$ | $\gamma$ | $\sigma$ | $e$ | $\alpha^2$ | $\alpha$ | $\alpha^3$ |
| $\gamma$ | $\gamma$ | $\sigma$ | $\beta$ | $\delta$ | $\alpha^2$ | $e$ | $\alpha^3$ | $\alpha$ |
| $\delta$ | $\delta$ | $\gamma$ | $\sigma$ | $\beta$ | $\alpha^3$ | $\alpha$ | $e$ | $\alpha^2$ |
| $\sigma$ | $\sigma$ | $\beta$ | $\delta$ | $\gamma$ | $\alpha$ | $\alpha^3$ | $\alpha^2$ | $e$ |

**Theorem 8.9**

$(D_4, \circ)$ is a group.

**PROOF** See Exercise 39. □

We can construct the group of rigid motions of other geometric figures (see, for example, Exercise 40).

## EXERCISES 8.4

**38.** **a)** Which elements of $S_3$ are their own inverses?

**b)** Which elements of $D_4$ are their own inverses?

**c)** How many elements of $S_3$ map 1 into 1?

**d)** How many elements of $S_n$ map 1 into 1?

**39.** Prove Theorem 8.9.

**40.** Construct the group of rigid motions of a rectangle that is not a square.

## 8.5 Subgroups

We now study subsets of a group that are themselves groups.

**Definition**

Let $(G, \circ)$ be a group, and let $H$ be a subset of $G$. If $\circ | H \times H$ is an operation on $H$ and $H$ is a group with respect to this operation, then $(H, \circ | H \times H)$ is called a **subgroup** of $G$.

Note that if $a$ and $b$ belong to $H$, then $a \circ b$ is $a \circ | H \times H\, b$. Obviously the notation $a \circ b$ is preferable.

If $(G, \circ)$ is a group and $e$ is the identity of $G$, then $G$ has two trivial subgroups: $\{e\}$ and $G$. Any other subgroup of $G$ is called a *proper subgroup* of $G$.

**EXAMPLE 18** The set of even integers is a subgroup of $(\mathbb{Z}, +)$. ❑

Note that if $(G, \circ)$ is a group, $H$ is a subgroup of $G$, and $a, b \in H$, then $a \circ b \in H$. That is, $H$ must be closed with respect to the operation $\circ$.

**EXAMPLE 19** The set $\{-1, 1\}$ is a subgroup of $(\mathbb{R}, \cdot)$. ❑

**Theorem 8.10**

Let $(G, \circ)$ be a group, and let $H$ be a subgroup of $G$.

**a)** The identity of $G$ is the identity of $H$.

**b)** If $a \in H$, then the inverse of $a$ in $G$ is the inverse of $a$ in $H$.

**PROOF**

**a)** Let $e$ denote the identity of $G$ and $e'$ the identity of $H$. If $(e')^{-1}$ denotes the inverse of $e'$ in $G$, then

$$(e')^{-1} \circ (e' \circ e') = (e')^{-1} \circ e'$$

$$[(e')^{-1} \circ e'] \circ e' = e$$

$$e \circ e' = e$$

$$e' = e$$

The proof of part (b) is left as an exercise. □

The following theorem provides a convenient method for deciding whether a subset of a group is a subgroup.

**Theorem 8.11**

Let $(G, \circ)$ be a group. A nonempty subset $H$ of $G$ is a subgroup of $G$ if and only if

**a)** $a, b \in H$ implies that $a \circ b \in H$, and

**b)** $a \in H$ implies that $a^{-1} \in H$.

**PROOF** See Exercise 45. □

**EXAMPLE 20** The set $\{(1), (123), (132)\}$ is a subgroup of $S_3$. The table for this subgroup is as follows:

| $\circ$ | (1) | (123) | (132) |
|---|---|---|---|
| (1) | (1) | (123) | (132) |
| (123) | (123) | (132) | (1) |
| (132) | (132) | (1) | (123) |

Once we have the table for the set $S = \{(1), (123), (132)\}$, it is easy to use Theorem 8.11 in order to see that $(S, \circ)$ is indeed a subgroup. ❑

**Theorem 8.12** Let $(G, \circ)$ be a group, and let $H$ denote a subgroup of $G$. If a relation $\sim$ is defined on $G$ by saying that $a \sim b$ if and only if $a \circ b^{-1} \in H$, then $\sim$ is an equivalence relation on $G$.

**PROOF** Let $a \in G$. Then $a \circ a^{-1} = e \in H$, so $a \sim a$. Thus $\sim$ is reflexive on $G$. Suppose that $a, b \in G$ and $a \sim b$. Then $a \circ b^{-1} \in H$. Hence, $(a \circ b^{-1})^{-1} \in H$, but, by Theorem 8.6(f), $(a \circ b^{-1})^{-1} = (b^{-1})^{-1} \circ a^{-1}$. Therefore, $b \circ a^{-1} \in H$ and hence $b \sim a$. So $\sim$ is symmetric.

Suppose that $a, b, c \in G$, $a \sim b$, and $b \sim c$. Then $a \circ b^{-1} \in H$ and $b \circ c^{-1} \in H$. So $(a \circ b^{-1}) \circ (b \circ c^{-1}) \in H$, but

$$(a \circ b^{-1}) \circ (b \circ c^{-1}) = a \circ [b^{-1} \circ (b \circ c^{-1})] = a \circ [(b^{-1} \circ b) \circ c^{-1}]$$
$$= a \circ [e \circ c^{-1}] = a \circ c^{-1}$$

Hence, $a \circ c^{-1} \in H$ and therefore $a \sim c$. Thus $\sim$ is transitive. Therefore, $\sim$ is an equivalence relation on $G$. □

**Definition** The equivalence classes that result from the equivalence relation defined in Theorem 8.12 are called the **right cosets of $H$ with respect to $G$.**

**EXAMPLE 21** If $H$ is the subgroup of $(\mathbb{Z}, +)$ consisting of the even integers, then there are two right cosets of $H$ with respect to $\mathbb{Z}$: (a) the set of even integers and (b) the set of odd integers. ❑

**Notation** If $H$ is a subgroup of a group $(G, \circ)$ and $a \in G$, let $H \circ a = \{h \circ a : h \in H\}$.

**EXAMPLE 22** If $H = \{[0], [4]\}$, then $H$ is a subgroup of $\mathbb{Z}_8$. The right cosets of $H$ with respect to $\mathbb{Z}_8$ are $H$, $H \oplus [1] = \{[1], [5]\}$, $H \oplus [2] = \{[2], [6]\}$, and $H \oplus [3] = \{[3], [7]\}$. ❑

**Theorem 8.13**

If $H$ is a subgroup of a group $(G, \circ)$ and $a, b \in G$, then the following conditions are equivalent:

**a)** $a \circ b^{-1} \in H$

**b)** $a = h \circ b$ for some $h \in H$

**c)** $a \in H \circ b$

**d)** $H \circ a = H \circ b$

**PROOF** We prove that (a)→(b)→(c)→(d)→(a). Let $a, b \in G$.

**(a)→(b)** Suppose that $a \circ b^{-1} \in H$. Then there exists $h \in H$ such that $a \circ b^{-1} = h$. Thus, $(a \circ b^{-1}) \circ b = h \circ b$. But $(a \circ b^{-1}) \circ b = a \circ (b^{-1} \circ b) = a \circ e = a$. Hence there is an element $h$ of $H$ such that $a = h \circ b$.

**(b)→(c)** Suppose that $a = h \circ b$ for some $h \in H$. Then $a \in H \circ b$.

**(c)→(d)** Suppose that $a \in H \circ b$. Then there exists $h \in H$ such that $a = h \circ b$. So $a \circ b^{-1} = (h \circ b) \circ b^{-1} = h \circ (b \circ b^{-1}) = h \circ e = h$. Therefore, $a \sim b$. Let $y \in H \circ a$. Then, using the same proof as $a \in H \circ b$ implies $a \sim b$, we can show that $y \sim a$. Since $\sim$ is transitive, $y \sim b$. Therefore, $y \circ b^{-1} \in H$. So there exists $k \in H$ such that $y \circ b^{-1} = k$. Thus, $(y \circ b^{-1}) \circ b = k \circ b$. But $(y \circ b^{-1}) \circ b = y \circ (b^{-1} \circ b) = y \circ e = y$. Therefore, $y \in H \circ b$ and hence $H \circ a \subseteq H \circ b$. The same proof shows that $H \circ b \subseteq H \circ a$.

**(d)→(a)** Suppose that $H \circ a = H \circ b$. Then there exist $h, k \in H$ such that $h \circ a = k \circ b$. Thus, $(h \circ a) \circ b^{-1} = (k \circ b) \circ b^{-1} = k \circ (b \circ b^{-1}) = k \circ e = k$. Therefore, $h^{-1} \circ [h \circ (a \circ b^{-1})] = h^{-1} \circ k$. But

$$h^{-1} \circ [h \circ (a \circ b^{-1})] = (h^{-1} \circ h) \circ (a \circ b^{-1}) = e \circ (a \circ b^{-1}) = a \circ b^{-1}$$

Since $H$ is a subgroup, $h^{-1} \circ k \in H$. Therefore, $a \circ b^{-1} \in H$.

□

In Example 22, $H$ has order 2, and each of the right cosets also has two elements. The following theorem generalizes this observation. In stating the theorem, we extend the notation $|G|$ to include sets, so that for a finite set $X$, $|X|$ will denote the number of members of $X$. If $Y$ is also a finite set, then $|X| = |Y|$ if and only if there is a one-to-one mapping of $X$ onto $Y$.

**Theorem 8.14**

If $H$ is a finite subgroup of a group $(G, \circ)$ and $a \in G$, then $|H| = |H \circ a|$.

**PROOF** See Exercise 46. □

We also have the notion of *left cosets*. Let $H$ be a subgroup of a group $(G, \circ)$, and define a relation $\sim$ on $G$ by $a \sim b$ if and only if $a^{-1} \circ b \in H$. Then $\sim$ is an equivalence relation on $G$, and the equivalence classes are called the *left cosets of $H$ with respect to $G$*. The left cosets are of the form $a \circ H = \{a \circ h: h \in H\}$. The analogue of Theorem 8.13 also holds.

The following theorem is useful in finding the subgroups of a finite group.

**Lagrange's Theorem**

If $H$ is a subgroup of a finite group $(G, \circ)$, then the order of $H$ is a divisor of the order of $G$.

**PROOF** Each right coset of $H$ with respect to $G$ is an equivalence class. Therefore, by Theorem 4.4, the right cosets of $H$ with respect to $G$ form a partition of $G$. So any two right cosets of $H$ with respect to $G$ are either equal or disjoint. Since $G$ is finite, there can be only finitely many of these cosets. Thus there exist distinct elements $a_1, a_2, \ldots, a_k$ of $G$ such that

$$G = H \circ a_1 \cup H \circ a_2 \cup \cdots \cup H \circ a_k \quad \text{and} \quad H \circ a_i \cap H \circ a_j = \varnothing \qquad \text{if } a_i \neq a_j$$

By Theorem 8.14, for each $i = 1, 2, \ldots, k$, $|H \circ a_i| = |H|$. Therefore, $|G| = k \cdot |H|$, and hence the order of $H$ is a divisor of the order of $G$. □

**Definition**

If $H$ is a subgroup of a finite group $(G, \circ)$, then the **index** of $H$ in $G$, denoted by $[G: H]$, is the number of right cosets of $H$ in $G$.

Lagrange's Theorem simplifies the problem of determining all the subgroups of a finite group. For example, if $(G, \circ)$ is a group of order 15, then the order of any subgroup of $(G, \circ)$ is either 1, 3, 5, or 15.

## EXERCISES 8.5

**41.** In each case, decide whether the given subset is a subgroup of $S_5$. Justify your answer.

**a)** $\{(1), (13)(45)\}$ **b)** $\{(1), (234), (345)\}$ **c)** $\{(1), (145), (154)\}$

**42.** Find all the subgroups of $S_3$.

**43.** Find a proper subgroup of $(\mathbb{Q}, +)$ that contains $\mathbb{Z}$ but is different from $\mathbb{Z}$.

**44. a)** What, if anything, is wrong with the following proof of Theorem 8.10(a)? Since $e$ and $e'$ are both identities, $e \circ e' = e'$ and $e \circ e' = e$. Therefore, $e = e'$.

**b)** Prove Theorem 8.10(b).

**45.** Prove Theorem 8.11.

**46.** Prove that if $H$ is a finite subgroup of a group $(G, \circ)$ and $a \in G$, then $|H| = |H \circ a|$.

**47.** Prove that if $H$ and $K$ are subgroups of a group $(G, \circ)$, then $H \cap K$ is a subgroup of $(G, \circ)$.

48. Find a group $(G, \circ)$ and subgroups $H$ and $K$ of $(G, \circ)$ such that $H \cup K$ is not a subgroup of $(G, \circ)$.
49. If $H$ is a subgroup of a finite group $(G, \circ)$, show that the number of right cosets of $H$ with respect to $G$ equals the number of left cosets with $H$ with respect to $G$.
50. If $H$ is a subgroup of a finite group $(G, \circ)$ of index 2, show that $a \circ H = H \circ a$ for every $a \in G$.
51. If $H$ is a nonempty finite subset of a group $(G, \circ)$ that is closed with respect to $\circ$, show that $H$ is a subgroup of $(G, \circ)$.
52. Find all left and right cosets of $\{(1), (23)\}$ with respect to $S_3$.
53. Let $(G, \circ)$ be an abelian group. Show that the set $\{x \in G: x \circ x = e\}$ is a subgroup of $(G, \circ)$.
54. If $H$ is a subgroup of $(\mathbb{Z}, +)$, show that there is a nonnegative integer $n$ such that $H = \{na: a \in \mathbb{Z}\}$.
55. Let $S$ be a nonempty set, let $a \in S$, and let $H = \{f \in \text{Sym}(S): f(a) = a\}$. Prove that $(H, \circ)$ is a subgroup of $(\text{Sym}(S), \circ)$.
56. Let $(G, \circ)$ be a group, and let $H$ be a nonempty finite subset of $G$ such that $\circ$ is an operation on $H$. Prove that $(H, \circ)$ is a subgroup of $(G, \circ)$.
57. Let $X$ be a nonempty finite subset of a set $S$, and let $H = \{f \in \text{Sym}(S): f(X) \subseteq X\}$. Prove that $(H, \circ)$ is a subgroup of $(G, \circ)$.
58. Let $(H, \circ)$ and $(K, \circ)$ be subgroups of an abelian group $(G, \circ)$, and let $HK = \{h \circ k: h \in H \text{ and } k \in K\}$. Prove that $(HK, \circ)$ is a subgroup of $(G, \circ)$.

## 8.6 Homomorphisms and Isomorphisms

Let $(\mathbb{Z}, +)$ denote the group of integers in which $+$ is the usual operation of addition. Let $(G, \circ)$ denote the group given by the following table:

| $\circ$ | $a$ | $b$ |
|---|---|---|
| $a$ | $a$ | $b$ |
| $b$ | $b$ | $a$ |

Define a function $h: \mathbb{Z} \to G$ by $h(n) = a$ if $n$ is even and $h(n) = b$ if $n$ is odd. It can be verified that $h(m + n) = h(m) \circ h(n)$ for all $m, n \in \mathbb{Z}$. This fact is often expressed by saying that the function $h$ preserves the operation. A function that preserves the operation is called a *homomorphism*. The precise definition follows.

**Definition**

If $(G, \circ)$ and $(H, *)$ are groups, then a function $h: G \to H$ is a **homomorphism** if $h(a \circ b) = h(a) * h(b)$ for all $a, b \in G$.

**EXAMPLE 23**

If $(\mathbb{Z}, +)$ denotes the group of integers with the usual operation of addition, then the function $h: \mathbb{Z} \to \mathbb{Z}$ defined by $h(n) = 2n$ for each $n \in \mathbb{Z}$ is a homomorphism, since $h(m + n) = 2(m + n) = 2m + 2n = h(m) + h(n)$ for all $m, n \in \mathbb{Z}$. ❑

**EXAMPLE 24** If $(\mathbb{Z}_n, \oplus)$ denotes the group defined in Section 8.3 (the integers modulo $n$ under addition), then the function $h: \mathbb{Z} \to \mathbb{Z}_n$ defined by $h(n) = [n]$ for each $n \in \mathbb{Z}$ is a homomorphism, since $h(m+n) = [m+n] = [m] \oplus [n] = h(m) \oplus h(n)$. ❑

**EXAMPLE 25** The function $h: \mathbb{Z} \to \mathbb{Z}$ defined by $h(n) = 0$ for each $n \in \mathbb{Z}$ is a homomorphism. ❑

It should be observed that the homomorphism given in Example 23 is a one-to-one function that does not map $\mathbb{Z}$ onto $\mathbb{Z}$. The homomorphism given in Example 24 maps $\mathbb{Z}$ onto $\mathbb{Z}_n$, but it is not one-to-one. The homomorphism given in Example 25 is not one-to-one, and it does not map $\mathbb{Z}$ onto $\mathbb{Z}$.

**Definition** If $(G, \circ)$ and $(H, *)$ are groups, then a homomorphism of $G$ onto $H$ that is also one-to-one is called an **isomorphism**, and we say that the two groups are isomorphic.

The notion of isomorphism allows us to treat certain groups as being alike. Consider the subgroup $H$ of $S_3$ given in Example 20 and $\mathbb{Z}_3$. The function $h: H \to \mathbb{Z}_3$ defined by $h((1)) = [0]$, $h((123)) = [1]$, and $h((132)) = [2]$ is a homomorphism that is a one-to-one function from $H$ onto $\mathbb{Z}_3$. So, even though the underlying sets and operations arise in different ways, the groups are isomorphic, and we could complete one table just by knowing the other.

However, it is somewhat surprising to observe that the set $E$ of even integers is a subgroup of $(\mathbb{Z}, +)$ and that the homomorphism given in Example 23 (now considered as a function from $\mathbb{Z}$ to $E$) is an isomorphism. This shows that if we consider these sets as groups, with the usual operation of addition, there will be no essential difference between them.

Let $f$ be the mapping from the additive group of real numbers into the multiplicative group of positive real numbers defined by $f(x) = e^x$. Since

$$f(x+y) = e^{x+y} = e^x e^y = f(x)f(y)$$

$f$ is a homomorphism. By Theorem 1.10, $f$ is one-to-one. If $y$ is any positive real number, then $\ln y$ is a real number such that $e^{\ln y} = y$. Hence $f$ maps onto the multiplicative group of positive reals. Therefore, $f$ is an isomorphism.

How many groups are there of order 3? Consider two groups to be the same if they are isomorphic. We have already seen that one subgroup of order 3 of $S_3$ is the same as $\mathbb{Z}_3$. Is there a group of order 3 that is different from $\mathbb{Z}_3$? Before attempting to answer this question, let us prove the following theorem.

**Theorem 8.15**

> Let $(G, \circ)$ and $(H, *)$ be groups, and let $h: G \to H$ be a homomorphism. If $e_G$ and $e_H$ denote the identity elements of $G$ and $H$, respectively, then $h(e_G) = e_H$.

**PROOF** Since $h$ is a homomorphism and $e_G$ is the identity of $G$, $h(e_G)*h(e_G) = h(e_G \circ e_G) = h(e_G)$. Since $h(e_G) \in H$ and $e_H$ is the identity of $H$, $h(e_G)*e_H = h(e_G)$. Therefore,

$$h(e_G)*h(e_G) = h(e_G)*e_H$$

Hence, by Theorem 8.6(b), $h(e_G) = e_H$. □

Now, let $(G, \circ)$ be any group of order 3. We know that there is an identity element $e$ of $G$. Let $a$ and $b$ denote the other two elements of $G$. Is there an isomorphsim $h: G \to \mathbb{Z}_3$? We know that any homomorphism from $G$ into $\mathbb{Z}_3$ must map $e$ into [0]. By Theorem 8.6, $a \circ b = e$, and it follows that the table for $G$ is

| $\circ$ | $e$ | $a$ | $b$ |
|---|---|---|---|
| $e$ | $e$ | $a$ | $b$ |
| $a$ | $a$ | $b$ | $e$ |
| $b$ | $b$ | $e$ | $a$ |

So the function $h: G \to \mathbb{Z}_3$ defined by $h(e) = [0]$, $h(a) = [1]$, and $h(b) = [2]$ is an isomorphism. Therefore, there is only one group of order 3.

**Theorem 8.16**

> Let $(G, \circ)$ and $(H, *)$ be groups, and let $h: G \to H$ be a homomorphism. If $a \in G$, then $h(a^{-1}) = (h(a))^{-1}$.

**PROOF** See Exercise 61. □

**Theorem 8.17**

> Let $(G, \circ)$ and $(H, *)$ be groups, and let $h: G \to H$ be a homomorphism. If $K$ is a subgroup of $G$, then $h(K)$ is a subgroup of $H$.

**PROOF** See Exercise 62. □

**Theorem 8.18**

> If $(G, \circ)$ and $(H, *)$ are isomorphic groups and $(G, \circ)$ is abelian, then so is $(H, *)$.

**PROOF** See Exercise 63. □

**Theorem 8.19**

Let $(G, \circ)$ and $(H, *)$ be groups, and let $h: G \to H$ be a homomorphism that is also one-to-one. Then $(G, \circ)$ and $(h(G), *)$ are isomorphic.

**PROOF** The function $k: G \to h(G)$ defined by $k(g) = h(g)$ for each $g \in G$ is an isomorphism. □

**EXAMPLE 26**

Let $H$ denote the group of permutations on $\mathbb{Z}_4$ and define $h: \mathbb{Z}_4 \to H$ by $h([0]) = i_{\mathbb{Z}_4}$, $h([1]) = ([0]\,[1]\,[2]\,[3])$, $h([2]) = ([0]\,[2])([1]\,[3])$, and $h([3]) = ([0]\,[3]\,[2]\,[1])$. Then the function $k: \mathbb{Z}_4 \to h(H)$ defined by $k(g) = h(g)$ for each $g \in \mathbb{Z}_4$ is an isomorphism (see Exercise 71). ❑

The function $h$ in Example 26 is not defined randomly. It was defined by looking at the table for $\mathbb{Z}_4$. The function $h$ maps the element $[a]$ (for each $[a] \in \mathbb{Z}_4$) into the permutation that maps $[b]$ (for each $[b] \in \mathbb{Z}_4$) into the entry in the $[a]$th row and $[b]$th column.

Example 26 shows that $(\mathbb{Z}_4, \oplus)$ is isomorphic to a subgroup of the group of permutations on $\mathbb{Z}_4$. The following theorem, known as Cayley's Theorem, asserts that every group is isomorphic to a group of permutations.

**Cayley's Theorem**

Let $(G, *)$ be a group. Then $(G, *)$ is isomorphic to a subgroup of the group of permutations on $G$.

**PROOF** For each $a \in G$, define $\lambda_a: G \to G$ by $\lambda_a(x) = a * x$ for each $x \in G$. By Theorem 8.6(e), for each $b \in G$ the equation $a * x = b$ has a unique solution in $G$. Therefore, $\lambda_a \in \text{Sym}\,(G)$ for each $a \in G$.

Define $h: G \to \text{Sym}\,(G)$ by $h(a) = \lambda_a$ for each $a \in G$. To prove that $h$ is a homomorphism, let $a, b \in G$. Then $h(a * b) = \lambda_{a*b}$ and $h(a) \circ h(b) = \lambda_a \circ \lambda_b$, so we need to show that $\lambda_{a*b} = \lambda_a \circ \lambda_b$. Let $x \in G$. Then

$$\lambda_{a*b}(x) = (a*b)*x = a*(b*x) = \lambda_a(b*x) = \lambda_a(\lambda_b(x)) = (\lambda_a \circ \lambda_b)(x)$$

Therefore, $h$ is a homomorphism.

To prove that $h$ is one-to-one, let $e$ denote the identity of $G$ and suppose that $a, b \in G$ and $h(a) = h(b)$. Then $\lambda_a = \lambda_b$, and in particular $\lambda_a(e) = \lambda_b(e)$. But $\lambda_a(e) = a * e = a$ and $\lambda_b(e) = b * e = b$, so $a = b$. Therefore, $h$ is one-to-one.

Thus by Theorem 8.18, $(G, *)$ and $(h(G), \circ)$ are isomorphic. □

**Definition**

Let $(G, \circ)$ and $(H, *)$ be groups, and let $h: G \to H$ be a homomorphism. The **kernel** of $h$, denoted by $\ker\,(h)$, is $\{a \in G: h(a) = e_H\}$.

**EXAMPLE 27** Let $h: \mathbb{Z} \to \mathbb{Z}_5$ be the homomorphism defined by $h(n) = [n]$ for each $n \in \mathbb{Z}$. Then $h$ is a homomorphism and ker $(h)$ is the set of all multiples of 5 (see Exercise 73). ❑

**Theorem 8.20** Let $(G, \circ)$ and $(H, *)$ be groups, and let $h: G \to H$ be a homomorphism. Then ker $(h)$ is a subgroup of $G$.

**PROOF** See Exercise 74. □

**Definition** Let $(G, \circ)$ be a group. A subgroup $(N, \circ)$ of $(G, \circ)$ is **normal** provided that $g \circ a \circ g^{-1} \in N$ for all $a \in N$ and all $g \in G$.

**Theorem 8.21** Let $(H, \circ)$ be a subgroup of an abelian group $(G, \circ)$. Then $(H, \circ)$ is normal.

**PROOF** See Exercise 76. □

In Exercise 77, you are asked to find a subgroup of $(S_3, \circ)$ that is not normal.

**Theorem 8.22** Let $(G, \circ)$ and $(H, *)$ be groups, and let $h: G \to H$ be a homomorphism. Then (ker $(h), \circ)$ is a normal subgroup of $(G, \circ)$.

**PROOF** See Exercise 78. □

## EXERCISES 8.6

**59. a)** Complete the following table to obtain a group that is isomorphic to $\mathbb{Z}_4$.

| $*$ | $a$ | $b$ | $c$ | $d$ |
|---|---|---|---|---|
| $a$ | | | | |
| $b$ | | | | |
| $c$ | | | | |
| $d$ | | | | |

**b)** What is an isomorphism?

**60. a)** Complete the following table to obtain a group that is not isomorphic to $\mathbb{Z}_4$.

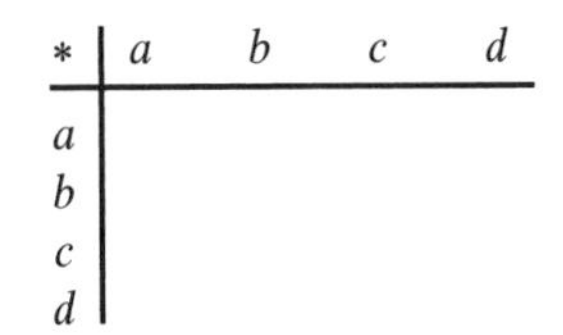

**b)** Prove that the two groups are not isomorphic.

**61.** Prove Theorem 8.16.

**62.** Let $(G, \circ)$ and $(H, *)$ be groups, and let $K$ be a subgroup of $G$. Prove that if $h: G \to H$ is a homomorphism, then $h(K)$ is a subgroup of $H$.

**63.** Prove that if $(G, \circ)$ is isomorphic to an abelian group, then $(G, \circ)$ is abelian.

**64.** Define an isomorphism between the following two groups:

| $\circ$ | $a$ | $b$ | $c$ |
|---|---|---|---|
| $a$ | $a$ | $b$ | $c$ |
| $b$ | $b$ | $c$ | $a$ |
| $c$ | $c$ | $a$ | $b$ |

| $*$ | $d$ | $e$ | $f$ |
|---|---|---|---|
| $d$ | $f$ | $d$ | $e$ |
| $e$ | $d$ | $e$ | $f$ |
| $f$ | $e$ | $f$ | $d$ |

**65.** Let $G$ be the set of all rational numbers of the form $2^m$, where $m$ is an integer, and let $\times$ denote usual multiplication. Prove that $(G, \times)$ is isomorphic to $(\mathbb{Z}, +)$.

**66.** Is $\mathbb{Z}_6$ isomorphic to $S_3$? Prove your answer.

**67.** Let $g$ be an element of a group $(G, \circ)$. Show that the function $f: G \to G$ defined by $f(x) = g^{-1} \circ x \circ g$ is an isomorphism.

**68.** Let $(G, \circ)$ be a group such that the mapping $f$ from $G$ into $G$ defined by $f(a) = a^{-1}$ is a homomorphism. Show that $(G, \circ)$ is abelian.

**69.** Let $(G, \circ)$ be an abelian group. Prove that the function $f: G \to G$ defined by $f(a) = a^{-1}$ for each $a \in G$ is an isomorphism.

**70.** Let $(G, *)$ be a group. Let $\hat{G}$ be the set of all isomorphisms on $G$, and let $\circ$ denote the usual composition of functions.

**a)** Prove that $(\hat{G}, \circ)$ is a group.

**b)** Let $S = \{f \in \hat{G}$: there is a $g \in G$ such that $f(x) = g^{-1} * x * g$ for each $x \in G\}$. Prove that $(S, \circ)$ is a subgroup of $(\hat{G}, \circ)$.

**71.** Prove that the function $k$ in Example 26 is an isomorphism.

**72.** Write the permutation associated with each element of $\mathbb{Z}_5$ by the function $h$ in the proof of Cayley's Theorem.

**73.** Prove that the function $h$ in Example 27 is a homomorphism and that its kernel is the set of all multiples of 5.

**74.** Let $(G, \circ)$ and $(H, *)$ be groups, and let $h: G \to H$ be a homomorphism. Prove that $\ker(h)$ is a subgroup of $G$.

**75.** Let $n, k \in \mathbb{N}$ and define $h: \mathbb{Z} \to \mathbb{Z}_n$ by $h(a) = [ka]$ for each $a \in G$.

**a)** Prove that $h$ is a homomorphism.

**b)** Describe $\ker(h)$.

**76.** Prove that every subgroup of an abelian group is normal.

77. Find a subgroup of $(S_3, \circ)$ that is not normal.
78. Let $(G, \circ)$ and $(H, *)$ be groups, and let $h: G \to H$ be a homomorphism. Prove that $(\ker(h), \circ)$ is a normal subgroup of $(G, \circ)$.
79. Let $(G, \circ)$ and $(H, *)$ be groups, let $h: G \to H$ be a homomorphism of $G$ onto $H$, and let $(N, \circ)$ be a normal subgroup of $(G, \circ)$. Prove that $(h(N), *)$ is a normal subgroup of $(H, *)$.
80. Let $\mathscr{A}$ denote a nonempty collection of normal subgroups of a group $(G, \circ)$. Prove that $(\cap_{N \in \mathscr{A}} N, \circ)$ is a normal subgroup of $(G, \circ)$.
81. Let $(H, \circ)$ be a subgroup of a group $(G, \circ)$. Prove that $H$ is normal if and only if $g \circ H = H \circ g$ for each $g \in G$.

## 8.7 Cyclic Groups

If $(G, \circ)$ is a group, $e$ is the identity element of $G$, $n \in \mathbb{N}$, and $a \in G$, then the integral powers of $a$ are defined inductively as follows:

$$a^0 = e,\ a^1 = a,\ a^2 = a \circ a, \ldots,\ a^{n+1} = a \circ a^n,\ a^{-n} = (a^{-1})^n$$

Observe that if $n \in \mathbb{Z}$ and $n < 0$, then $(a^{-1})^n = [(a^{-1})^{-1}]^{-n} = a^{-n}$. Therefore, $a^{-n} = (a^{-1})^n$ for all $n \in \mathbb{Z}$.

**Theorem 8.23**

If $(G, \circ)$ is a group, $m, n \in \mathbb{Z}$, and $a \in G$, then

**a)** $a^m \circ a^n = a^{m+n}$

**b)** $(a^m)^n = a^{mn}$

**PROOF** We prove only part (a). The proof of part (b), which is similar, is left as Exercise 82.

Let

$$S = \{m \in \mathbb{N}: \text{if } n \in \mathbb{Z} \text{ then } a^m \circ a^n = a^{m+n}\}$$

We first show that $1 \in S$.

***Case I.*** $n > 0$. Then $a^1 \circ a^n = a \circ a^n = a^{n+1} = a^{1+n}$.

***Case II.*** $n = 0$. Then $a^1 \circ a^0 = a \circ e = a = a^1 = a^{1+0}$.

***Case III.*** $n = -1$. Then $a^1 \circ a^{-1} = a \circ a^{-1} = e = a^{1+(-1)}$.

***Case IV.*** $n \leqslant -2$. Then $-n - 1 \geqslant 1$. So

$$\begin{aligned} a^1 \circ a^n &= a \circ (a^{-1})^{-n} \\ &= a \circ (a^{-1})^{-n-1+1} \end{aligned}$$

$$
\begin{aligned}
&= a \circ [(a^{-1}) \circ (a^{-1})^{-n-1}] \\
&= (a \circ a^{-1}) \circ (a^{-1})^{-n-1} \\
&= e \circ (a^{-1})^{-n-1} \\
&= (a^{-1})^{-(n+1)} \\
&= a^{n+1} \\
&= a^{1+n}
\end{aligned}
$$

Therefore, $1 \in S$.

Suppose that $m \in S$. Then $a^m \circ a^n = a^{m+n}$ for all $n \in \mathbb{Z}$. So

$$
\begin{aligned}
a^{m+1} \circ a^n &= (a \circ a^m) \circ a^n \\
&= a \circ (a^m \circ a^n) \\
&= a \circ a^{m+n} \\
&= a^1 \circ a^{m+n} \\
&= a^{1+m+n} \\
&= a^{m+1+n}
\end{aligned}
$$

Therefore, $m + 1 \in S$.

By the Principle of Mathematical Induction, $S = \mathbb{N}$. We have proved that $a^m \circ a^n = a^{m+n}$ for all $m \in \mathbb{N}$ and all $n \in \mathbb{Z}$.

Suppose that $m = 0$ and $n \in \mathbb{Z}$. Then

$$a^0 \circ a^n = e \circ a^n = a^n = a^{0+n}$$

Now suppose that $m < 0$. Then for all $n \in \mathbb{Z}$,

$$
\begin{aligned}
a^m \circ a^n &= (a^{-1})^{-m} \circ a^n \\
&= (a^{-1})^{-m} \circ [(a^{-1})^{-1}]^n \\
&= (a^{-1})^{-m} \circ (a^{-1})^{-n} \\
&= (a^{-1})^{-m-n} \qquad \text{(since } -m > 0\text{)} \\
&= (a^{-1})^{-(m+n)} \\
&= a^{m+n}
\end{aligned}
$$

We have now proved that $a^m \circ a^n = a^{m+n}$ for all $m, n \in \mathbb{Z}$. This completes the proof of part (a). □

If $(G, \circ)$ is a group and $a \in G$, let $\langle a \rangle = \{a^n : n \in \mathbb{Z}\}$.

**Theorem 8.24**

If $(G, \circ)$ is a group and $a \in G$, then $\langle a \rangle$ is a subgroup of $G$.

**PROOF** See Exercise 83. □

**Definition**

If $(G, \circ)$ is a group and $a \in G$, the subgroup $\langle a \rangle$ is called the **subgroup generated by *a***. If $H$ is a subgroup of $G$ and there exists $a \in G$ such that $H = \langle a \rangle$, then $H$ is a **cyclic subgroup** and $a$ is a **generator of *H***.

Note that $(\mathbb{Z}, +)$ is cyclic, since $\mathbb{Z} = \langle 1 \rangle$ (or $\mathbb{Z} = \langle -1 \rangle$).

**EXAMPLE 28** Observe that $(\mathbb{Z}_5 - \{[0]\}, \odot)$ is a group. This group is cyclic since it is generated by $[2]$. ❑

**EXAMPLE 29** The group $(\mathbb{Z}_{10}, \oplus)$ is generated by $[1]$. Is there an element of $\mathbb{Z}_{10}$ (different from $[1]$) that generates $\mathbb{Z}_{10}$? See Exercise 84. ❑

**Definition**

If $(G, \circ)$ is a group and $a \in G$, then the smallest positive integer $n$ such that $a^n = e$, if it exists, is called the **order of *a***, and it is denoted by $o(a)$. If there is no such integer, then $a$ has **infinite order**.

In Example 28, the order of $[2]$ is 4.

It would be natural to hope that the order of an element of a group would be the same as the order of the cyclic group generated by that element. The desired result is an immediate corollary of part (d) of the following theorem.

**Theorem 8.25**

If $(G, \circ)$ is a group, $e$ is the identity of $G$, $a \in G$, and there exist distinct integers $r$ and $s$ such that $a^r = a^s$, then

**a)** There is a smallest natural number $n$ such that $a^n = e$.

**b)** For each integer $m$, $a^m = e$ if and only if $n$ is a divisor of $m$.

**c)** If $t, u \in \mathbb{Z}$, $0 \leqslant t \leqslant n-1$, and $0 \leqslant u \leqslant n-1$, then $a^t \neq a^u$.

**d)** $\langle a \rangle = \{a^t : t \in \mathbb{Z} \text{ and } 0 \leqslant t \leqslant n-1\}$.

**PROOF**

**a)** We can assume that $r > s$. Since $a^r = a^s$, $a^{r-s} = a^r \circ a^{-s} = a^s \circ a^{-s} = a^0 = e$. Thus there is a positive integer $r - s$ such that $a^{r-s} = e$. By the Least-Natural-Number Principle, there is a smallest such integer $n$.

**b)** Suppose that $n$ is a divisor of $m$. Then there is an integer $q$ such that $m = nq$. Hence $a^m = a^{nq} = (a^n)^q = e^q = e$. Suppose that $a^m = e$. By the Division Algorithm, there are integers $q$ and $r$ such that $m = nq + r$, where $0 \leqslant r < n$. Hence, $e = a^m = a^{nq+r} = a^{nq} \circ a^r = (a^n)^q \circ a^r = e^q \circ a^r = e \circ a^r = a^r$. Therefore, since $0 \leqslant r < n$ and $n$ is the smallest positive integer such that $a^n = e$, $r = 0$. So $m = nq$ and $n$ is a divisor of $m$.

**c)** Suppose that there exist $t, u \in \mathbb{Z}$ with $0 \leqslant t \leqslant n-1$ and $0 \leqslant u \leqslant n-1$ such that $a^t = a^u$. We can assume that $t \geqslant u$. Then $a^{t-u} = a^t \circ a^{-u} = a^u \circ a^{-u} = a^0 = e$. By part (b), $n$ is a divisor of $t - u$. But $t - u$ is less than $n$, and therefore $t - u = 0$. So $t = u$.

**d)** By part (c), $\{a^t : t \in \mathbb{Z} \text{ and } 0 \leqslant t \leqslant n-1\}$ consists of $n$ distinct elements. In order to complete the proof, it is sufficient to show that if $m$ is an integer, then $a^m \in \{a^t : t \in \mathbb{Z} \text{ and } 0 \leqslant t \leqslant n-1\}$. By the Division Algorithm, there are integers $q$ and $r$ such that $m = nq + r$, where $0 \leqslant r < n$. Therefore, $a^m = a^{nq+r} = a^{nq} \circ a^r = (a^n)^q \circ a^r = e^q \circ a^r = e \circ a^r = a^r$. □

**Theorem 8.26**

Every group of prime order is cyclic.

**PROOF** See Exercise 87. □

If $(G, \circ)$ is a group, $a \in G$, $H = \langle a \rangle$, and $b \in G$ such that $H = \langle b \rangle$, then it is clear that $o(a) = o(b)$.

**Theorem 8.27**

Every cyclic group of infinite order is isomorphic to $(\mathbb{Z}, +)$.

**PROOF** See Exercise 89. □

**Theorem 8.28**

If $n \in \mathbb{N}$, then every cyclic group of order $n$ is isomorphic to $(\mathbb{Z}_n, \oplus)$.

**PROOF** Let $(G, \circ)$ be a cyclic group of order $n$, and let $a$ be a generator of $G$. Then $G = \{a^t : t \in G \text{ and } 0 \leqslant t \leqslant n-1\}$. Define $f: (G, \circ) \to (\mathbb{Z}_n, \oplus)$ by $f(a^m) = [m]$. Since $f(a^m \circ a^p) = f(a^{m+p}) = [m+p] = [m] \oplus [p] = f(a^m) \oplus f(a^p)$, $f$ is a homomorphism. Suppose that $f(a^m) = f(a^p)$. Then $[m] = [p]$. Since $0 \leqslant m \leqslant n-1$ and $0 \leqslant p \leqslant n-1$, $m = p$. Therefore, $a^m = a^p$ and hence $f$ is one-to-one. Suppose that $[m] \in \mathbb{Z}_n$. We can assume that $0 \leqslant m \leqslant n-1$ and hence $f(a^m) = [m]$. Therefore, $f$ maps $G$ onto $\mathbb{Z}_n$. □

**Corollary 8.29**

Every cyclic group is abelian.

**Theorem 8.30**

Every subgroup of a cyclic group is itself a cyclic group.

**PROOF** Let $(G, \circ)$ be a cyclic group with generator $a$, and let $H$ be a subgroup of $G$. If $H = \{e\}$, then $H$ is cyclic. Suppose that $H \neq \{e\}$. By the Least-Natural-Number Principle, there is a smallest positive integer $n$ such that $a^n \in H$. Suppose that $a^m \in H$. By the Division Algorithm, there are integers $q$ and $r$ such that $m = nq + r$, where $0 \leqslant r < n$. Now $a^m = a^{nq+r} = a^{nq} \circ a^r = (a^n)^q \circ a^r$. Clearly, $(a^n)^q \in H$, and therefore $[(a^n)^q]^{-1} \in H$. Consequently,

$$a^r = e \circ a^r = \{[(a^n)^q]^{-1} \circ (a^n)^q\} \circ a^r$$

$$= [(a^n)^q]^{-1} \circ [(a^n)^q \circ a^r] \in H$$

Therefore, $r = 0$ and hence $m$ is a multiple of $n$. So $H = \langle a^n \rangle$. □

## EXERCISES 8.7

**82.** Prove Theorem 8.23(b).

**83.** Prove Theorem 8.24.

**84.** Either find an element of $\mathbb{Z}_{10}$ different from $[1]$ that generates $\mathbb{Z}_{10}$ or prove that no such element exists.

**85.** Find the subgroup of $S_3$ generated by $(123)$.

**86.** Prove that every group of prime order is cyclic.

**87.** If $(G, \circ)$ is a group and $a \in G$, prove that $o(a) = |\langle a \rangle|$.

**88.** Prove that every cyclic group of infinite order is isomorphic to $(\mathbb{Z}, +)$.

**89.** Prove that if $n \in \mathbb{N}$, $(G, \circ)$ is a cyclic group of order $n$, and $k \in \mathbb{N}$ such that $k$ divides $n$, then there exists a subgroup of $G$ of order $k$.

**90.** Let $k, n \in \mathbb{N}$ and $(G, \circ)$ be a cyclic group of order $n$ with generator $a$. Prove that $G = \langle a^k \rangle$ if and only if $\gcd(k, n) = 1$.

**91.** Find all subgroups of $(\mathbb{Z}_{18}, \oplus)$.

**92.** Let $(G, \circ)$ be a group, $e$ be the identity element of $G$, and $a, b \in G$ such that $a$, $b$, and $a \circ b$ have order 2. Show that

**a)** $a \circ b = b \circ a$

**b)** $\{e, a, b, a \circ b\}$ is a subgroup of $G$.

**93.** Show that $\mathbb{Z}_m \times \mathbb{Z}_n$ is cyclic if and only if $\gcd(m, n) = 1$.

**94.** Give an example to show that an infinite group need not be cyclic.

**95.** Let $(G, \circ)$ be an abelian group, and let $a, b \in G$. Prove that for each integer $n$, $(a \circ b)^n = a^n \circ b^n$.

**96.** Let $(G, \circ)$ be a group such that $(a \circ b)^2 = a^2 \circ b^2$ for all $a, b \in G$. Prove that $(G, \circ)$ is abelian.

**97.** Let $(G, \circ)$ be a finite group. Show that there is a natural number $n$ such that $a^n = e$ for all $a \in G$.

**98.** Let $(G, \circ)$ be a finite group of even order. Prove that there is a member $a$ of $G$ such that $a \neq e$ and $a^2 = e$.

## WRITING EXERCISES

1. On page 250, we considered a square and all rigid motions of it into itself and obtained a group $(D_4, \circ)$. With pictures and words, describe all rigid motions of an equilateral triangle into itself and show that you have a group. Is the group abelian? What is the order of the group?

2. Let $a$ and $n$ be natural numbers and let $d = \gcd(a, n)$. Show that the equation $ax \equiv 1 \pmod{n}$ has a solution if and only if $d = 1$. Thus it follows that a natural number $a$ has a multiplicative inverse modulo $n$ if and only if $a$ and $n$ are relatively prime. For each natural number $n > 1$, we define $U(n)$ to be the set of all natural numbers less than $n$ and relatively prime to $n$. Show that if $\odot_n$ denotes multiplication modulo $n$, then $(U(n), \odot_n)$ is a group. Explain how to use the Euclidean algorithm to find the inverse of an arbitrary member of $U(17)$.

CHAPTER

# 9

# FOUNDATIONS OF ADVANCED CALCULUS

Because they were the first to understand its fundamental theorem, Isaac Newton (1642–1727) and Gottfried Wilhelm Leibniz (1646–1716) are credited with the discovery of calculus. Their calculus was "infinitesmal" calculus. That is, it depended upon the intuitive notion of an arbitrarily small but positive number, which they called an "infinitesmal," but which Bishop Berkeley called "the ghost of a departed quantity." Newton, in some of the murkiest mathematics ever written, tried to defend his concept of an infinitesmal against Berkeley's attacks, but Newton never came up with anything more sound than Berkeley's ironic definition. Leibniz, for his part, admitted that the notion of "infinitesmal" could not be defended, but he believed that one day a system of logic would be developed in which the concept could be given a rigorous interpretation. Indeed, around the middle of the 20th century, Abraham Robinson developed nonstandard analysis and vindicated Leibniz's position. But the notion of a limit, due to Augustin Cauchy (1789–1857), can be used to replace the notion of an infinitesmal in developing calculus, and it is the use of Cauchy's limit that we study in this chapter. Thus, the chapter is not really about calculus, about which the reader knows far more than we cover here; rather, the chapter is about the use of limits to describe calculus. The challenge is to learn how to use limits to provide a solid foundation to calculus, without losing the intuition, which for all of us is secretly based on the ghosts of departed quantities.

Personal Note

## Isaac Newton (1642–1727) and Gottfried Wilhelm Leibniz (1646–1716)

Newton and Leibniz are generally considered to be the cofounders of calculus. Of course, many others contributed to the groundwork of this subject, and as Newton himself put it, "If I have seen farther than others, it is because I have stood on the shoulders of giants." Still, given the preeminence of Newton and Leibniz, it is disconcerting to see how differently their lives turned out. Honors flowed to Newton, who was knighted by Queen Anne in 1705 and was buried in Westminster Abbey with such pride, pomp, and circumstance that Voltaire remarked, "I have seen a professor of mathematics buried like a king only because he was great in his vocation." The last years of Leibniz's life were embittered by accusations that Leibniz had plagiarized Newton's early work on calculus. Naturally, those in continental Europe backed Leibniz while the English backed Newton. When Queen Anne died childless, Leibniz's employer and benefactor, George of Hanover, became the first German King of England, and Leibniz was left in the lurch. It is said that when Leibniz died two years later, his funeral was attended only by the man who had once been Leibniz's secretary.

*Personal Note*

## Maria Gaetana Agnesi (1718–1799) and Marquis de L'Hospital (1661–1704)

The first two influential calculus textbooks were *Analyse des infiniment petits* by the French mathematician, Marquis de L'Hospital, and *Analytical Intuitions* by the Italian mathematician, Maria Agnesi. A curious tale accompanies each text. L'Hospital made a deal with Jean Bernoulli in which L'Hospital paid Bernoulli for a supply of mathematical results that L'Hospital took as his own. Consequently, much of L'Hospital's text is due to Bernoulli; in particular, Bernoulli supplied the result, published in that text, now known as L'Hospital's Rule.

Maria Agnesi's *Analytical Intuitions* was more scholarly and comprehensive than L'Hospital's text. Agnesi spent ten years writing her two-volume work; because she read Latin and Greek and several modern European languages, she was able to incorporate mathematics from disparate sources, and she included results of both Leibniz and Newton. When her text was translated into English shortly after Agnesi's death, the translator mixed up the word *versiera* (from the Latin *versare*, "to turn often") with the word *avversiera*. From then on, the curve $xy^2 = a^2(a - x)$ became known in English as the *witch of Agnesi*, and today the correct English word for this curve, *versiera*, is more useful to Scrabble players than to mathematicians.

# 9.1 Sequences and Convergence

We begin by recalling properties of the absolute value function.

**Definition**

If $x$ is any real number, the *absolute value* of $x$, denoted by $|x|$, is defined as follows:

$$|x| = \max\{x, -x\}$$

The proof of the following theorem is left as Exercise 1.

**Theorem 9.1**

Let $p$ and $q$ be real numbers. Then

**a)** $|p-q|=0$ if and only if $p=q$.

**b)** $|p-q|=|q-p|$.

**c)** $|pq|=|p|\,|q|$.

**d)** $|p+q|\leqslant|p|+|q|$.

**e)** $\big||p|-|q|\big|\leqslant|p-q|$.

**f)** For each $\varepsilon>0$, $|p|<\varepsilon$ if and only if $-\varepsilon<p<\varepsilon$.

Notice that part (f) implies that for any $\varepsilon>0$ the following statements are equivalent:

**a)** $|p-q|<\varepsilon$.

**b)** $-\varepsilon<p-q<\varepsilon$.

**c)** $q-\varepsilon<p<q+\varepsilon$.

**d)** $p-\varepsilon<q<p+\varepsilon$.

If $p$ and $q$ are real numbers, then $|p-q|$ is the *distance from $p$ to $q$*, and

$$\{q\in\mathbb{R}:|p-q|<\varepsilon\}=\{q\in\mathbb{R}:p-\varepsilon<q<p+\varepsilon\}$$

is the set of points that are within $\varepsilon$ of the point $p$, namely the open interval $(p-\varepsilon,p+\varepsilon)$.

In Section 5.1, we defined a sequence to be a function whose domain is the set of natural numbers, and if $x:\mathbb{N}\to\mathbb{R}$, we denoted the sequence $x$ by $\langle x(n)\rangle$ or by $\langle x_n\rangle$. Prior to defining a convergent sequence, we consider an example.

**EXAMPLE 1**

Let $\varepsilon=1/10^{17}$. For sufficiently large $n$, $1/n<\varepsilon$. However, although there are infinitely many $n$ for which $1-(-1)^n=0$, no matter how large $n$ is, either $1-(-1)^n$ or $1-(-1)^{n+1}$ is 2. Thus we cannot say that for sufficiently large $n$, $1-(-1)^n$ is within $\varepsilon$ of 0. There is a major difference between the two sequences $\langle 1/n\rangle$ and $\langle 1-(-1)^n\rangle$, and as we shall see, the first one converges to zero, whereas the second does not converge to any number. ❑

It is customary to denote a sequence $x:\mathbb{N}\to\mathbb{R}$ by $\langle x_n\rangle$. Thus the $n$th term of the sequence is denoted by $(n, x_n)$ rather than $(n, x(n))$.

**Definition**

Let $A$ be a real number. A sequence $\langle x_n\rangle$ *converges to $A$*, denoted by $\langle x_n\rangle\to A$, provided that for each positive number $\varepsilon$ there is a natural number $N$ such that if $n>N$, then $|x_n-A|<\varepsilon$. A sequence $\langle x_n\rangle$ *converges* if there is a real number $A$ such that $\langle x_n\rangle\to A$. If $\langle x_n\rangle$ does not converge, then it is said to *diverge*.

Thus a sequence $\langle x_n \rangle$ converges to $A$ provided that for each positive number $\varepsilon$ there is a natural number $N$ such that if $n > N$, then $x_n$ lies in the open interval shown in Figure 9.1.

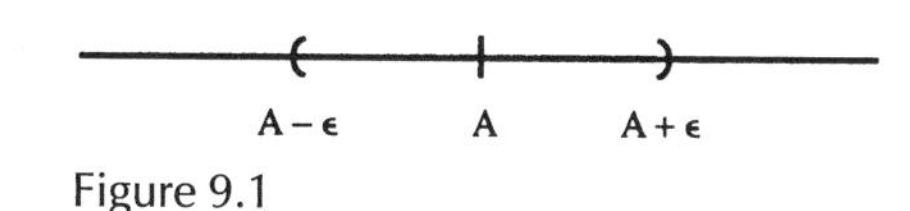


Figure 9.1

Notice that this definition says that given any positive number $\varepsilon$ we can find a natural number $N$ with specified properties. The choice of $N$ may indeed depend upon $\varepsilon$. We illustrate this dependence by returning to Example 1. Let $\varepsilon > 0$. In order to show that $\langle 1/n \rangle$ converges to 0, we must find a natural number $N$ such that if $n > N$, then $|(1/n) - 0| < \varepsilon$. In this example, we choose $N$ to be $\lceil 1/\varepsilon \rceil$. Then $N \geqslant 1/\varepsilon$, so $N\varepsilon \geqslant 1$ and hence $1/N \leqslant \varepsilon$. Therefore, if $n > N$, $|(1/n) - 0| = 1/n < 1/N \leqslant \varepsilon$, and so $\langle 1/n \rangle$ converges to 0.

In order to show that a sequence $\langle x_n \rangle$ does not converge to a real number, we must find, for each real number $A$, a positive number $\varepsilon$ such that if $N$ is any natural number then there is an $n > N$ such that $|x_n - A| \geqslant \varepsilon$. Again, let us return to Example 1 and consider the sequence $\langle 1 - (-1)^n \rangle$. Notice that the first term of this sequence is (1, 2), the second term is (2, 0), the third term is (3, 2), the fourth term is (4, 0), and in general the $n$th term is $(n, 2)$ if $n$ is odd and $(n, 0)$ if $n$ is even. As we said, in order to show that this sequence does not converge to 0, we seek a suitable positive number $\varepsilon$. Let $\varepsilon = 1$ (you will see shortly that any $\varepsilon$ satisfying $0 < \varepsilon \leqslant 2$ will suffice). Let $N$ be any natural number. Then if $n$ is odd and $n > N$, $|(1 - (-1)^n) - 0| = 2 > \varepsilon$, so the sequence does not converge to 0. Let $A$ be a nonzero real number and set $\varepsilon = |A|/2$. Then for each natural number $N$ there is an $n > N$ such that $|x_n - A| = |A| > \varepsilon$. So $\langle x_n \rangle$ does not converge to $A$.

Suppose a sequence $\langle x_n \rangle$ converges to a real number $A$, and let $\varepsilon > 0$. Then there is a natural number $N$ such that if $n > N$ then $|x_n - A| < \varepsilon$. This means that if $x_p$ is not in the open interval $(A - \varepsilon, A + \varepsilon)$, then $p \leqslant N$. In other words, for all but a finitely many $n \in \mathbb{N}$, $x_n$ lies in the interval $(A - \varepsilon, A + \varepsilon)$. We use this observation to characterize the convergence of a sequence, but first we give a definition.

**Definition**

A set $Q$ of real numbers is a *neighborhood* of a real number $A$ provided there is a positive number $\varepsilon$ such that the open interval $(A - \varepsilon, A + \varepsilon)$ is a subset of $Q$.

**EXAMPLE 2**

Let $Q = (0, 1]$. Then if $A$ belongs to the open interval (0, 1), $Q$ is a neighborhood of $A$ because if $\varepsilon = \min\{1 - A, A - 0\}$, then $(A - \varepsilon, A + \varepsilon) \subseteq Q$. However, $Q$ is not a neighborhood of 1, because if $\varepsilon$ is any positive number then $(1 - \varepsilon, 1 + \varepsilon)$ contains $1 + \varepsilon/2$, but $1 + \varepsilon/2 \notin Q$. ❑

**Theorem 9.2**

A sequence $\langle x_n \rangle$ converges to a real number $A$ if and only if every neighborhood of $A$ contains $x_n$ for all but finitely many $n \in \mathbb{N}$.

**PROOF** Suppose $\langle x_n \rangle$ converges to $A$, and let $Q$ be a neighborhood of $A$. Then there is an $\varepsilon > 0$ such that $(A - \varepsilon, A + \varepsilon) \subseteq Q$. Since $\langle x_n \rangle$ converges to $A$, there is a natural number $N$ such that if $n > N$, then $|x_n - A| < \varepsilon$. Thus if $n > N$, then $x_n \in (A - \varepsilon, A + \varepsilon)$, so if $x_p \notin Q$ then $1 \leqslant p \leqslant N$. Therefore $Q$ contains $x_n$ for all but finitely many $n \in \mathbb{N}$.

Suppose each neighborhood of $A$ contains $x_n$ for all but finitely many $n \in \mathbb{N}$. Let $\varepsilon > 0$. Then $Q = (A - \varepsilon, A + \varepsilon)$ is a neighborhood of $A$, so it contains $x_n$ for all but finitely many $n \in \mathbb{N}$. If $\{p \in \mathbb{N} : x_p \notin Q\} = \varnothing$, let $N = 1$; otherwise, let $N = \max\{p \in \mathbb{N} : x_p \notin Q\}$. Let $n > N$. Then $x_n \in Q$, so $|x_n - A| < \varepsilon$. Therefore, $\langle x_n \rangle$ converges to $A$. □

Now we consider whether a sequence can converge to more than one number. Figure 9.2 provides a visualization of the proof of Theorem 9.3.

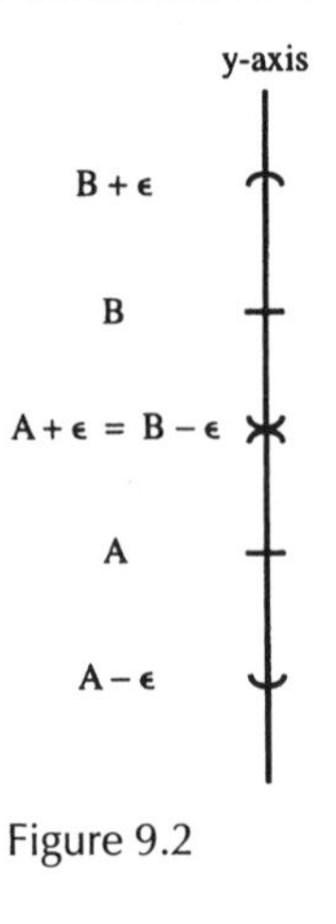


Figure 9.2

**Theorem 9.3**

If $\langle x_n \rangle$ is a sequence that converges to both $A$ and to $B$, then $A = B$.

**PROOF** The proof is by contradiction. Suppose $\langle x_n \rangle$ converges to $A$ and to $B$ and $A \neq B$. We can assume that $A < B$. Let $\varepsilon = \frac{1}{2}(B - A)$. Note that $A + \varepsilon = B - \varepsilon$. Since $\langle x_n \rangle \to A$, there is a natural number $N_1$ such that if $n > N_1$ then $|x_n - A| < \varepsilon$. Since $\langle x_n \rangle \to B$, there is a natural number $N_2$ such that if $n > N_2$ then $|x_n - B| < \varepsilon$. Let $N = \max\{N_1, N_2\}$. Then if $n > N$, $|x_n - A| < \varepsilon$ and $|x_n - B| < \varepsilon$, so $x_n < A + \varepsilon = B - \varepsilon < x_n$. This is a contradiction. □

**Definition**

If a sequence converges, then the unique number to which it converges is called the *limit* of the sequence.

The uniqueness is given by Theorem 9.3.

It seems intuitively clear that the sequence, $\langle n \rangle$, whose $n$th term is $(n, n)$, cannot converge because its values get larger and larger without bound.

**Definition**

Let $S$ be a subset of $\mathbb{R}$. A real number $b$ is an *upper bound* of $S$ provided that for each $x \in S$, $x \leqslant b$, and a real number $b$ is a *lower bound* of $S$ provided that for each $x \in S$, $b \leqslant x$. The set $S$ is *bounded* provided there is a real number $B$ such that $|x| < B$ for all $x \in S$. A sequence $\langle x_n \rangle$ is *bounded* provided that its range is bounded.

In Exercise 2 you are asked to prove that a subset $S$ of $\mathbb{R}$ is bounded if and only if there are real numbers $P$ and $Q$ such that $P < x < Q$ for each $x \in S$. We prove that every convergent sequence is bounded.

**Theorem 9.4**

If the sequence $\langle x_n \rangle$ converges, then it is bounded.

**PROOF** Let $\langle x_n \rangle$ be a convergent sequence, and let $A$ be the real number such that $\langle x_n \rangle \to A$. By choosing $\varepsilon$ to be 1, we can find a natural number $N$ such that if $n > N$, then $A - 1 < x_n < A + 1$. Let $P = \min \{A - 1, x_i: 1 \leqslant i \leqslant N\}$ and let $Q = \max \{A + 1, x_i: 1 \leqslant i \leqslant N\}$. Then for each natural number $n$, $P - 1 < x_n < Q + 1$. Thus $\langle x_n \rangle$ is bounded. □

In order to prove that a sequence is convergent, we must first guess the real number to which it converges and then prove that this number is the limit of the sequence. We have seen that the sequence $\langle 1/n \rangle$ converges to 0. Thus, we would guess that the sequence $\langle 1 + 1/n \rangle$ converges to 1. In the following example, we prove that this is the case. The proof is the same as the one we used to show that $\langle 1/n \rangle \to 0$.

**EXAMPLE 3**

Let $\varepsilon > 0$ and let $N = \lceil 1/\varepsilon \rceil$. Then if $n > N$, $|(1 + 1/n) - 1| = 1/n < 1/N \leqslant \varepsilon$. ❑

If a sequence $\langle x_n \rangle$ converges to $A$, then for each $\varepsilon > 0$ there is an integer $N$ such that for all $m, n > N$, $x_m$ and $x_n$ are each $(\varepsilon/2)$-close to $A$. But if $x_m$

and $x_n$ are $(\varepsilon/2)$-close to $A$ they must be $\varepsilon$-close to each other. The following definition, named in honor of the French mathematician Augustin Cauchy (1789–1857), formalizes this notion.

**Definition**

A sequence $\langle x_n \rangle$ is *Cauchy* provided that for each positive number $\varepsilon$ there is a natural number $N$ such that if $m, n > N$, then $|x_m - x_n| < \varepsilon$.

We show that every convergent sequence is a Cauchy sequence by proving the following theorem.

**Theorem 9.5**

A sequence $\langle x_n \rangle$ is Cauchy if and only if for each positive number $\varepsilon$ there is a real number $L$ and a natural number $N$ such that if $n > N$, then $|x_n - L| < \varepsilon$.

**PROOF** Suppose $\langle x_n \rangle$ is a Cauchy sequence and let $\varepsilon > 0$. Then there exists a natural number $N$ such that if $m, n > N$, then $|x_m - x_n| < \varepsilon$. Let $L = x_{N+1}$ and suppose $p > N$. Since $p$ and $N+1$ are greater than $N$, $|x_p - L| = |x_p - x_{N+1}| < \varepsilon$.

Suppose $\langle x_n \rangle$ is a sequence with the property that if $\varepsilon$ is a positive number, then there is a real number $L$ and a natural number $N$ such that if $n > N$, then $|x_n - L| < \varepsilon$. Let $\varepsilon > 0$. Since $\varepsilon/2 > 0$, there is a real number $L$ and a natural number $N$ such that if $n > N$, then $|x_n - L| < \varepsilon/2$. Suppose $m, n > N$. Then $|x_m - L| < \varepsilon/2$ and $|x_n - L| < \varepsilon/2$. Thus $|x_m - x_n| = |(x_m - L) + (L - x_m)| \leqslant |x_m - L| + |L - x_n| < \varepsilon/2 + \varepsilon/2 = \varepsilon$. □

**Corollary 9.6**

Every convergent sequence is a Cauchy sequence.

In Exercise 3 you are asked to prove the following theorem.

**Theorem 9.7**

Every Cauchy sequence is bounded.

**Definition**

Let $S \subseteq \mathbb{R}$ and let $x \in \mathbb{R}$. Then $x$ is said to be an *accumulation point* of $S$ provided every neighborhood of $x$ contains infinitely many members of $S$.

In Exercise 4, you are asked to give an example of a set $S$ that has an accumulation point that does not belong to $S$.

**Theorem 9.8**

Let $S \subseteq \mathbb{R}$ and let $x \in \mathbb{R}$. Then $x$ is an accumulation point of $S$ if and only if every neighborhood of $x$ contains a member of $S$ that is different from $x$.

**PROOF** Suppose $x$ is an accumulation point of $S$. Then every neighborhood of $x$ contains a member of $S$ that is different from $x$.

Suppose $x$ is not an accumulation point of $S$. Then there is a neighborhood $Q$ of $x$ such that $Q \cap S$ is finite. Since we seek a neighborhood of $x$ that contains no members of $S$ different from $x$, we may assume that $Q \cap S$ contains only a finite number of points $x_1, x_2, \ldots, x_n$ of $S$ that are different from $x$. For each $i = 1, 2, \ldots, n$, let $\varepsilon_i = |x_i - x|$. Then, for each $i = 1, 2, \ldots, n$, $(x - \varepsilon_i, x + \varepsilon_i)$ is a neighborhood of $x$ that does not contain $x_i$, so $Q \cap (\cap_{i=1}^{n} (x - \varepsilon_i, x + \varepsilon_i))$ is a neighborhood of $x$ that does not contain any member of $S$ that is different from $x$ (see Exercise 5). □

**EXAMPLE 4**

Let $S = \{1/n\colon n \text{ is a natural number}\}$. At the beginning of this section, we showed that the sequence $\langle 1/n \rangle$ converges to 0. Thus, every neighborhood of 0 contains infinitely many members of $S$. Therefore, 0 is an accumulation point of $S$. ❑

One might think that if the sequence $\langle x_n \rangle$ converges to $x$, then $x$ is an accumulation point of the range of $\langle x_n \rangle$. However, the sequence $\langle x_n \rangle$, where $x_n = 1$ for each $n$, converges to 1, but its range consists of only one point and therefore its range can have no accumulation point.

Which subsets of $\mathbb{R}$ have at least one accumulation point? No finite set does, and $\mathbb{Z}$ is an infinite subset of $\mathbb{R}$ without an accumulation point. The question is settled by a famous result, which is now commonly named the Bolzano–Weierstrass Theorem even though it was first proved by Bernard Bolzano in 1817.

The proof of Bolzano's theorem uses the Least-Upper-Bound Axiom. For completeness, we also state the Greatest-Lower-Bound Axiom.

**Least-Upper-Bound Axiom** Every nonempty subset of $\mathbb{R}$ that has an upper bound has a least upper bound.

**Greatest-Lower-Bound Axiom** Every nonempty subset of $\mathbb{R}$ that has a lower bound has a greatest lower bound.

**Theorem 9.9**

(The Bolzano–Weierstrass Theorem). Every bounded infinite subset of $\mathbb{R}$ has an accumulation point.

**PROOF** Let $S$ be a bounded infinite set. There are integers $a$ and $b$ such that $S \subseteq [a, b]$ and, by the Pigeonhole Principle, there is an integer $j$ such that $S \cap [j, j+1]$ is infinite. Divide $[j, j+1]$ into 10 intervals each of length 1/10 in the obvious way:

$$[j, j+1/10], [j+1/10, j+2/10], \ldots, [j+9/10, j+1]$$

Choose the first such interval whose intersection with $S$ is infinite and label this interval $[x_1, y_1]$. Divide $[x_1, y_1]$ into 10 intervals each of length 1/100 in the obvious way:

$$[x_1, x_1+1/100], [x_1+1/100, x_1+2/100], \ldots, [x_1+9/100, y_1]$$

Choose the first such interval whose intersection with $S$ is infinite and label this interval $[x_2, y_2]$. By induction, we can define sequences $\langle x_n \rangle$ and $\langle y_n \rangle$ with the following properties: For each natural number $n$,

**a)** $j \leqslant x_n < y_n \leqslant j+1$

**b)** $x_n \leqslant x_{n+1}$

**c)** $y_{n+1} \leqslant y_n$

**d)** $[x_n, y_n] \cap S$ is infinite

**e)** $y_n - x_n = 1/10^n$

Let $R$ denote the range of the sequence $\langle x_n \rangle$. Then $R$ is a nonempty set and $j+1$ is an upper bound of $R$. Let $x$ denote the least upper bound of $R$. We show that $x$ is an accumulation point of $S$. Let $\varepsilon > 0$. There is an $x_n \in R$ such that $x - \varepsilon < x_n \leqslant x$ (why?). For each $m > n$, $x - \varepsilon < x_n \leqslant x_m < y_m$ because of properties (a) and (b) given earlier in the proof. Can we find a natural number $m > n$ such that $y_m < x + \varepsilon$? Yes. (The Archimedian property of $\mathbb{R}$ promises an $m > n$ such that $1/10^m < \varepsilon$, and since $y_m - x_m = 1/10^m < \varepsilon$, $y_m < x_m + \varepsilon < x + \varepsilon$.) For this natural number $m$ we have

$$x - \varepsilon < x_n \leqslant x_m < y_m < x_m + \varepsilon < x + \varepsilon$$

and so $(x - \varepsilon, x + \varepsilon) \cap S$ is infinite. (Indeed even $[x_m, y_m] \cap S$ is infinite.) □

**Corollary 9.10**

A subset $S$ of $\mathbb{R}$ has an accumulation point if and only if $S$ has a bounded infinite subset.

**Theorem 9.11**

Every Cauchy sequence is convergent.

**PROOF** Let $\langle x_n \rangle$ be a Cauchy sequence, and let $S$ denote the range of $\langle x_n \rangle$. We consider two cases.

***Case I.*** Suppose $S$ is finite. Then we can write $S = \{s_1, s_2, \ldots, s_r\}$, and we may assume that $S$ has at least two members. (Why?) Let $\varepsilon = \min\{|s_i - s_j| : i \neq j, i, j = 1, 2, \ldots, r\}$. Since $\langle x_n \rangle$ is a Cauchy sequence, there is a natural number $N$ such that if $m, n > N$, $|x_m - x_n| < \varepsilon$. Thus, by the choice of $\varepsilon$, $x_m = x_n$ for all $m, n > N$. So there is a real number $A$ such that $x_n = A$ for all $n > N$. Thus, by Exercise 6, $\langle x_n \rangle \to A$.

***Case II.*** Suppose $S$ is infinite. By Theorem 9.7, $S$ is bounded. By the Bolzano–Weierstrass Theorem, $S$ has an accumulation point $A$. We complete the proof by showing that $\langle x_n \rangle \to A$. Let $\varepsilon > 0$. Then $(A - \varepsilon/2, A + \varepsilon/2)$ is a neighborhood of $A$, so it contains infinitely many points of $S$. Since $\langle x_n \rangle$ is a Cauchy sequence, there is a natural number $N'$ such that if $m, n > N'$, then $|x_m - x_n| < \varepsilon/2$. Since $(A - \varepsilon/2, A + \varepsilon/2)$ contains infinitely many points of $S$, there is a natural number $N > N'$ such that $x_N \in (A - \varepsilon/2, A + \varepsilon/2)$. Then if $n > N$, $|x_n - A| \leqslant |x_n - x_N| + |x_N - A| < \varepsilon/2 + \varepsilon/2 = \varepsilon$. Thus $\langle x_n \rangle \to A$. □

The proof of Theorem 9.11 establishes that $\langle x_n \rangle$ is bounded. This fact reflects a general principle. In order to show that a given sequence with certain properties converges, it is usually necessary to show first that the sequence is bounded.

## EXERCISES 9.1

1. Prove Theorem 9.1.
2. Prove that a subset $S$ of $\mathbb{R}$ is bounded if and only if there are real numbers $P$ and $Q$ such that $P < x < Q$ for each $x \in S$.
3. Prove Theorem 9.7.
4. Give an example of a set $S$ that has an accumulation point that does not belong to $S$.
5. Let $n \in \mathbb{N}$ and for each $i = 1, 2, \ldots, n$, let $\varepsilon_i > 0$. Prove that for each $x \in \mathbb{R}$, $\cap_{i=1}^{n} (x - \varepsilon_i, x + \varepsilon_i)$ is a neighborhood of $x$.
6. Let $\langle x_n \rangle$ be a sequence with the property that there is a real number $A$ and a natural number $N$ such that $x_n = A$ for all $n > N$. Prove that $\langle x_n \rangle \to A$.
7. Show that $[0, 1]$ is a neighborhood of $7/8$.
8. Let $x \in \mathbb{R}$ and let $\varepsilon > 0$. Prove that if $y \in (x - \varepsilon, x + \varepsilon)$, then $(x - \varepsilon, x + \varepsilon)$ is a neighborhood of $y$.
9. Let $x, y \in \mathbb{R}$ with $x \neq y$. Prove that there is a neighborhood $U$ of $x$ and a neighborhood $V$ of $y$ such that $U \cap V = \varnothing$.
10. Prove that the sequence $\langle 5n/(3n + 1) \rangle$ converges.
11. **a)** Suppose $\langle x_n \rangle \to A$. Prove that $\langle |x_n| \rangle \to |A|$. Is the converse true? Justify your answer.

    **b)** Suppose that $\langle |y_n| \rangle \to |0|$. Does it follow that $\langle y_n \rangle \to 0$? Justify your answer.
12. Let $\langle x_n \rangle$ and $\langle y_n \rangle$ be Cauchy sequences. Prove that

    **a)** $\langle x_n + y_n \rangle$ is a Cauchy sequence.

    **b)** $\langle x_n y_n \rangle$ is a Cauchy sequence.

**13.** Give an example of a set that has exactly two accumulation points.

**14.** Let $\langle x_n \rangle$ be a sequence and suppose that the range of $\langle x_n \rangle$ has two accumulation points. Prove that $\langle x_n \rangle$ does not converge.

**15.** Prove or disprove each of the following statements:

**a)** If $\langle x_n \rangle \to x$, then $x$ is an accumulation point of the range of $\langle x_n \rangle$.

**b)** If $\langle x_n \rangle$ is a sequence whose range is bounded and $x$ is an accumulation point of the range of $\langle x_n \rangle$, then $\langle x_n \rangle \to x$.

**c)** If $\langle x_n \rangle$ is a Cauchy sequence and $x$ is an accumulation point of the range of $\langle x_n \rangle$, then $\langle x_n \rangle \to x$.

**16.** Let $x_1, x_2 \in \mathbb{R}$ with $x_1 \neq x_2$, and for each $n \geqslant 3$, let $x_n = (x_{n-1} + x_{n-2})/2$. Prove that

**a)** For each $n \geqslant 3$, $x_{n+1} - x_n = (-\frac{1}{2})^{n-1}(x_2 - x_1)$.

**b)** $\langle x_n \rangle$ is a Cauchy sequence.

**17.** Let $\langle x_n \rangle$ be a sequence with the following property. If $\varepsilon > 0$, there is a natural number $N$ such that if $n > N$, $|x_N - x_n| < \varepsilon$. Is $\langle x_n \rangle$ necessarily a Cauchy sequence? Give a proof or counterexample.

***18.** Let $\langle x_n \rangle$ be a sequence with the following property. If $\varepsilon > 0$, there is a natural number $N$ such that if $n > N$, $|x_n - x_{n+1}| < \varepsilon$. Is $\langle x_n \rangle$ necessarily a Cauchy sequence? Give a proof or counterexample.

## 9.2 *Subsequences and Arithmetic Operations on Sequences*

In this section we show that convergent sequences relate in a friendly way with addition, multiplication, division, and order on $\mathbb{R}$. Then we examine the convergence of subsequences.

**Theorem 9.12**

If $\langle x_n \rangle \to x$ and $\langle y_n \rangle \to y$, then $\langle x_n + y_n \rangle \to x + y$.

**PROOF** Let $\varepsilon > 0$. There are natural numbers $N_1$ and $N_2$ such that if $n > N_1$ then $|x_n - x| < \varepsilon/2$ and if $n > N_2$ then $|y_n - y| < \varepsilon/2$. Let $N = \max\{N_1, N_2\}$ and let $n > N$. Then $|(x_n + y_n) - (x + y)| = |(x_n - x) + (y_n - y)| \leqslant |x_n - x| + |y_n - y| < \varepsilon/2 + \varepsilon/2 = \varepsilon$. Therefore, $\langle x_n + y_n \rangle \to x + y$. □

**Theorem 9.13**

If $\langle x_n \rangle \to x$ and $\langle y_n \rangle \to y$, then $\langle x_n y_n \rangle \to xy$.

**PROOF** By Corollary 9.6, $\langle x_n \rangle$ is a Cauchy sequence, so by Theorem 9.7, it is bounded. Therefore, there is a real number $B$ such that $|x_n| < B$ for all $n \in \mathbb{N}$. Let $\varepsilon > 0$. There is a natural number $N_1$ such that if $n > N_1$ then

$|x_n - x| < \varepsilon/(B + |y|)$, and there is a natural number $N_2$ such that if $n > N_2$ then $|y_n - y| < \varepsilon/(B + |y|)$. Let $N = \max\{N_1, N_2\}$. Then if $n > N$,

$$\begin{aligned} |x_n y_n - xy| &= |x_n y_n - x_n y + x_n y - xy| \\ &\leqslant |x_n|\,|y_n - y| + |y|\,|x_n - x| \\ &\leqslant [\varepsilon/(B + |y|)][|x_n| + |y|] \end{aligned}$$

Since $|x_n| < B$, $(|x_n| + |y|)/(B + |y|) < 1$. Therefore, $|x_n y_n - xy| < \varepsilon$. □

**Theorem 9.14**

If $\langle x_n \rangle \to x$ and there is a natural number $N$ such that $x_n \geqslant 0$ for all $n > N$, then $x \geqslant 0$.

**PROOF** See Exercise 19. □

**Theorem 9.15**

If $\langle x_n \rangle \to x$ and $\langle y_n \rangle \to y$, then $\langle x_n - y_n \rangle \to x - y$.

**PROOF** See Exercise 20. □

**Theorem 9.16**

If $\langle x_n \rangle \to x$ and $\langle y_n \rangle \to y$ and there is a natural number $N$ such that for all $n > N$, $x_n \geqslant y_n$, then $x \geqslant y$.

**PROOF** See Exercise 21. □

**Theorem 9.17**

Let $\langle x_n \rangle \to x$ and $\langle z_n \rangle \to x$, and let $\langle y_n \rangle$ be a sequence for which there is a natural number $N$ such that if $n > N$ then $x_n \leqslant y_n \leqslant z_n$. Then $\langle y_n \rangle$ is a convergent sequence and in fact $\langle y_n \rangle \to x$.

**PROOF** See Exercise 22. □

**Theorem 9.18**

Let $\langle x_n \rangle \to x$ and suppose that $x \neq 0$. Then there is a natural number $N$ and a positive real number $B$ such that for all $n > N$, $|x_n| > B$.

**PROOF** See Exercise 23. □

**Theorem 9.19**

Let $\langle x_n \rangle \to x$ and suppose that for each $n \in \mathbb{N}$, $x_n \neq 0$. If $x \neq 0$, then $\langle 1/x_n \rangle \to 1/x$.

**PROOF** See Exercise 24. □

**Theorem 9.20**

Let $x$ be a real number and let $y$ be a nonzero real number. Let $\langle x_n \rangle \to x$ and $\langle y_n \rangle \to y$, and suppose that for each $n \in \mathbb{N}$, $y_n \neq 0$. Then $\langle x_n/y_n \rangle \to x/y$.

**PROOF** See Exercise 25. □

**Theorem 9.21**

Let $S$ be a subset of $\mathbb{R}$ and let $x$ be a real number. The following statements are equivalent:

**a)** For each positive number $\varepsilon$, $\{y \in \mathbb{R}: |y - x| < \varepsilon\} \cap S$ is infinite.

**b)** For each positive number $\varepsilon$, $\{y \in \mathbb{R}: |y - x| < \varepsilon\} \cap S$ has at least two members.

**c)** For each positive number $\varepsilon$, $\{y \in \mathbb{R}: |y - x| < \varepsilon\} \cap S$ has some member other than $x$ (more formally, $(\{y \in \mathbb{R}: |y - x| < \varepsilon\} - \{x\}) \cap S \neq \varnothing$).

**PROOF** Only the implication (c)⇒(a) requires proof. Assume that condition (c) holds and suppose that $\varepsilon$ is a positive number such that $\{y \in \mathbb{R}: |y - x| < \varepsilon\} \cap S$ is finite. Let

$$D = \{|y - x|: y \in S - \{x\} \quad \text{and} \quad |y - x| < \varepsilon\}$$

Condition (c) implies that $D$ is a nonempty finite set. Let $\delta$ be the smallest member of $D$. Then $\{y \in \mathbb{R}: |y - x| < \delta\} \cap S$ has no member other than $x$, which contradicts condition (c). □

We recall the definition of increasing and decreasing for real-valued functions.

**Definition**

A real-valued function $f$ is *increasing* (*decreasing*) provided that whenever $x$ and $y$ belong to the domain $f$ and $x \leqslant y$, then $f(x) \leqslant f(y) [f(y) \leqslant f(x)]$. A sequence $f: \mathbb{N} \to \mathbb{R}$ is *monotone* if either it is an increasing function or it is a decreasing function.

**Theorem 9.22**

Let $\langle a_n \rangle$ be a monotone sequence. Then $\langle a_n \rangle$ converges if and only if it is bounded.

**PROOF** Suppose that $\langle a_n \rangle$ is increasing and $\{a_n: n \in \mathbb{N}\}$ is bounded. Let $p$ be the least upper bound of $\{a_n: n \in \mathbb{N}\}$. Let $\varepsilon > 0$. Then there is an $N \in \mathbb{N}$ such that $p - \varepsilon < a_N \leqslant p$. Since $\langle a_n \rangle$ is increasing, for each $n > N$ we have $p - \varepsilon < a_N \leqslant a_n \leqslant p < p + \varepsilon$. Therefore $\langle a_n \rangle$ converges to $p$.

Now suppose that $\langle a_n \rangle$ is decreasing and its range is bounded. Then $\langle -a_n \rangle$ is a bounded increasing sequence and so $\langle -a_n \rangle$ converges say to $p$. Then $\langle a_n \rangle \to -p$. (Why?) □

We recall the definition of a subsequence.

**Definition**

If $\langle x_n \rangle$ is a sequence and $g: \mathbb{N} \to \mathbb{N}$ is a sequence such that $g(n+1) > g(n)$ for each $n \in \mathbb{N}$ then $\langle x \circ g(n) \rangle$, which we denote by $\langle x_{g(n)} \rangle$, is a *subsequence* of $\langle x_n \rangle$.

**Theorem 9.23**

A sequence converges if and only if each of its subsequences converges. In fact, if every subsequence converges, then they all converge to the same limit.

**PROOF** If every subsequence of a sequence converges, then the sequence converges, because every sequence is a subsequence of itself.

Suppose that $\langle x_n \rangle$ is a sequence that converges to $x$ and that $\langle x_{g(n)} \rangle$ is a subsequence of $\langle x_n \rangle$. Let $\varepsilon > 0$. Then there is a natural number $N$ such that if $n > N$, then $|x_n - x| < \varepsilon$. By Exercise 28, if $n > N$ then $g(n) > N$. Thus for all $n > N$, $|x_{g(n)} - x| < \varepsilon$. Therefore $\langle x_{g(n)} \rangle$ converges to $x$. □

In Exercise 26, you are asked to prove that every bounded sequence of real numbers has a convergent subsequence. We use this fact to prove the following theorem.

**Theorem 9.24**

Let $\langle x_n \rangle$ be a bounded sequence. If all its convergent subsequences have the same limit, then the sequence is convergent.

**PROOF** Suppose all the convergent subsequences of $\langle x_n \rangle$ have the same limit, say $x$. By Exercise 26, there is a convergent subsequence of $\langle x_n \rangle$.

Suppose $\langle x_n \rangle$ does not converge to $x$. Then there is a positive number $\varepsilon$ such that if $N$ is any natural number then there is an $n > N$ such that $|x_n - x| \geqslant \varepsilon$. We define a function $g: \mathbb{N} \to \mathbb{N}$ such that $g(n+1) > g(n)$ for all $n \in \mathbb{N}$. This function is defined recursively. There is a natural number, call it $g(1)$, greater than 1 such that $|x_{g(1)} - x| \geqslant \varepsilon$. Likewise, there is a natural number, call it $g(2)$, greater than $g(1)$ such that $|x_{g(2)} - x| \geqslant \varepsilon$. Suppose $k \in \mathbb{N}$, $k > 2$, and $g(1), g(2), \ldots, g(k-1)$ have been defined so that for each $i = 2, 3, \ldots, k-1$, $g(i) > g(i-1)$ and for each $i = 1, 2, \ldots, k-1$, $|x_{g(i)} - x| \geqslant \varepsilon$. As above there is a natural number, call it $g(k)$, greater than $g(k-1)$ such that $|x_{g(k)} - x| \geqslant \varepsilon$. The range of $\langle x_{g(n)} \rangle$ is a subset of the range of $\langle x_n \rangle$, so $\langle x_{g(n)} \rangle$ is bounded. Therefore, by Exercise 26, $\langle x_{g(n)} \rangle$ has a convergent subsequence.

This convergent subsequence is a subsequence of $\langle x_n \rangle$ that does not converge to $x$ (see Exercise 29). This contradicts the hypothesis. □

**EXAMPLE 5** Let $x_1 = \sqrt{2}$, and for each $n > 1$, let $x_n = \sqrt{2 + \sqrt{x_{n-1}}}$. We prove, using the Principle of Mathematical Induction, that $\langle x_n \rangle$ is increasing and that for each $n \in \mathbb{N}$, $x_n \leqslant 2$. Thus, by Theorem 9.22, the sequence converges.

Let $S = \{n \in \mathbb{N}: x_{n+1} \geqslant x_n \text{ and } x_n \leqslant 2\}$. Since $x_2 = \sqrt{2 + \sqrt[4]{2}} \geqslant \sqrt{2} = x_1$ and $x_1 \leqslant 2$, $1 \in S$. Suppose $n \in S$. Since $x_{n+2} = \sqrt{2 + \sqrt{x_{n+1}}} \geqslant \sqrt{2 + \sqrt{x_n}} = x_{n+1}$ and $x_{n+1} = \sqrt{2 + \sqrt{x_n}} \leqslant \sqrt{2 + \sqrt{2}} \leqslant \sqrt{2 + 2} = 2$, $n + 1 \in S$. Therefore, by the Principle of Mathematical Induction, $S = \mathbb{N}$. □

**EXAMPLE 6** Let $x_1 = 1$ and for each $n > 1$, let $x_n = \sqrt{2x_{n-1}}$. The first three terms of the sequence are $x_1 = 1$, $x_2 = \sqrt{2}$, and $x_3 = \sqrt{2\sqrt{2}}$. Thus, it seems reasonable to expect that the sequence is increasing. It is not clear that it is bounded, but if it is, then, by Theorem 9.22, it converges. Let's assume for the moment that the sequence converges and attempt to find the number to which it converges. Suppose $\langle x_n \rangle$ converges to $x$. Then $\langle x_n^2 \rangle$ converges to $x^2$, but $x_{n+1}^2 = 2x_n$. Therefore, $\langle x_{n+1}^2 \rangle = \langle 2x_n \rangle$ converges to $2x$. Thus, the only candidates for $x$ must be the solutions of the equation $x^2 = 2x$. The only solutions of this equation are 0 and 2. It is clear that the sequence $\langle x_n \rangle$ does not converge to 0. So if $\langle x_n \rangle$ converges then it must converge to 2. It remains to show that the sequence converges. We prove this, using the Principle of Mathematical Induction, by showing that the sequence is increasing and that it is bounded.

Let $S = \{n \in \mathbb{N}: x_{n+1} \geqslant x_n \text{ and } x_n \leqslant 2\}$. It is clear that $1 \in S$. Suppose $n \in S$. Then $x_{n+2} = \sqrt{2x_{n+1}} \geqslant \sqrt{2x_n} = x_{n+1}$, and $x_{n+1} = \sqrt{2x_n} \leqslant \sqrt{2 \times 2} = 2$. Therefore, $n + 1 \in S$ and by the Principle of Mathematical Induction, $S = \mathbb{N}$. So, by Theorem 9.22, $\langle x_n \rangle$ converges. □

**Theorem 9.25** Let $X \subseteq \mathbb{R}$. Then $x_0 \in \mathbb{R}$ is an accumulation point of $X$ if and only if there is a sequence $\langle x_n \rangle$ of members of $X$ such that $x_n \neq x_0$ for any $n \in \mathbb{N}$ and $\langle x_n \rangle \to x_0$.

**PROOF** Suppose $x_0$ is an accumulation point of $X$. Then for each $n \in \mathbb{N}$, there is a member $x_n$ of $X$ such that $0 < |x_n - x_0| < 1/n$. Since $|x_n - x_0| > 0$, $x_n \neq x_0$ for any $n \in \mathbb{N}$. We show that $\langle x_n \rangle \to x_0$. Let $\varepsilon > 0$. Then there exists $N \in \mathbb{N}$ such that $1/\varepsilon < N$. Thus if $n > N$, $|x_n - x_0| < 1/n < 1/N < \varepsilon$. Therefore, $\langle x_n \rangle \to x_0$. □

Suppose there is a sequence $\langle x_n \rangle$ of members of $X$ such that $x_n \neq x_0$ for any $n \in \mathbb{N}$ and $\langle x_n \rangle \to x_0$. Since every neighborhood of $x_0$ contains $x_n$ for all but finitely many $n$, every neighborhood of $x_0$ must contain at least one member of $X$ that is different from $x_0$. Thus, by Theorem 9.21, $x_0$ is an accumulation point of $X$.

## EXERCISES 9.2

19. Prove Theorem 9.14.
20. Prove Theorem 9.15.
21. Prove Theorem 9.16.
22. Prove Theorem 9.17.
23. Prove Theorem 9.18.
24. Prove Theorem 9.19.
25. Prove Theorem 9.20.
26. Let $g: \mathbb{N} \to \mathbb{N}$ be a function such that $g(n+1) > g(n)$ for each $n$. Prove that for each $n \in \mathbb{N}$, $g(n) \geqslant n$.
27. Find a convergent subsequence of the sequence $\langle (-1)^n(1-1/n) \rangle$.
28. Prove that every bounded sequence of real numbers has a convergent subsequence.
29. Let $\langle x_{g(n)} \rangle$ be a subsequence of a sequence $\langle x_n \rangle$. Prove that any subsequence of $\langle x_{g(n)} \rangle$ is also a subsequence of $\langle x_n \rangle$.
30. Let $x_1 = \sqrt{6}$, and for $n > 1$ let $x_n = \sqrt{x_{n-1} + 6}$. Prove that $\langle x_n \rangle$ converges and find the limit.
*31. Prove that if $a$ is a real number that is greater than 1, then $\langle \sqrt[n]{a} \rangle$ converges to 1.
32. Suppose $\langle x_n \rangle$ and $\langle y_n \rangle$ are sequences such that $\langle x_n \rangle$ and $\langle x_n + y_n \rangle$ converge. Prove that $\langle y_n \rangle$ converges.
33. Let $x$ be a nonzero real number, and let $\langle x_n \rangle$ and $\langle y_n \rangle$ be sequences such that $\langle x_n y_n \rangle$ converges and $\langle x_n \rangle \to x$. Prove that $\langle y_n \rangle$ converges.
34. Suppose $\langle x_n \rangle \to x$ and, for each $n \in \mathbb{N}$, $x_n \geqslant 0$. Prove that $\langle \sqrt{x_n} \rangle \to \sqrt{x}$.
35. Assume that the Least-Upper-Bound Axiom holds. Prove that the Greatest-Lower-Bound Axiom is a consequence.

## 9.3 Limits of Functions

Calculus is based on the concept of the existence of the limit of a function at a point. Thus, the introductory material that we present in this section is the basis for understanding calculus.

**Definition**

Let $X \subseteq \mathbb{R}$, let $f: X \to \mathbb{R}$, and let $x_0$ be an accumulation point of $X$. Then *$f$ has a limit $L$ at $x_0$* if for each positive number $\varepsilon$ there is a positive number $\delta$ such that if $x \in X$ and $0 < |x - x_0| < \delta$ then $|f(x) - L| < \varepsilon$.

Notice that the definition does not specify that if $L$ exists then it is unique. The proof of this fact is left as Exercise 36. If $f$ has a limit $L$ at $x_0$, we write $\lim_{x \to x_0} f(x) = L$.

We begin by considering the geometric interpretation of limit (see Figure 9.3). Let $\varepsilon > 0$, and draw the horizontal lines whose equations are $y = L - \varepsilon$ and $y = L + \varepsilon$. In order for $\lim_{x \to x_0} f(x) = L$, we must be able to find $\delta$ such that if the point $(x, f(x))$, where $x \neq x_0$, on the graph of $f$ lies between the vertical lines whose equations are $x = x_0 - \delta$ and $x = x_0 + \delta$, then $(x, f(x))$ lies between $y = L - \varepsilon$ and $y = L + \varepsilon$.

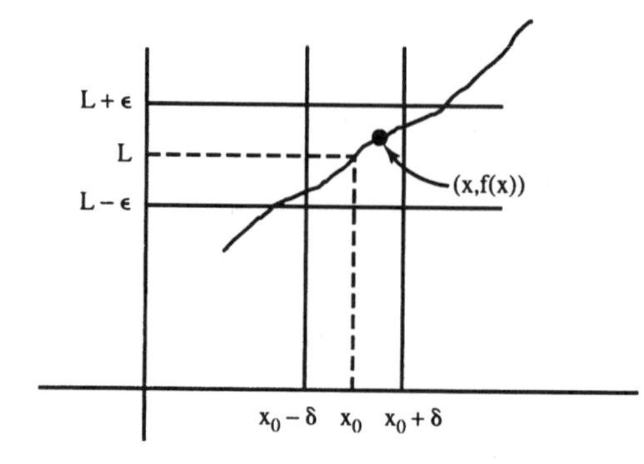

Figure 9.3

Now we consider two examples.

**EXAMPLE 7** Define $f: \mathbb{R} \to \mathbb{R}$ by $f(x) = (2x^2 - x - 1)/(x - 1)$ if $x \neq 1$ and $f(1) = 4$. If $x \neq 1$, $f(x) = 2x + 1$. Thus if $x \neq 1$, $f$ is a linear function with slope 2. The value of $f$ at 1 (or even whether 1 is in the domain of $f$) has nothing to do with the existence of the limit at $x = 1$. All we need from the number 1 is that it is an accumulation point of the domain of $f$, and this is obvious. The graph of $f$ has a gap at $x = 1$. It seems reasonable that as $x$ gets closer and closer to 1, the value of $f$ at $x$ gets closer and closer to the second coordinate of

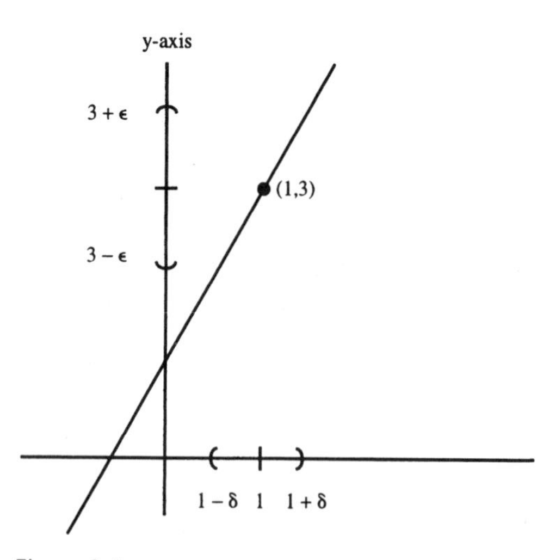


Figure 9.4

the point that fills in this gap. Thus we conjecture that $\lim_{x\to 1} f(x) = 3$, and we prove this fact. Let $\varepsilon > 0$. As illustrated in Figure 9.4, we want to find $\delta$ such that if $x \in (1-\delta, 1) \cup (1, 1+\delta)$, then $f(x) \in (3-\varepsilon, 3+\varepsilon)$.

In view of the geometric interpretation of limit and the fact that the slope of $f$ is 2, it seems reasonable to try $\delta = \varepsilon/2$. Suppose $\delta = \varepsilon/2$ and $0 < |x-1| < \delta$. Then $|f(x) - 3| = |(2x^2 - x - 1)/(x-1) - 3| = |(2x+1) - 3| = |2x - 2| = 2|x-1| < 2\delta = 2(\varepsilon/2) = \varepsilon$. Therefore, $\lim_{x\to 1} f(x) = 3$. ❑

Now we give an example of a function that fails to have a limit at a point. In order to show that a function $f$ does not have a limit at a point $x_0$, we must show that no number $L$ will meet the requirements of the definition. The strategy is to assume that $f$ has a limit at $x_0$ and reach a contradiction.

**EXAMPLE 8** Define $f: \mathbb{R} \to \mathbb{R}$ by $f(x) = |x|/x$ if $x \neq 0$ and $f(0) = 0$. Then if $x < 0$, $f(x) = -1$ and if $x > 0$, $f(x) = 1$. The graph of $f$ is shown in Figure 9.5.

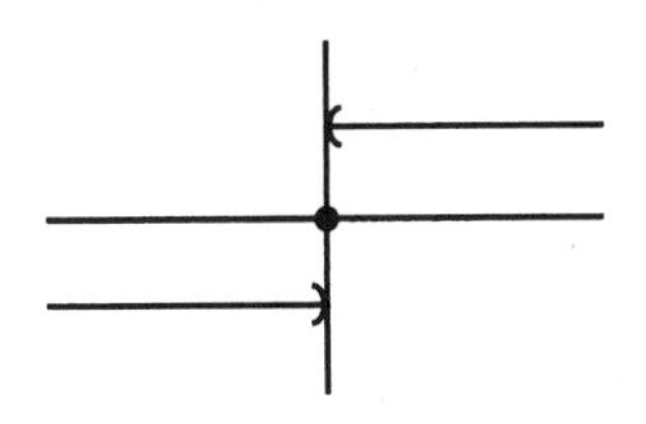

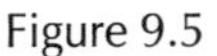

Figure 9.5

This suggests that $f$ does not have a limit at 0. In order to show that the requirements of the definition are not met, we must let $L$ be any number and find a positive number $\varepsilon$ such that if $\delta$ is any positive number, then there is an $x$ such that $0 < |x - 0| < \delta$ but $|f(x) - L| \geq \varepsilon$. Let $L$ be any number, let $\varepsilon = 1$, and let $\delta$ be any positive number. We consider two cases:

***Case I.*** Suppose $L \geq 0$. Then there is a negative number $x$ such that $0 < |x - 0| < \delta$, so $|f(x) - L| = |-1 - L| \geq 1 = \varepsilon$.

***Case II.*** Suppose $L < 0$. Then there is a positive number $x$ such that $0 < |x - 0| < \delta$, so $|f(x) - L| = |1 - L| \geq 1 = \varepsilon$.

Therefore $\lim_{x\to 0} f(x)$ does not exist. ❑

There is a strong relationship between the limits of sequences and limits of functions. Suppose $X \subseteq \mathbb{R}$, $x_0$ is an accumulation point of $X$, and $f: X \to \mathbb{R}$. We say that a sequence $\langle x_n \rangle$ is *good for $f$ at $x_0$* provided that

**a)** for each $n \in \mathbb{N}$, $x_n$ is in the domain of $f$,

**b)** for each $n \in \mathbb{N}$, $x_n \neq x_0$, and

**c)** $\langle x_n \rangle \to x_0$.

The connection between limits of sequences and limits of functions is given by the following theorem, which is easier to understand and remember when it is stated informally: If $\lim_{x \to a} f(x) = L$, then the image sequence $\langle f(x_n) \rangle$ of any good sequence $\langle x_n \rangle$ converges to $L$. Conversely, if for each good sequence, the corresponding image sequence converges, then the image sequences of all good sequences converge to the same real number $L = \lim_{x \to a} f(x)$.

**Theorem 9.26**

Suppose $X \subseteq \mathbb{R}$, $x_0$ is an accumulation point of $X$, and $f: X \to \mathbb{R}$. Then $f$ has a limit at $x_0$ if and only if whenever $\langle x_n \rangle$ is a good sequence for $f$ at $x_0$, $\langle f(x_n) \rangle$ is a convergent sequence. Moreover, when this condition is met and $\langle x_n \rangle$ is a good sequence for $f$ at $x_0$, $\langle f(x_n) \rangle \to \lim_{x \to x_0} f(x)$.

In order to facilitate the proof of this theorem, we first prove the following lemma.

**Lemma 9.27**

Suppose $X \subseteq \mathbb{R}$, $x_0$ is an accumulation point of $X$, and $f: X \to \mathbb{R}$ has the property that whenever $\langle x_n \rangle$ is a good sequence for $f$ at $x_0$, then $\langle f(x_n) \rangle$ converges. If $\langle x_n \rangle$ and $\langle y_n \rangle$ are two good sequences for $f$ at $x_0$ and $\langle f(x_n) \rangle \to L_1$ and $\langle f(y_n) \rangle \to L_2$, then $L_1 = L_2$.

**PROOF** Let $\langle x_n \rangle$ and $\langle y_n \rangle$ be two sequences that are good for $f$ at $x_0$ such that $\langle f(x_n) \rangle \to L_1$ and $\langle f(y_n) \rangle \to L_2$. Define a sequence $\langle z_n \rangle$ by $z_{2n-1} = y_n$ and $z_{2n} = x_n$ for each $n \in \mathbb{N}$. Then $\langle z_n \rangle$ is a good sequence for $f$ at $x_0$, so by hypothesis, $\langle f(z_n) \rangle$ converges, say, to $L$. By Theorem 9.23, $L_1 = L_2 = L$. □

**PROOF OF THEOREM 9.26** Suppose $f$ has a limit $L$ at $x_0$, and let $\langle x_n \rangle$ be a good sequence for $f$ at $x_0$. Let $\varepsilon > 0$. There is a positive number $\delta$ such that if $x \in X$ and $0 < |x - x_0| < \delta$, then $|f(x) - L| < \varepsilon$. There is a natural number $N$ such that if $n > N$, then $|x_n - x_0| < \delta$. Thus, for $n > N$, $|f(x_n) - L| < \varepsilon$, so $\langle f(x_n) \rangle$ converges to $L$.

Now suppose that whenever $\langle x_n \rangle$ is a good sequence for $f$ at $x_0$, the sequence $\langle f(x_n) \rangle$ converges. By Theorem 9.25, there is a good sequence $\langle a_n \rangle$ for $f$ at $x_0$. By hypothesis there is a number $L$ such that $\langle f(a_n) \rangle \to L$. Suppose $L$ is not a limit of $f$ at $x_0$. Then there is a positive number $\varepsilon$ such that if $\delta$ is any positive number, there is an $x \in X$ such that $0 < |x - x_0| < \delta$ but $|f(x) - L| \geqslant \varepsilon$. In particular, for each natural number $n$, there is an $x_n \in X$ such that $0 < |x_n - x_0| < 1/n$ but $|f(x_n) - L| \geqslant \varepsilon$. Then $\langle x_n \rangle$ is a good sequence for $f$ at $x_0$, and so by hypothesis there is a number $L_1$ such that $\langle f(x_n) \rangle \to L_1$. By Lemma 9.27, $L_1 = L$. But $\langle f(x_n) \rangle$ cannot converge to $L$ because for each $n \in \mathbb{N}$, $|f(x_n) - L| \geqslant \varepsilon$. □

**Theorem 9.28**

Suppose $X \subseteq \mathbb{R}$, $x_0$ is an accumulation point of $X$, and $f: X \to \mathbb{R}$. If for each sequence in $X - \{x_0\}$ converging to $x_0$ the sequence $\langle f(x_n) \rangle$ is Cauchy, then $f$ has a limit at $x_0$.

**PROOF** See Exercise 37. □

**Theorem 9.29**

Let $X \subseteq \mathbb{R}$, let $x_0$ be an accumulation point of $X$, and let $f: X \to \mathbb{R}$. If $f$ has a limit at $x_0$, then there is a neighborhood $Q$ of $x_0$ and a positive number $B$ such that $|f(x)| \leqslant B$ for each $x \in X \cap Q$.

**PROOF** Let $\varepsilon = 1$ and let $L = \lim_{x \to x_0} f(x)$. Then there is a $\delta > 0$ such that if $x \in X$ and $0 < |x - x_0| < \delta$ then $|f(x) - L| < \varepsilon = 1$. If $x_0 \in X$, let $B = \max\{|L-1|, |L+1|, |f(x_0)|\}$; otherwise let $B = \max\{|L-1|, |L+1|\}$. Then if $Q = (x_0 - \delta, x_0 + \delta)$ and $x \in X \cap Q$, $|f(x)| \leqslant B$. □

**Theorem 9.30**

Let $X \subseteq \mathbb{R}$, let $f, g: X \to \mathbb{R}$, and let $x_0$ be an accumulation point of $X$. If $\lim_{x \to x_0} f(x) = L_1$ and $\lim_{x \to x_0} g(x) = L_2$, then

**a)** $\lim_{x \to x_0} (f+g)(x) = L_1 + L_2$.

**b)** $\lim_{x \to x_0} (fg)(x) = L_1 L_2$.

**c)** If $g(x) \neq 0$ for any $x \in X$ and $L_2 \neq 0$, then $\lim_{x \to x_0} (f/g)(x) = L_1/L_2$.

**PROOF**

**a)** We use Theorem 9.26. Let $\langle x_n \rangle$ be a good sequence for $f+g$ at $x_0$. Clearly $\langle x_n \rangle$ is a good sequence for both $f$ and $g$ at $x_0$, so $\langle f(x_n) \rangle \to L_1$ and $\langle g(x_n) \rangle \to L_2$. By Theorem 9.12, $\langle (f+g)(x_n) \rangle = \langle f(x_n) + g(x_n) \rangle$ converges to $L_1 + L_2$. Therefore by Theorem 9.26, $\lim_{x \to x_0} (f+g)(x) = L_1 + L_2$.

**b)** See Exercise 38.

**c)** See Exercise 39. □

**Theorem 9.31**

Let $X \subseteq \mathbb{R}$, let $f, g: X \to \mathbb{R}$, let $x_0$ be an accumulation point of $X$, and suppose $f$ and $g$ have limits at $x_0$. If $f(x) \leqslant g(x)$ for all $x \in X$, then $\lim_{x \to x_0} f(x) \leqslant \lim_{x \to x_0} g(x)$.

**PROOF** See Exercise 40. □

**Theorem 9.32**

Let $X \subseteq \mathbb{R}$, let $f, g\colon X \to \mathbb{R}$, and let $x_0$ be an accumulation point of $X$. If $f$ is bounded in a neighborhood of $x_0$ and $\lim_{x \to x_0} g(x) = 0$, then $\lim_{x \to x_0} (fg)(x) = 0$.

**PROOF** Let $\varepsilon > 0$. There are positive numbers $\delta_1$ and $B$ such that if $x \in X$ and $|x - x_0| < \delta_1$, then $|f(x)| \leqslant B$. Let $\varepsilon_1 = \varepsilon/B$. There is a positive number $\delta_2$ such that if $x \in X$ and $0 < |x - x_0| < \delta_2$, then $|g(x) - 0| < \varepsilon_1$. Let $\delta = \min\{\delta_1, \delta_2\}$. Then if $x \in X$ and $0 < |x - x_0| < \delta$, $|(fg)(x) - 0| = |f(x)g(x)| = |f(x)|\,|g(x)| \leqslant B\varepsilon_1 = \varepsilon$. Therefore, $\lim_{x \to x_0} (fg)(x) = 0$. □

## EXERCISES 9.3

**36.** Let $X \subseteq \mathbb{R}$, let $f\colon X \to \mathbb{R}$, and let $x_0$ be an accumulation point of $X$. Using the definition of a limit of a function at $x_0$, show that if $L_1$ and $L_2$ are limits of $f$ at $x_0$, then $L_1 = L_2$.

**37.** Prove Theorem 9.28.

**38.** Prove Theorem 9.30, part (b).

**39.** Prove Theorem 9.30, part (c).

**40.** Prove Theorem 9.31.

**41.** Define $f\colon (0, 1) \to \mathbb{R}$ by $f(x) = (x^2 - 1)/(x - 1)$. Prove that $f$ has a limit at 1 and find it.

**42.** Give an example of a function $f\colon (-2, 0) \to \mathbb{R}$ that has a limit at every point of $(-2, 0)$ except $-1$. Use the definition of limit to show that your function has the desired properties.

**43.** Suppose $X \subseteq \mathbb{R}$ and $f\colon X \to \mathbb{R}$ has a limit $L$ at $x_0$. Define $|f|\colon X \to \mathbb{R}$ by $|f|(x) = |f(x)|$ for each $x \in X$. Prove that $|f|$ has limit $|L|$ at $x_0$.

**44.** Let $X \subseteq \mathbb{R}$, let $f, g, h\colon X \to \mathbb{R}$, and let $x_0$ be an accumulation point of $X$. Suppose $f(x) \leqslant h(x) \leqslant g(x)$ for all $x \in X$ and $f$ and $g$ have the same limit $L$ at $x_0$. Show that $h$ has limit $L$ at $x_0$.

**45.** Define $f\colon (0, 1) \to \mathbb{R}$ by $f(x) = (x^3 + 7x^2 + 2x)/(x^2 - 7x)$. Prove that $f$ has a limit at 0 and find that limit.

**46.** Define $f\colon \mathbb{R} \to \mathbb{R}$ by $f(x) = x - \lfloor x \rfloor$. Find those points at which $f$ has a limit, and justify your answer.

**47.** Let $f\colon [a, b] \to \mathbb{R}$ be an increasing function. Show that if $A \neq \varnothing$ and $A \subseteq [a, b]$, then $\{f(x)\colon x \in A\}$ has a least upper bound and a greatest lower bound.

**48.** Let $f\colon [a, b] \to \mathbb{R}$ be an increasing function. For each $x \in (a, b)$, let $U(x) = \text{glb}\,\{f(y)\colon x < y\}$ and $L(x) = \text{lub}\,\{f(y)\colon y < x\}$. Show that $f$ has a limit at $x_0 \in (a, b)$ if and only if $U(x_0) = L(x_0)$.

**49.** Let $f\colon [a, b] \to \mathbb{R}$ be an increasing function. Show that $f$ has a limit at each of $a$ and $b$.

**50.** Suppose $f\colon [a, b] \to \mathbb{R}$ and for all $x \in [a, b]$, $f([a, x])$ has an upper bound. For each $x \in [a, b]$, set $g(x) = \text{lub}(f([a, x]))$. Does $g$ have a limit at $a$ and $b$? Give an example of a function $f$ as above such that $\lim_{x \to a} g(x) \neq g(a)$ and $\lim_{x \to b} g(x) \neq g(b)$.

**51.** Suppose $X$ and $Y$ are disjoint subsets of $\mathbb{R}$ and that $x_0$ is an accumulation point of both $X$ and $Y$, and suppose $f: X \to \mathbb{R}$ and $g: Y \to \mathbb{R}$. Define $h: X \cup Y \to \mathbb{R}$ by $h(x) = f(x)$ if $x \in X$ and $h(x) = g(x)$ if $x \in Y$. Show that $h$ has a limit at $x_0$ if and only if each of $f$ and $g$ has a limit at $x_0$ and $\lim_{x \to x_0} f(x) = \lim_{x \to x_0} g(x)$.

**52.** Show that Exercises 41 and 45 are consequences of Exercise 44.

## 9.4 Continuity

Whereas the value of a function $f$ at $x_0$, even if it exists, bears no relationship to the limit of $f$ at $x_0$, the concept of continuity brings $f(x_0)$ back into the picture. Intuitively, a function $f$ is continuous at $x_0$ provided that $f(x)$ is close to $f(x_0)$ whenever $x$ is close to $x_0$. The precise definition is stated in terms of epsilons and deltas.

**Definition**

Let $X \subseteq \mathbb{R}$, let $x_0 \in X$, and let $f: X \to \mathbb{R}$. Then $f$ is *continuous at* $x_0$ provided that for each $\varepsilon > 0$ there is a $\delta > 0$ such that if $x \in X$ and $|x - x_0| < \delta$, then $|f(x) - f(x_0)| < \varepsilon$. We say that $f$ is *continuous* provided it is continuous at each point of $X$.

Compare this definition with the definition of limit of a function at a point. For continuity at $x_0$, $x_0$ must belong to $X$ but need not be an accumulation point of $X$. Notice that if $x_0$ is not an accumulation point of $X$, then there is a $\delta > 0$ such that if $x \in X$ and $|x - x_0| < \delta$, then $x = x_0$. Thus, if $x_0$ is not an accumulation point of $X$, then $f$ is automatically continuous at $x_0$. So the only interesting case is when $x_0$ is an accumulation point of $X$. In this case, in order for $f$ to be continuous at $x_0$, $f$ must have a limit at $x_0$ and this limit must be $f(x_0)$. Hence we have the following theorem.

**Theorem 9.33**

Let $X \subseteq \mathbb{R}$, let $x_0$ be an accumulation point of $X$, and let $f: X \to \mathbb{R}$. Then the following statements are equivalent:

**a)** $f$ is continuous at $x_0$.

**b)** If $\langle x_n \rangle$ is a sequence in $X$ that converges to $x_0$, then $\langle f(x_n) \rangle$ converges to $f(x_0)$.

**c)** $f$ has a limit at $x_0$ and $\lim_{x \to x_0} f(x) = f(x_0)$.

**PROOF** We prove that (a) implies (b), (b) implies (c), and (c) implies (a).

**(a)→(b)** Suppose $f$ is continuous at $x_0$ and $\langle x_n \rangle$ is a sequence in $X$ that converges to $x_0$. Let $\varepsilon > 0$. Since $f$ is continuous at $x_0$, there is a $\delta > 0$ such that if $x \in X$ and $|x - x_0| < \delta$, then

$|f(x) - f(x_0)| < \varepsilon$. Since $\langle x_n \rangle \to x_0$, there is a natural number $N$ such that if $n > N$, then $|x_n - x_0| < \delta$. Therefore, for $n > N$, $|f(x_n) - f(x_0)| < \varepsilon$ and $\langle f(x_n) \rangle \to f(x_0)$.

**(b)→(c)** Suppose (b) holds. Then by Theorem 9.26, $f$ has a limit at $x_0$ and this limit must be $f(x_0)$.

**(c)→(a)** Suppose $f$ has a limit at $x_0$ and $\lim_{x \to x_0} f(x) = f(x_0)$. Let $\varepsilon > 0$. Then there is a $\delta > 0$ such that if $x \in X$ and $0 < |x - x_0| < \delta$, then $|f(x) - f(x_0)| < \varepsilon$. The only thing necessary to fulfill the definition of continuity is to observe that if $|x - x_0| = 0$ then $x = x_0$, and hence $f(x) = f(x_0)$. □

We prove that the sum, product, and quotient of continuous functions are continuous, but first we establish the following lemma, which we use in proving that the quotient of continuous functions is continuous. Intuitively this lemma promises that if $g$ is continuous at $x_0$ and $g(x_0) \neq 0$, then $g$ is bounded away from 0 in some neighborhood of $x_0$.

**Lemma 9.34**

Let $X \subseteq \mathbb{R}$, let $x_0 \in X$, and let $g: X \to \mathbb{R}$ be continuous at $x_0$. If $g(x_0) \neq 0$, then there are positive numbers $B$ and $\delta$ such that if $x \in X$ and $|x - x_0| < \delta$ then $|g(x)| \geqslant B$.

**PROOF** Let $B = |g(x_0)|/2$. Then $B > 0$ and, since $g$ is continuous at $x_0$, there is a positive number $\delta$ such that if $x \in X$ and $|x - x_0| < \delta$, then $|g(x) - g(x_0)| < B$. Thus, for $x \in X$,

$$|g(x)| = |g(x_0) - (g(x_0) - g(x))| \geqslant |g(x_0)| - |g(x) - g(x_0)| > |g(x_0)| - B$$
$$= |g(x_0)|/2 = B.$$
□

**Theorem 9.35**

Let $X \subseteq \mathbb{R}$, let $x_0 \in X$, and suppose $f, g: X \to \mathbb{R}$ are continuous at $x_0$. Then

**a)** $f + g$ is continuous at $x_0$.

**b)** $fg$ is continuous at $x_0$.

**c)** If $g(x_0) \neq 0$, then $f/g$ is continuous at $x_0$.

**PROOF** The proofs of parts (a) and (b) are left as Exercises 53 and 54.

**c)** Suppose $f$ and $g$ are continuous at $x_0 \in X$ and $g(x_0) \neq 0$. By Lemma 9.34, there are positive numbers $B$ and $\delta$ such that if $x \in X$ and $|x - x_0| < \delta$, then $|g(x)| \geqslant B$. Let $\varepsilon > 0$ and let $\varepsilon_1 = B|g(x_0)|\varepsilon$. Then $\varepsilon_1 > 0$, and since $g$ is continuous at $x_0$, there is a $\delta_1 > 0$ such that if $x \in X$ and $|x - x_0| < \delta_1$, then $|g(x) - g(x_0)| < \varepsilon_1$. Let $\delta_2 =$

$\min\{\delta, \delta_1\}$. Then $\delta_2 > 0$, and if $x \in X$ and $|x - x_0| < \delta_2$, then

$$|(1/g)(x) - (1/g)(x_0)| = |(g(x_0) - g(x))/g(x)g(x_0)| < \varepsilon_1/|g(x)|\,|g(x_0)|$$

$$\leqslant \varepsilon_1/B|g(x_0)| = \varepsilon.$$

Therefore, $1/g$ is continuous at $x_0$. By part (b), $f/g$ is continuous at $x_0$. □

Now we prove that the composition of continuous functions is continuous. The proof is illustrated in Figure 9.6.

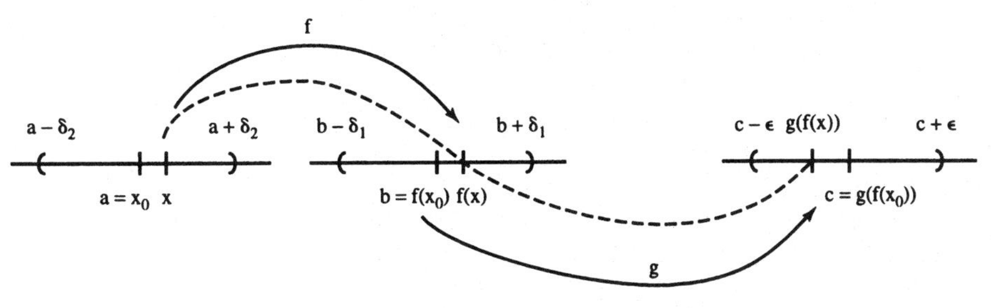

Figure 9.6

**Theorem 9.36**

> Let $X$ and $Y$ be subsets of $\mathbb{R}$, let $x_0 \in X$, and let $f: X \to \mathbb{R}$ be continuous at $x_0$. Suppose the range of $f$ is a subset of $Y$ and $g: Y \to \mathbb{R}$ is continuous at $f(x_0)$. Then $g \circ f$ is continuous at $x_0$.

**PROOF** Let $\varepsilon > 0$. Since $g$ is continuous at $f(x_0)$, there is a $\delta_1 > 0$ such that if $y \in Y$ and $|y - f(x_0)| < \delta_1$, then $|g(y) - g(f(x_0))| < \varepsilon$. Since $f$ is continuous at $x_0$, there is a $\delta_2 > 0$ such that if $x \in X$ and $|x - x_0| < \delta_2$ then $|f(x) - f(x_0)| < \delta_1$. Since the range of $f$ is a subset of $Y$, if $x \in X$ and $|x - x_0| < \delta_2$, then $f(x), f(x_0) \in Y$ and $|f(x) - f(x_0)| < \delta_2$. Therefore, $|(g \circ f)(x) - (g \circ f)(x_0)| = |g(f(x)) - g(f(x_0))| < \varepsilon$, and so $g \circ f$ is continuous at $x_0$. □

Suppose $f: [0, 1] \to \mathbb{R}$ is continuous, $f(0) < 0$, and $f(1) > 0$. Then we would guess that the graph of $f$ must cross the $x$-axis; that is, there is a member $x$ of $(0, 1)$ such that $f(x) = 0$. This is indeed the case, and we have the following theorem.

**Theorem 9.37**

> (Intermediate-Value Theorem) Suppose $f: [a, b] \to \mathbb{R}$ is continuous and $f(a) < y < f(b)$ [or $f(b) < y < f(a)$]. Then there is a $c \in (a, b)$ such that $f(c) = y$.

**PROOF** We assume that $f(a) < y < f(b)$. Let $A = \{x \in [a, b]: f(x) \leqslant y\}$. Then $A \neq \varnothing$ because $a \in A$. Also, $A$ is bounded because $A \subseteq [a, b]$. Therefore, $A$

has a least upper bound; call it $c$. If $c \in A$, then $f(c) \leqslant y$. Suppose $c \notin A$. Then by Exercise 55, $c$ is an accumulation point of $A$. So by Theorem 9.25, there is a sequence $\langle x_n \rangle$ of members of $A$ such that $\langle x_n \rangle \to c$. Since $f$ is continuous at $c$, by Theorem 9.33, $\langle f(x_n) \rangle \to f(c)$. Since $x_n \in A$ for each $n \in \mathbb{N}$, $f(x_n) \leqslant y$ for each $n \in \mathbb{N}$. Therefore, by Theorem 9.16, $f(c) \leqslant y$. Thus in any case, $f(c) \leqslant y$.

Now suppose $f(c) < y$. Let $\varepsilon = (y - f(c))/2$. Then $\varepsilon > 0$, and since $f$ is continuous at $c$, there is a $\delta > 0$ such that if $x \in [a, b]$ and $|x - c| < \delta$ then $|f(x) - f(c)| < \varepsilon$. Since $f(c) < y$ and $f(b) > y$, $c < b$. Thus, there is an $x_1 \in [a, b]$ such that $c < x_1 < c + \delta$. Therefore, $f(x_1) < f(c) + \varepsilon = f(c) + (y - f(c))/2 = (f(c) + y)/2 < y$. This is a contradiction, because $c$ is the least upper bound of $A$. Therefore, $f(c) = y$. □

**Conclusion:** As is easily verified, Theorem 9.37 is the last theorem in this book. Of course, there's lots more analysis to be done, and some will be disappointed that this chapter has not discussed compactness, uniform continuity, and differentiation, to name just a few important omissions. But this book is intended as an introduction to methods of proof, and although we have tried to discuss some mathematical matters that are considered folklore and therefore often given short shrift, our intent has always been to emphasize the shaping of mathematics rather than mathematics itself. We are reminded of the ending of Tom Sawyer:

> So endeth this chronicle. It being strictly a history of a boy, it must stop here; the story could not go much further without becoming the history of a man.

## EXERCISES 9.4

**53.** Prove part (a) of Theorem 9.35.

**54.** Prove part (b) of Theorem 9.35.

**55.** Suppose $f: [a, b] \to \mathbb{R}$ is continuous and $f(a) < y < f(b)$. Let $A = \{x \in [a, b]: f(x) \leqslant y\}$, and let $c$ be the least upper bound of $A$. Show that if $c \notin A$, then $c$ is an accumulation point of $A$.

**56.** Let $f: \mathbb{R} - \{0\} \to \mathbb{R}$ be defined by $f(x) = 1/x$. Is $f$ continuous? Is $f$ continuous at 0?

***57.** Let $f: [0, 1] \to [0, 1]$ be a continuous function. Prove that there is a number $x$ such that $f(x) = x$. (Such a number is called a fixed point of $f$). [*Hint:* Use the Intermediate-Value Theorem.]

**58.** Let $f: [a, b] \to \mathbb{R}$ be a continuous one-to-one function such that $f(a) < f(b)$. Prove that if $a \leqslant x < y \leqslant b$, then $f(x) < f(y)$.

## Writing Exercises

**1.** There are two common intuitive explanations of what it means to say that a function $f$ is continuous:

**a)** You can draw the function without lifting your pencil.

**b)** For any $a$, $f(x)$ is near $f(a)$ whenever $x$ is near $a$.

What is the basis for the first of these explanations? In your opinion, which explanation comes closer to the spirit of the definition of continuity? Justify your answer with examples.

**2.** The Parthian cavalry invented the military tactic that is now known as the parting shot. A parting shot: Review the text and find some mathematical explanation that strikes you as unclear. Write your own explanation, which clarifies the offending paragraph. In the untoward event that everything in the text seems perfectly clear, write an essay in which you defend the use of this text in future classes.

# REFERENCES

The following list, which includes references cited in the text, is intended to suggest suitable sources for further reading. Many of the references are available in paperback and should be in every interested reader's mathematical library; some references are, alas, out of print (we have indicated this by an asterisk next to the reference number). The notation "G" indicates that the reference is of general interest or pertains to many topics from our text; the notation "W" indicates that the reference concerns writing or discovering mathematical arguments. Bracketed numbers indicate those chapters from our text to which a reference is closely related.


**1.** Allenby, R. B. (1991) *Rings, Fields and Groups: An Introduction to Abstract Algebra* 2nd ed. UK: E. Arnold: Routledge Chapman & Hall. [8]

**2.** Cajori, Florian (1993) *A History of Mathematical Notations,* Vols 1–2. New York: Dover Publications. [5, G]

***3.** Devlin, K. J. (1977) "The Axiom of Constructibility: A Guide for the Mathematician." *Lecture Notes in Mathematics,* Vol. 617. New York: Springer-Verlag. [2, 7]

**4.** Dodge, Clayton W. (1975) *Numbers & Mathematics.* Boston: Prindle, Weber & Schmidt. [1–7, G]

**5.** Durbin, John R. (1991) *Modern Algebra: An Introduction* 3rd ed. New York: Wiley. [8]

**6.** Euclid (1926) *The Elements.* Heath, Thomas L., ed. New York: Dover Publications. [3]

**7.** Eves, Howard (1969) *In Mathematical Circles.* Boston: PWS Publishers. [G]

**8.** Eves, Howard (1972) *Mathematical Circles Revisited* Boston: PWS Publishers. [G]

**9.** Eves, Howard (1972) *Mathematical Circles Squared.* Boston: PWS Publishers. [G]

**10.** Eves, Howard (1988) *Return to Mathematical Circles.* Boston: PWS Publishers. [G]

***11.** Fadiman, Clifton. ed. (1985) *Fantasia Mathematica: An Introduction.* New York: Simon & Schuster. [G]

**12.** Gardner, Martin (1982) *Aha! Gotcha: Paradoxes to Puzzle & Delight.* New York: W. H. Freeman. [1, G]

**13.** Gaughan, Edward (1993) *Introduction to Analysis.* 4th ed. Brooks-Cole. [9]

**14.** Hadamard, Jacques (1945) *Psychology of Invention in the Mathematical Field.* New York: Dover Publications. [W]

**15.** Halmos, Paul R. (1983) "How to Write Mathematics." *Selecta: Expository Writing.,* Vol. 2, New York: Springer-Verlag, 157–190. [W]

**16.** Halmos, Paul R. (1991) *Naive Set Theory.* (Undergraduate Texts in Mathematics Series.) New York: Springer-Verlag. [2, 3, 7]

17. Herstein, Israel N. (1989) *Abstract Algebra.* 2nd ed. Macmillan. [8]

18. Hilbert, David. (1902) "Mathematical Problems." (Lecture delivered before the International Congress at Paris in 1900.) Translated by Mary Winston Newsom. *Bull. Amer. Math. Soc.,* No. 8, 437–79. [7]

*19. Kazarinoff, Nicholas D. (1970) *Ruler and the Round.* Boston: Prindle, Weber & Schmidt. [8]

*20. Kline, M. (1969) *Mathematics in the Modern World: Readings from Scientific American.* New York: W. H. Freeman. [G]

*21. Lakatos, Imre. (1977) *Proofs and Refutations.* Cambridge, England: Cambridge University Press. [W]

22. Larson, L. C. (1993) *Problem-Solving Through Problems.* ed. by P. R. Halmos. New York: Springer-Verlag. [1, 3, W]

23. Ore, Oystein. (1990) *Graphs & Their Uses.* rev. ed. Wilson, Robin New Mathematical Library: No. 34 Math Assn. New York: Random House. [6]

*24. Pinter, Charles C. (1971) *Set Theory.* Reading. Mass.: Addison-Wesley. [7]

25. Polya, George (1981) *Mathematical Discovery: On Understanding, Learning, and Teaching Problem Solving,* Vol. 1 New York: Wiley. [W]

26. Polya, George (1981) *Mathematical Discovery: On Understanding, Learning, and Teaching Problem Solving.* Vol. 2 New York: Wiley. [W]

27. Polya, George (1971) *How to Solve It.* Princeton, N.J.: Princeton University Press. [1, W]

28. Polya, George et al (1990) *Notes on Introductory Combinatorics.* Progress in Computer Science Series, Vol 4. Boston: Birkhauser. [4, 6]

29. Rothman, Tony (1982) *Genius and biographies, The fictionalization of Evariste Galois.* Amer. Math. Monthly 89 No. 2, 225–240. [8]

30. Smullyan, Raymond (1986) *What Is the Name of This Book? The Riddle of Dracula & Other Logical Puzzles.* S & S Trade. [1, G]

31. Solow, Daniel (1990) *How to Read & Do Proofs: An Introduction to Mathematical Thought Processes.* 2nd ed. New York: Wiley. [1, W]

32. Stewart, Ian and David O. Tall (1977) *The Foundation of Mathematics.* New York: Oxford University Press. [1–7, G]

33. Vaught, Robert L. (1994) *Set Theory: An Introduction.* Cambridge, Mass.: Birkhauser. [2]

*34. Vilenkin, N. Ya. (1968) *Stories About Sets.* Orlando, Fla: Academic Press. [7]

35. Weeks, Jeffrey R. (1985) "The Shape of Space." *Pure and Applied Mathematics,* No. 96. New York: Marcel Dekker. [4]

# SELECTED ANSWERS AND HINTS

## CHAPTER 1

**1.** (a) and (c) are propositions. In (e), "I" acts as a variable.

**3. a)** The hypothesis is: Mary is 24 years old. The conclusion is: I am a monkey's uncle.

**c)** The hypothesis is: $r$ is a rational number. The conclusion is: $r^2$ is rational.

**e)** The hypothesis is: $a$ is rational and $b$ is irrational. The conclusion is: $a+b$ is irrational.

**g)** The hypothesis is: The candidate passes the driver's test. The conclusion is: The candidate must be able to parallel park.

**5.** (b) is true.

**7. a)** The converse is: If $2<5$, then $\sqrt{2}<\sqrt{5}$. The contrapositive is: If $2 \geqslant 5$, then $\sqrt{2} \geqslant \sqrt{5}$.

**11.** Yes

**13. a)** If $3 \leqslant 1$ then $5 \leqslant 1$.

**16.** (c)

**17. a)** $\neg P \vee Q$

**19.** Let $Q$ be the proposition $3=4$. Then $P \vee Q$ is false and $P \wedge Q$ is false, so $P \vee Q \leftrightarrow P \wedge Q$ is true. Let $Q$ be the proposition $3<4$. Then $P \vee Q$ is true and $P \wedge Q$ is false, so $P \vee Q \leftrightarrow P \wedge Q$ is false.

**22.** (a) and (c)

**25. a)** converse: $(Q \vee R) \rightarrow P$ or $(Q \rightarrow P) \wedge (R \rightarrow P)$
contrapositive: $\neg(Q \vee R) \rightarrow \neg P$ or $\neg Q \wedge \neg R \rightarrow \neg P$
negation: $P \wedge \neg(Q \vee R)$ or $P \wedge \neg Q \wedge \neg R$

**27. b)** Suppose John is not ten years old. Then Mary is ten years old. Thus John is not smart, and we have a contradiction. So the statement is true.

**29.** *Hint*: If 1.1 (c) holds for all propositions $P$ and $Q$, then it holds for $\neg P$ and $\neg Q$.

**31.** By Proposition 1.2(e), $P \rightarrow R$, since $P \rightarrow Q$ and $Q \rightarrow R$. By Proposition 1.2(d), $R$ is equivalent to $P$, since $R \rightarrow P$ and $P \rightarrow R$. In the same manner, it can be shown that $Q$ is equivalent to $P$ and that $R$ is equivalent to $Q$.

**35. a)** $2 \neq 4$, but $f(2)=f(4)$.

**37. a)** If $a=0$ or $b=0$, then $ab \neq 0$.

**39.** negation: Either $x$ is an even integer or $x>17$, and either $x$ is not a multiple of 4 or $x<5$.

contrapositive: If $x$ is not a multiple of 4 or $x < 5$, then $x$ is not an even integer and $x \leqslant 17$.

**42.** An integer $x$ does not have property $P$ provided that there exist $a$ and $b$ such that $x$ divides $ab$ but $x$ does not divide $a$ and $x$ does not divide $b$.

**44.** There is a pair of real numbers $a$ and $b$ with $a < b$ such that if $r$ is any rational number then either $r \leqslant a$ or $r \geqslant b$.

**46.** The limit of $f(x)$, as $x$ approaches $a$, is not $L$, provided that there is an $\varepsilon > 0$ such that if $d$ is any positive number, then there is an $x \neq a$ such that $|x - a| < d$ but $|f(x) - L| \geqslant \varepsilon$.

**48.** The negation is: $x$ is a positive number, and for each $\varepsilon > 0$, $x \geqslant \varepsilon$ or $1/\varepsilon \geqslant x$. The contrapositive is: If for each $\varepsilon > 0$ there exists an $x$ such that $x \geqslant \varepsilon$ or $1/\varepsilon \geqslant x$, then $x$ is either negative or zero.

**50. a)** $\{3\}$

**53. a)** There is a tax return that can be postmarked after April 15.

**54. a)** is true.

**58.** Since $A$ divides $B$, there is an integer $P$ such that $B = A \times P$. Since $B$ divides $C$, there is an integer $Q$ such that $C = B \times Q$. Therefore $C = B \times Q = (A \times P) \times Q = A \times (P \times Q)$. Since $P \times Q$ is an integer, $A$ divides $C$.

**63.** Since $A$ is an even integer, there is an integer $M$ such that $A = 2M$. Since $B$ is an odd integer, there is an integer $N$ such that $B = 2N + 1$. So $A + B = 2M + 2N + 1 = 2(M + N) + 1$, and hence $A + B$ is odd.

**66.** Suppose $x = -1$. Then $x^3 + x^2 + x + 1 = (-1)^3 + (-1)^2 + (-1) + 1 = 0$. Suppose $x^3 + x^2 + x + 1 = 0$. Now $x^3 + x^2 + x + 1 = (x+1)(x^2+1)$, so $(x+1)(x^2+1) = 0$. Hence either $x + 1 = 0$ or $x^2 + 1 = 0$. Since $x$ is a real number, $x^2 + 1 \neq 0$. Therefore $x + 1 = 0$, and hence $x = -1$.

**70.** *Hint*: Consider Theorem 1.6.

**71.** Recall that $(n-1)! = (n-1)(n-2)\cdots(3)(2)(1)$.

**78.** Suppose $x$ is a positive real number and $x/(x+1) \geqslant (x+1)/(x+2)$. Then $x + 1$ and $x + 2$ are positive real numbers, so $(x+1)(x+2)$ is a positive real number. Multiplying the above inequality by $(x+1)(x+2)$, we obtain $x(x+2) \geqslant (x+1)^2$. So $x^2 + 2x \geqslant x^2 + 2x + 1$, and hence $0 \geqslant 1$. This is a contradiction.

**83.** *Hint*: Show that $B$ divides $R - S$ and compare the sizes of $|B|$ and $|R - S|$.

**85.** *Hint*: Assume that $P$ is the largest prime and consider any prime $Q$ that divides $P! + 1$. How are $P$ and $Q$ related?

**88.** No, it is not true: $n = 3$ is a counterexample because 3 is prime but $2^3 + 1 = 9$ is not prime.

**93.** Let $n$ be a natural number greater than 1. Then 2 divides $n!$, so $n!$ is even; that is, there is an integer $k$ such that $n! = 2k$. Therefore $n! + 1 = 2k + 1$, so $n! + 1$ is odd.

**99.** Only one. If $q - p = 3$, then one of $p$ and $q$ must be even and the other odd. Since the only even prime is 2, $p$ must be 2 and $q$ must be 5.

## CHAPTER 2

**5.** The members of $\mathscr{P}(A)$ are: $\varnothing$, $\{1\}$, $\{2\}$, and $A$.

**7.** No

**9.** 128.

**11.** (a), (b), (d), and (h) are true.

**15. a)** This set is empty because no real number $x$ can be between $-\sqrt{5}$ and $\sqrt{5}$ and greater than 3.

**21. a)** $A \cup B = \{1, 3, 4, 5, 6\}$

**c)** $A \cap B = \{3, 6\}$

**e)** $(A \cup B) \cap C = \{3, 6\}$

**g)** $(A')' = \{1, 3, 6\}$

**i)** $A' \cup B' = \{1, 2, 4, 5, 7, 8, 9, 10\}$

**23.** Suppose $A \cup B = A$, and let $x \in B$. Then $x \in A \cup B$, and so $x \in A$. Therefore $B \subseteq A$. Thus we have proved that if $A \cup B = A$ then $B \subseteq A$.

**28. a)**

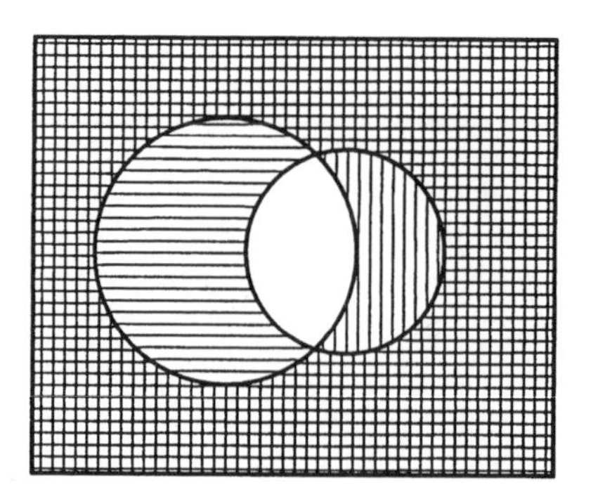

**29. a)** Suppose $\emptyset \cap A \neq \emptyset$. Then there exists $x \in \emptyset \cap A$. Hence $x \in \emptyset$, and we have a contradiction. Therefore $\emptyset \cap A = \emptyset$. If $x \in A$ then $x \in \emptyset$ or $x \in A$, so $A \subseteq \emptyset \cup A$. Suppose $x \in \emptyset \cup A$. Then $x \in \emptyset$ or $x \in A$. Since $x \notin \emptyset$, $x \in A$. So $\emptyset \cup A \subseteq A$ and $\emptyset \cup A = A$.

**39.** Suppose $A \cup B = B$. By Proposition 2.5(b), $A \cap B \subseteq A$. Let $x \in A$. By Proposition 2.5(c), $A \subseteq A \cup B$. Therefore $x \in A \cup B$. Since $A \cup B = B$, $x \in B$. Therefore $x \in A$ and $x \in B$, so $x \in A \cap B$. Hence $A \subseteq A \cap B$, and thus $A \cap B = A$.

**43.** Suppose $A - B \subseteq C$, and let $x \in A - C$. Then $x \in A$ and $x \notin C$. Since $x \notin C$ and $A - B \subseteq C$, $x \notin A - B$. Therefore $x \notin A$ or $x \in B$. But $x \in A$, so $x \in B$. Therefore $A - C \subseteq B$.

**54.** $\cup_{n \in \mathbb{N}} A_n = \mathbb{N}$ and $\cap_{n \in \mathbb{N}} A_n = \emptyset$

**56.** $\cup_{n \in \mathbb{N}} A_n = [0, 2]$ and $\cap_{n \in \mathbb{N}} A_n = \{1\}$

**60.** $\cup_{x \in \mathbb{R}} A_x = \mathbb{R}$ and $\cap_{x \in \mathbb{R}} A_x = \{3\}$

**71.** *Hint*: Take, as obvious, that any subset of a finite set is finite.

**73.** $\emptyset, \{\emptyset\}, \{\emptyset, \{\emptyset\}\}, \{\emptyset, \{\emptyset\}, \{\emptyset, \{\emptyset\}\}\}$

**75. a)** $\cup \emptyset = \cup \{\emptyset\} = \emptyset$

**c)** Let $S = \{\emptyset\}$. Then $\cup S = \emptyset$, so $\cup S \in S$.

**77. a)** Of course $76 < 122$. The question is, in terms of sets, what does it mean to say that $76 < 122$?

**79. a)** *Hint*: Consider a singleton set.

**b)** Try a proof by contradiction. Assume that there is a family $S = \{A_n : n \in \mathbb{N}\}$ such that for each $n \in \mathbb{N}$, $A_{n+1} \in A_n$ and show that no member of $S$ is disjoint from $S$.

## CHAPTER 3

**1. e)** Let $S = \{n \in \mathbb{N}: 2 + 5 + 8 + \cdots + (3n - 1) = n(3n + 1)/2\}$. Since $1[3(1) + 1]/2 = 2$, $1 \in S$. Suppose $n \in S$. Then $2 + 5 + 8 + \cdots + (3n - 1) = n(3n + 1)/2$. So $2 + 5 + 8 + \cdots + (3n - 1) + [3(n + 1) - 1] = n(3n + 1)/2 + 3(n + 1) - 1 = [n(3n + 1) + 6(n + 1) - 2]/2 = (3n^2 + 7n + 4)/2 = (3n + 4)(n + 1)/2 = (n + 1)[3(n + 1) + 1]/2$. Therefore $n + 1 \in S$. By the Principle of Mathematical Induction, $S = \mathbb{N}$.

**l)** Let $S = \{n \in \mathbb{N}: 3^n - 1$ is divisible by $2\}$. Since $3^1 - 1 = 2$ is divisible by 2, $1 \in S$. Suppose $n \in S$. Then $3^n - 1$ is divisible by 2. Now $3^{n+1} - 1 = 3 \cdot 3^n - 3 + 3 - 1 = 3(3^n - 1) + 2$. Therefore by Exercise 62 of Chapter 1, $3^{n+1} - 1$ is divisible by 2. So $n + 1 \in S$. By the Principle of Mathematical Induction, $S = \mathbb{N}$.

**3.** *Hint*: Why is $2 \times 64$ the answer to Problem 9 in Chapter 2?

**9.** If $n = 1$, $A_3$ does not exist.

**13.** *Hint*: Consider $n = 40$ or $n = 41$.

**24.** Let $S = \{n \in \mathbb{N}: n$ can be written as the sum of natural numbers, each of which is a 2 or a $3\}$. Since $5 = 2 + 3$, $5 \in S$. Since $6 = 3 + 3$, $6 \in S$. Suppose $n \geqslant 6$ and $\{5, 6, \ldots, n\} \subseteq S$. Then $n + 1 = (n - 1) + 2$, and since $n - 1 \in S$, it can be written as the sum of natural numbers, each of which is a 2 or a 3. Therefore $n+1 \in S$. By the Extended Second Principle of Mathematical Induction, $\{n \in \mathbb{N}: n \geqslant 5\} \subseteq S$.

**33. a)** It is a routine calculation to show that if $1<k<16$ then $k^4>2^k$, that $16^4=2^{16}$, and that $17^4<2^{17}$.

**35. a)** 16

**b)** 31

*Remarks*: The whole point of this exercise is to show that we cannot determine how things are going by considering any finite number of cases. Nevertheless, it is natural to ask, what is the real formula for cutting up a pie? The formula for the maximum number of pieces of pie is $f(n)=(n^4-6n^3+23n^2-18n+24)/24$. Although this formula can be proved by induction, the authors believe the exercise unreasonably difficult at this stage. It is not even obvious that $f(n)$ is always an integer, but this result, at least has been established in Exercise 1(p).

**39.** *Note*: We hope that in each case you assumed that you already knew how to do the previous case so that you did not write the description again, but merely stated that you would use previous knowledge in order to make use of the previous case.

**50.** *Hint*: For a given natural number $n$, which of the following numbers are divisible by 3: $6n+1$, $6n+2$, $6n+3$, $6n+4$, $6n+5$, $6n+6$? How about $x_{6n+1}, x_{6n+2}, \ldots, x_{6n+6}$?

**61. a)** For $n=1, 2, 3, 4, 5$, $f(n)=2, 3, 5, 8, 13$ respectively.

**65.** *Hint*: It is helpful to prove parts (a) and (b) simultaneously.

**71.** *Hint*: Pólya's Principle applies. Prove the general result that the second player has a winning strategy in the "name-blah" game as long as "blah" is a multiple of 4.

**75.** *Hint*: Let $S=\{i \in \mathbb{N}: \text{ if } j \leqslant i, \text{ then } f_j=a_j\}$. Explain why it is sufficient to show that $S=\mathbb{N}$. Show that $2 \in S$ and make clever use of the equalities $(1+\sqrt{5})/2+(1-\sqrt{5})/2=1$ and $[(1+\sqrt{5})/2]\times[1-\sqrt{5})/2]=-1$.

**79.** *Hint*: Let $P_n$ denote the probability that two consecutive heads do not appear in $n$ tosses. Consider two cases: one in which the first toss is heads and one in which it is tails. Show that if $n>2$, then $P_n=(1/2)P_{n-1}+(1/4)P_{n-2}$.

**86. a)** $72 \cap 12=\gcd(72, 12)=12$;
$(72 \cap 12)'=720/12=60$.
$30 \cap 9=\gcd(30, 9)=3$;
$(30 \cap 9)'=720/3=240$.
$1 \cap 720=\gcd(1, 720)=1$;
$(1 \cap 720)'=720/1=720$.

**92.** *Hint*: Pólya's Principle applies. You can prove that every common divisor of $a$ and $b$ is a common divisor of $b$ and $r$ and that every common divisor of $b$ and $r$ is a common divisor of $a$ and $b$.

**104.** *Hint*: Use Exercise 103.

**109.** *Hint*: gcd $(a, b)$ divides $12a-5b$.

**110.** *Hint*: If 3 does not divide $p$, then $p$ can be written as $3m+1$ or as $3m-1$.

**112.** *Hint*: Suppose that $p_1, p_2, \ldots, p_r$ are the only primes that can be written in this form. Let $N=p_1p_2\cdots p_r-1$ and show that there exists a prime of the form $4n+3$ that divides $N$.

**116.** Since $3=\gcd(6, 15)$ and 3 does not divide 83, by Theorem 3.16(a), $6x+15y=83$ does not have a solution.

**117.** *Hint*: One solution is $x=23$ and $y=1$.

## CHAPTER 4

**3.** *Hint*: List a member of $A \times (B \times C)$. Is the member you listed also a member of $(A \times B) \times C$?

**5. a)** $\{-\sqrt{15}, \sqrt{15}\}$

**e)** $\{y \in \mathbb{R}: y>3 \text{ or } y<-3\}$

**8.** There are of course many correct answers. For example, $S=\{(x, y) \in \mathbb{R} \times \mathbb{R}: y=x\}$ and $T=\{(x, y) \in \mathbb{R} \times \mathbb{R}: y=x+1\}$.

**13.** 7, 14, 21, 28, and 35 are members of $R[7]$.

**17.** (a) and (d) are symmetric.

**18. d)**

**19.**

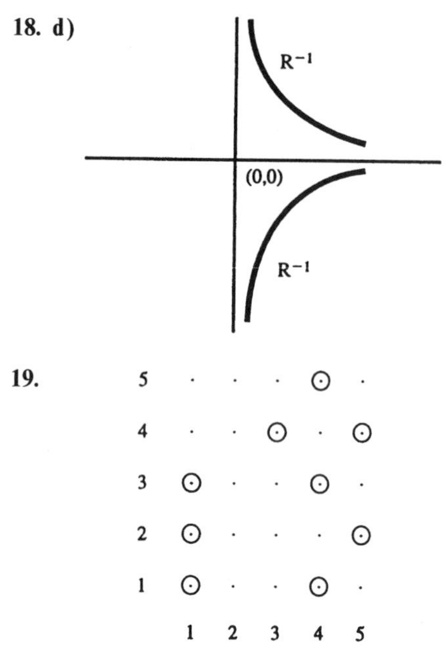


**22.** No

**25.** Yes

**26. c)** $1, 1 + 2\pi, 1 + 4\pi$

**29.** $R$ is not symmetric, but it is transitive.

**31.** $R$ is reflexive on $\mathbb{N}$ and transitive, but not symmetric.

**34. i)** (b) and (d) are reflexive on $\mathbb{R}$.

**36. b)** $[(0, 0)] = \{(0, 0)\}$

**38. a)** 1, 2, 3, 4, 5, 6, 7, and 8 are the members of $\cup\mathscr{A}$.

**39.** (3, 4), (6, 8), (9, 12), and (12, 16) are members of $[(6, 8)]$.

**40. b)** $f(x) = x^2 + 17x + 11$, $g(x) = 2x^2 + 34x$, $h(x) = 3x^2 + 51x$, $k(x) = 17x^2 + 289x$

**41.** We need to prove: (1) If $R = R^{-1}$, then $(y, x) \in R$ whenever $(x, y) \in R$, and (2) if $(y, x) \in R$ whenever $(x, y) \in R$, then $R = R^{-1}$.

**50.** (1, 1), (1, 2), (2, 1), (2, 2), (3, 3), (4, 4), (4, 5), (5, 4), (5, 5)

**51.** $\{1\}, \{2, 3\}, \{4, 5, 6\}$

**53.** The Möbius strip has only one edge and one side. The resulting strip is not a Möbius strip. It has 2 twists, and it has two edges and two sides. If someone is not convinced that the resulting strip is not a Möbius strip, tell him (her) to cut it again in the same manner. After it is cut the second time, it will be clear that it had to be different from the Möbius strip because the second cut will result in an object that consists of two pieces.

**57. a)** Yes

**b)** Yes

**61. a)**

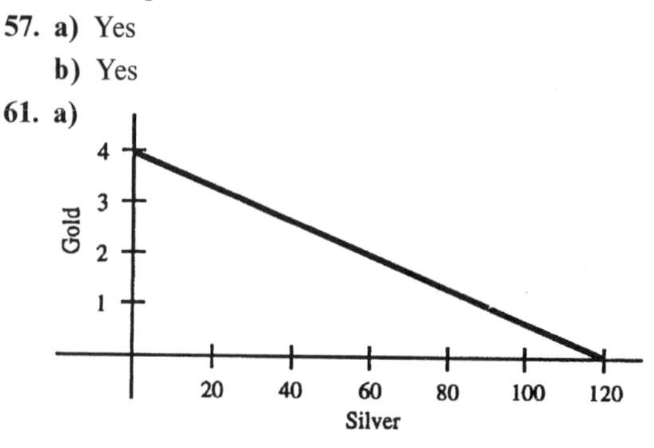


**69.** Let $n = 4$, $a = 0$, and $b = 2$.

**72.** If $b$ is an integer and $\gcd(a, n) = 1$, then we can prove that there is an integer $c$ such that $ac \equiv b \pmod{n}$. So we need to choose $a$ and $n$ such that $\gcd(a, n) \neq 1$. Let $n = 6$, $a = 4$, and $b = 1$. Then those integers that are congruent to $b$ mod 6 are integers of the form $6p + 1$ whereas $ac$ is a multiple of 4.

**73.** $[10] = [3]$, $[10^2] = [3^2] = [2]$, $[10^3] = [3^3] = [6]$, $[10^4] = [3^4] = [4]$, $[10^5] = [3^5] = [5]$, and $[10^6] = [3^6] = [1]$.
*Remark*: If $n$ is a natural number of the type discussed in this problem, then $n$ is odd (indeed $n$ is an odd prime) and the decimal representation of $1/n$ repeats in a block of $n - 1$ digits. The first $(n - 1)/2$ digits and the last $(n - 1)/2$ digits add, digit by digit, to 9. For example, $1/7 = .\overline{142857}$, where $1 + 8 = 4 + 5 = 2 + 7 = 9$, and $1/17 = .\overline{0588235294117647}$. Finally, we note that $2/7 = .\overline{285714}$, $3/7 = .\overline{428571}$, $4/7 = .\overline{571428}$, $5/7 = .\overline{714285}$, and $6/7 = .\overline{857142}$.

**77. a)** $8 \equiv 1 \pmod{7}$, so, by Exercise 70, $8^n \equiv 1 \pmod{7}$ for each natural number $n$.

**b)** Since $8^n \equiv 1 \pmod{7}$ for each natural number $n$, $(a_k \ldots a_1 a_0)_8 = a_k 8^k + \cdots + a_1 8^1 + a_0 \equiv a_k + \cdots + a_1 + a_0 \pmod{7}$. Therefore $(a_k \ldots a_1 a_0)_8$ is divisible by 7 if and only if $\Sigma_{i=0}^{k} a_i$ is divisible by 7. Since $1 + 1 + 4 + 2 + 6 + 5 = 19$, the base 8 number 114265 is not divisible by 7.

**78.** *Hint*: Consider the congruence classes mod 9 and use the Pigeonhole Principle.

**84. a)** Use Exercise 81 with $a = 2$.

**b)** Since $1024 = 3(341) + 1$, $2^{10} \equiv 1 \pmod{341}$. It follows that $(2^{10})^{34} \equiv 1 \pmod{341}$, and so $2^{341} \equiv 2 \pmod{341}$.

**90. a)** $R^{-1} = \{(2, 1), (5, 3), 2, 2)\}$

**c)** $\{(2, 2), (5, 5), (5, 2)\}$

**91. b)** Let $R = \{(1, 2), (2, 1)\}$ and $S = \{(1, 3), (3, 1)\}$. Then $R$ and $S$ are symmetric, but $R \circ S = \{(3, 2)\}$, so $R \circ S$ is not symmetric.

**d)** Let $R = \{(1, 2), (3, 4)\}$ and $S = \{(5, 6), (7, 8)\}$. Then $R \circ S = \varnothing$ and $S \circ R = \varnothing$, so $R \circ S = S \circ R$, but $R \neq S$. By adding $(9, 9)$ to both $R$ and $S$ we can avoid empty compositions.

**f)** Let $R = \{(1, 2), (3, 2)\}$, $S = \{(4, 1), (5, 1)\}$, and $T = \{(4, 3), (5, 3)\}$. Then $R \circ S = \{(4, 2), (5, 2)\}$ and $R \circ T = \{(4, 2), (5, 2)\}$. Hence $(R \circ S) \cap (R \circ T) = \{(4, 2), (5, 2)\}$. Now $S \cap T = \varnothing$, so $R \circ (S \cap T) = \varnothing$. Hence $(R \circ S) \cap (R \circ T) \neq R \circ (S \cap T)$.

**92.** Let $R = \{(1, 2), (3, 2)\}$ and $S = \{(4, 1), (5, 1)\}$.

**94.** *Hints*: (1) Let $B$, $C$, $D$, and $E$ be relations on $A$. Prove that if $B \subseteq C$ and $D \subseteq E$, then $B \circ D \subseteq C \circ E$. (2) Prove that $R^{\infty} \circ R = R^{\infty}$.

**95. b)** Note that $R = E$, because $R$ is symmetric. There are two equivalence classes, the set of all positive numbers and the set of all negative numbers.

**99. b)** Functions $f$ and $g$ are in the same equivalence class with respect to the equivalence relation $E$ if and only if they have the same integral on $[0, 1]$.

**101. a)** $S = \{1, 2\}$, $R = S \times S$

**102.** *Hint*: Consider the directed graph of Example 18(b).

**105. a)** Let $(\mathbb{N}, \leqslant)$ be the set of all natural numbers with the usual order. Then $(2, 3) = \varnothing$.

**107.** The statement is false. Let $<$ be the usual less than relation on $\mathbb{R}$. Then $<$ is a strict partial order, but since $3 \nless 3$, $<$ is not a partial order on $\mathbb{R}$.

## CHAPTER 5

**1. a)** $f$ is one-to-one: If $2m + 1 = 2n + 1$, then $m = n$; $f$ does not map onto $\mathbb{Z}$. Indeed $f(\mathbb{Z})$ is the set of all odd integers.

**3.**

```
⊙ · · ·
· · · ⊙
· · ⊙ ·
· ⊙ · ·
  f∘g
```

**6.** $f$ is not one-to-one, but $f$ maps $\mathbb{N}$ onto $\mathbb{N}$.

**7. a)** $f(n) = 2n$

**13. b)** The second term is $(2, 0)$, the ninth term is $(9, 7/5)$, and the tenth term is $(10, 8/5)$.

**15.** $g(n) = 2n - 1$, $\mathrm{Dom}\,(g) = \mathbb{N}$

**16.** *Hint*: First prove that if $g\colon \mathbb{N} \to \mathbb{N}$ and $g(n + 1) > g(n)$ for all $n \in \mathbb{N}$, then $g(n) \geqslant n$ for all $n \in \mathbb{N}$.

**18.** Suppose $f(a) = f(b)$. Then $\sqrt{17 - 3a} = \sqrt{17 - 3b}$, so $17 - 3a = 17 - 3b$. Therefore $a = b$, and hence $f$ is one-to-one. Let $y \in [0, \infty)$. Then $x = (17 - y^2)/3 \in (-\infty, \frac{17}{3}]$ and $f(x) = \sqrt{17 - 3((17 - y^2)/3)} = \sqrt{y^2} = y$. Therefore $f$ maps $(-\infty, \frac{17}{3}]$ onto $[0, \infty)$. Let $f^{-1}(x) = (17 - x^2)/3$.

**23.** Let $c \in C$. There exists $a \in A$ such that $f \circ g(a) = c$. Thus there is $b \in B$ such that $g(a) = b$ and $f(b) = c$. Therefore $f$ maps $B$ onto $C$. *Example*: Let $A = \{1\}$, $B = \{3, 4, 5\}$, and

$C = \{6, 7\}$. Define $f$ and $g$ by $g(1) = 3$, $f(3) = 6$, and $f(4) = f(5) = 7$. The example shows that in Theorem 6.7(a) it is necessary that $g$ maps onto $B$.

**32.** **a)** $(123)(4576)$

**34.** (a), (c), and (d)

**35.** **c)** *Hint*: Consider part (a).

**d)** *Hint*: Consider parts (a) and (b).

**36.** $(12)(34)(56)$

**38.** $\alpha \circ \beta = (265)(34)$

**40.** The partition is the collection of all singleton subsets of $X$.

**43.** Yes

**46.** **a)** $\mathbb{R} - \{3\}$

**c)** $\mathbb{R} - \{0\}$

**e)** $(3, \infty)$

**g)** $[3, 12) \cup (12, \infty)$

**i)** $(-\infty, -1/3] \cup (0, \infty)$

**47.** **a)** $f^{-1}(x) = 3 + 1/x$

**c)** $(f + g)(x) = 1/(x - 3) + \sqrt{x - 3}$

**e)** $f \circ g(x) = 1/(\sqrt{x - 3} - 3)$

**g)** $(f \circ g)^{-1}(x) = (3 + 1/x)^2 + 3$

**51.**

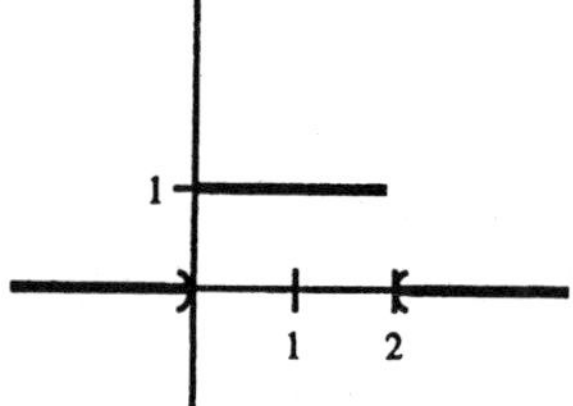


**53.** **a)** Dom $(f) = \mathbb{R} - \{3\}$; $f$ is one-to-one: Suppose that $f(a) = f(b)$. Then $7/(a - 3) = 7/(b - 3)$, so $b - 3 = a - 3$ and $a = b$. Dom $(f^{-1}) =$ range $(f) = \mathbb{R} - \{0\}$.

**57.** $f: \mathbb{R} \to \mathbb{R}$ defined by $f(x) = x^2$ and $g: \mathbb{R} \to \mathbb{R}$ defined by $g(x) = x$

**59.** **a)** 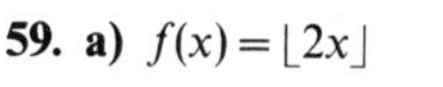 $f(x) = \lfloor 2x \rfloor$

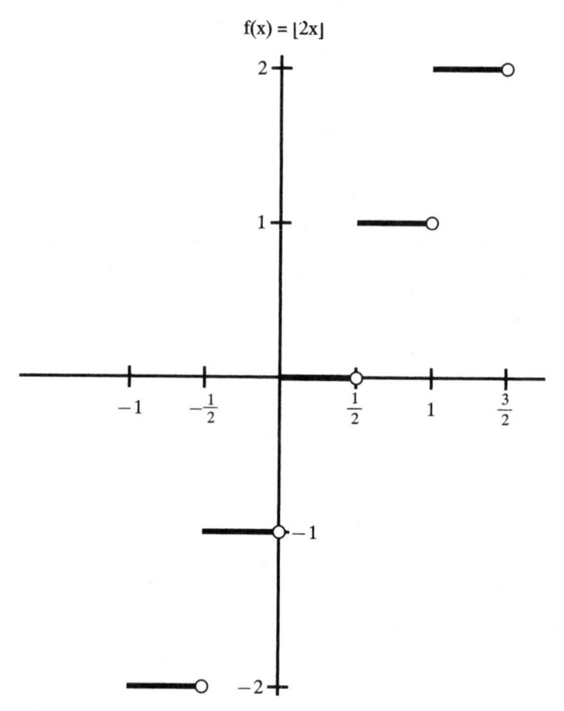


**61.** *Hint*: In the case in which $d \leqslant n$, the list of positive multiples of $d$ not exceeding $n$ is $d$, $2d$, $3d, \ldots, kd$, where $kd \leqslant n$ but $(k + 1)d > n$.

**63.** Since $f(0) = f(1) = 0$, $f \mid \mathbb{Z}$ is not one-to-one. Since $f(2n) = n$ for each $n \in \mathbb{Z}$, $f \mid \mathbb{Z}$ maps onto $\mathbb{Z}$.

**65.** **a)** $\lfloor \pi \rfloor ! = 2^{\alpha(2)} \cdot 3^{\alpha(3)}$. $\alpha(2) = \lfloor \pi/2 \rfloor + \lfloor \pi/4 \rfloor + \cdots$ and all remaining terms are, like $\lfloor \pi/4 \rfloor$, equal to zero. $\alpha(3) = \lfloor \pi/3 \rfloor + \lfloor \pi/9 \rfloor + \cdots = 1 + 0 + 0 + \cdots = 1$. Thus $\lfloor \pi \rfloor ! = 2^1 \cdot 3^1 = 6$.

**67.** **a)** $[14, 29]$

**c)** $\{0\}$

**e)** $\{5, 14, 29\}$

**69.** **a)** $E$

**c)** $E$

**e)** $(E \times \mathbb{N}) \cup (\mathbb{N} \times E)$

**71.** **a)**

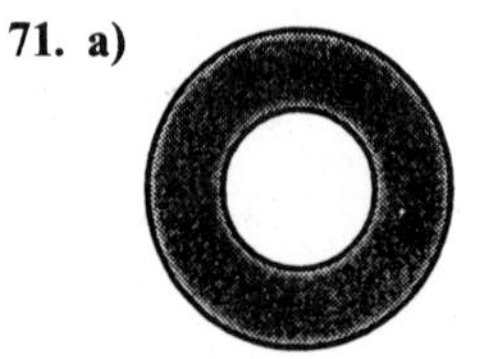

**76.** The argument needed to justify that $f(A)' = f(A')$ is the same argument needed to establish the original proposition.

**81. a)** Yes

**85.** Let $\Lambda = \{-1, 1\}$ and let $f: \mathbb{R} \to \mathbb{R}$ be defined by $f(x) = x^2$. For each $\alpha \in \Lambda$, let $A_\alpha = \{\alpha, 0\}$. Then $f(\cap\{A_\alpha : \alpha \in \Lambda\}) = f(\{0\}) = \{f(0)\} = \{0\}$, whereas $\cap\{f(A_\alpha): \alpha \in \Lambda\} = \{0, 1\}$.

## CHAPTER 6

**1. a)** 8

**b)** 16

**3. a)** $2^{10} = 1024$

**b)** $3^{10} = 59{,}049$

**5. b)** 300

**7. a)** 5040

**c)** 120

**e)** 2520

**9. a)** 24

**11.** 120

**15.** 0

**17.** 55,440

**19.** 616

**21.** 600

**25.** The number of subsets of an $n$-element set, which is $2^n$, is obviously the sum of the number of $i$-element subsets of an $n$-element set, which is $\Sigma_{i=0}^{n} C(n, i)$.

**28.** *Hint*: Imagine performing the long division $n\overline{)1.000\ldots}$ and keep in mind that whenever you divide $n$ into a natural number, the only possible remainders are $0, 1, 2, \ldots, n-1$.

**31.** *Hint*: Consider the different parity (odd–even) patterns of the coordinates of the 9 points.

**32.** *Hint*: If $1 \leqslant x \leqslant 2n$, then $x = 2^m y$, where $y$ is an odd integer and $m$ is a nonnegative integer.

**34.** $C(100, 50)$

**36. a)** $C(14, 7) = 3432$

**39.** 18

**46. a)** 32,768

**48. a)** $C(25, 12)$

**50.** You would be finding the number of two-element subsets of a seven-element set instead of finding the number of five-element subsets of a seven-element set. Notice that these are the same, because for each two-element subset you choose, the set that is left over is the five-element set.

**52. a)** $m_1 = 14$ and $m_4 = 0$

**54.** *Hint*: Since $10{,}000 = 7(1428) + 4$, there are 1428 numbers not exceeding 10,000 that are divisible by 7.

**55.** There are 100 perfect squares that do not exceed 10,000, and there are 21 perfect cubes that do not exceed 10,000. There are 4 numbers that do not exceed 10,000 that are perfect squares and perfect cubes. So the number of natural numbers not exceeding 10,000 that are perfect squares or perfect cubes in $100 + 21 - 4 = 117$. So the number of natural numbers not exceeding 10,000 that are neither perfect squares nor perfect cubes is $10{,}000 - 117 = 9883$.

**60. a)**

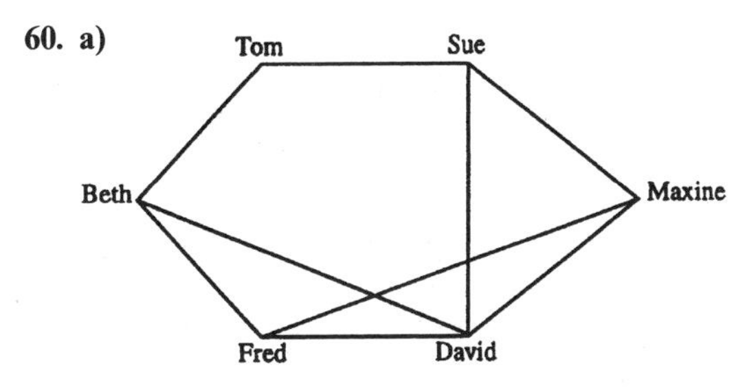


**b)** 2 days; either Sue or Beth

**c)** Yes; if Tom and Sue stopped talking to each other, it would take 3 days to pass a rumor from Tom to Maxine.

**64.** $\lceil p/2 \rceil$

**67.** The primes $p$ such that $2p > n$

**68.** The following graph models the situation; Mr. and Mrs. Smith each shook 3 hands.

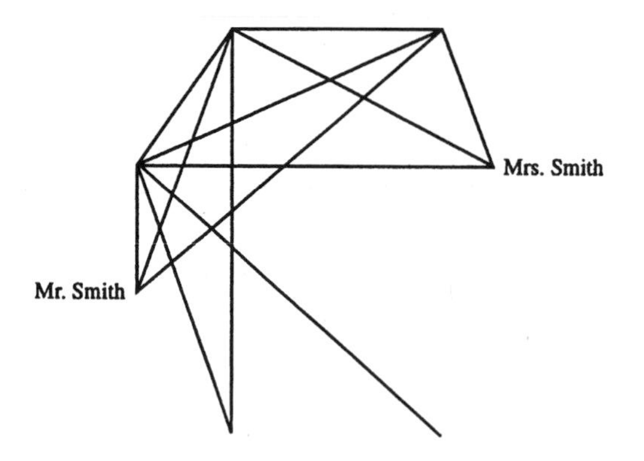


## CHAPTER 7

**1. a)** $i_A$ is a one-to-one function mapping $A$ onto $A$.

**b)** There is a one-to-one function $f$ mapping $A$ onto $B$. Then $f^{-1}$ is a one-to-one function mapping $B$ onto $A$.

**3.** Let $g = \{(n^2, n^3): n \in \mathbb{N}\}$. Then $g$ is a one-to-one mapping from the set of all perfect squares onto the set of all perfect cubes. (To see that $g$ is one-to-one, suppose that $n^3 = g(n^2) = g(m^2) = m^3$. Then $m = n$, so $m^2 = n^2$.)

**5.** $f: (a, b) \to (c, d)$ defined by $f(x) = x + c - a$ is clearly one-to-one. To see that $f$ maps onto $(c, d)$, let $y \in (c, d)$. Then $c < y < d = d - c + c = (b - a) + c$. Therefore $a + c < y + a < b + c$ and $a < y - c + a < b$. Thus $x = y - c + a$ is a member of $(a, b)$ and $f(x) = y$.

**9.** The function $f$ defined by $f(x) = x/(1 + |x|)$ discussed in Proposition 6.6 has the required properties.

**13. a)** *Hint*: Consider the function $f$ defined by $f((x, y)) = (x/2, y/2)$.

**17. a)** $f(n) = n/2$

**b)** *Hint*: By the Fundamental Theorem of Arithmetic (except for the order of appearance), no natural number can be written as the product of prime numbers in more than one way. Thus if $5^{11} \cdot 2^{13} \cdot 17^{19} = 2^x \cdot 17^y \cdot 5^z$, then $x = 13$, $y = 19$, and $z = 11$.

**19.** Apply the Fletcher–Patty Principle.

**20. a)** *Hint*: Pólya's Principle applies. Try insisting that the real numbers are all between 0 and 1 and have only 3's and 4's in their decimal representations.

**b)** *Hint*: Try using only subsets of $\mathbb{N}$ whose members are powers of 2, 3, 5, and 7.

**23. a)** Let $f: \mathscr{C} \to \cup\mathscr{C}$ be defined by $f((-\infty, a]) = a - 17$. Note that for each $C \in \mathscr{C}$, $f(C) \in C$.

**25.** In light of Proposition 7.12, it suffices to argue that the set of decimals between 0 and 1 that are of length $n$ is countable. But there are exactly $10^n$ members in this set.

**28.** Let $A$ be an infinite set. By Proposition 7.11, there is a one-to-one function $g: \mathbb{N} \to A$. Since we are given a one-to-one function $f: A \to \mathbb{N}$, it follows from the Schroeder–Bernstein Theorem that $A$ is countably infinite.

**30.** By Proposition 7.10 there is a one-to-one function $f: A \to \mathbb{N}$. Clearly $f|(A - \{x\})$ maps $A - \{x\}$ into $\mathbb{N}$. Therefore by Proposition 7.10, $A - \{x\}$ is countable.

**33.** By Proposition 7.10, there is a one-to-one

function $f: B \to \mathbb{N}$. Define $g: \{(a, b): b \in B\} \to \mathbb{N}$ by $g((a, b)) = f(b)$. Since $g$ is one-to-one, by Proposition 7.10, $\{(a, b): b \in B\}$ is countable.

**35.** *Hint*: Consider Example 17 from Chapter 6.

**38.** Suppose that $S - A$ is countable. Then, by Proposition 7.12, $S = (S - A) \cup A$ is countable — a contradiction.

**41.** *Hint*: Use the result of Exercise 19.

**43.** By Proposition 7.5, $\mathscr{P}(\mathbb{N})$ and $\mathscr{P}(\mathscr{P}(\mathbb{N}))$ do not have the same cardinality. Let $S = \{\{A\}: A \in \mathscr{P}(\mathbb{N})\}$. Then $S \sim \mathscr{P}(\mathbb{N})$ and both $S$ and $\mathscr{P}(\mathbb{N})$ are uncountable. It follows from Exercise 29 that $\mathscr{P}(\mathscr{P}(\mathbb{N}))$ is also uncountable.

**46.** There are one-to-one functions $\alpha$ mapping $A$ onto $A^*$ and $\beta$ mapping $B$ onto $B^*$. Then $\alpha \cup \beta$: $A \cup B \to A^* \cup B^*$ is a one-to-one function mapping onto $A^* \cup B^*$.

**48.** Let $A$ be a set such that $\alpha = |A|$. Then $\alpha + 0 = |A| + |\varnothing| = |A \cup \varnothing| = |A| = \alpha$.

**51.** Let $A = \{1/(n+1): n \in \mathbb{N}\}$. Then $\aleph_0 + \aleph_0 = |A| + |\mathbb{N}| = |A \cup \mathbb{N}|$. But by Proposition 7.12, $|A \cup \mathbb{N}| = |\mathbb{N}| = \aleph_0$.

**55.** $\aleph_0 \aleph_0 = |\mathbb{N}|\,|\mathbb{N}| = |\mathbb{N} \times \mathbb{N}| = |\cup_{n=1}^{\infty} \{n\} \times \mathbb{N}|$. Under the assumption that the countable union of countably infinite sets is countably infinite, we have that $|\cup_{n=1}^{\infty} \{n\} \times \mathbb{N}| = \aleph_0$.

## CHAPTER 8

**3. a)** [0]
**b)** No

**5.** Yes

**7.** No

**9.**

| * | a | b | c | d |
|---|---|---|---|---|
| a | b | a | d | c |
| b | a | b | c | d |
| c | d | c | b | a |
| d | c | d | a | b |

**14. a)** $[a] \odot [b] = [ab] = [ba] = [b] \odot [a]$.

**16.** Since $\oplus$ is commutative, it is sufficient to show that $[a] \oplus [-a] = [0]$ for each $a \in \mathbb{Z}$. If $a \in \mathbb{Z}$, then $[a] \oplus [-a] = [a + (-a)] = [0]$.

**20.** If $n$ is prime, then by Exercise 19, there are no integers $a$ such that $[a]$ is a divisor of zero in $\mathbb{Z}_n$. Suppose $n$ is not prime. If $a \in \mathbb{Z}$, $1 < a < n$, and $a$ divides $n$, then $n/a$ is an integer, $[a] \neq [0]$, and $[n/a] \neq [0]$, but $[a] \odot [n/a] = [n] = [0]$. Therefore $[a]$ is a divisor of zero. But if $a$ does not divide $n$, then $n/a$ is not an integer, so $[a]$ is not a divisor of zero in $\mathbb{Z}_n$. Therefore those integers $a$ such that $[a]$ is a divisor of zero are those integers which are in the same equivalence class of a divisor of $n$.

**23. a)** Yes, it is a group. The identity element is 1. If $a$ is a positive real number, then $1/a$ is the inverse of $a$.

**c)** Yes, it is a group. The identity element is 1. Each element is its own inverse.

**e)** Yes, it is a group. The identity element is $b$. Each element is its own inverse.

**g)** No, it is not a group. There is no element $e$ such that $a/e = e/a = a$ and 1 is not associative.

**i)** No, it is not a group. Matrix multiplication is not an operation on the given set, and not all elements have inverses.

**25.** No

**27.** Yes

**31.** *Hint*: Use the preceding exercise.

**34.** Associativity follows from Exercise 32. If $f_{1,a}, f_{1,b} \in K$, then $f_{1,a} \circ f_{1,b}(x) = f_{1\cdot 1, 1\cdot b + a}(x) = f_{1,b+a}(x)$, and $f_{1,b+a} \in K$. Therefore $\circ$ is an operation on $K$. Since $f_{1,0} \in K$, $K \neq \varnothing$ and $K$ has an identity. If $f_{1,a} \in K$, then $f_{1,-a} \in K$, $f_{1,-a} \circ f_{1,a}(x) = f_{1\cdot 1, a-a}(x) = f_{1,0}(x)$, and $f_{1,a} \circ f_{1,-a}(x) = f_{1\cdot 1, -a+a}(x) = f_{1,0}(x)$. Therefore $(K, \circ)$ is a group. Finally, $(K, \circ)$ is abelian, since for any $f_{1,a}, f_{1,b} \in K$, $f_{1,a} \circ f_{1,b}(x) = f_{1,b+a}(x) = f_{1,a+b}(x) = f_{1,b} \circ f_{1,a}(x)$.

**38. a)** (12), (13), (23), and (1)

**c)** Two

**41. a)** Yes

**c)** Yes

**43.** Let $S = \{q \in \mathbb{Q}$: when $q$ is reduced to lowest terms, the denominator is a power of 2$\}$. If $n \in \mathbb{Z}$, then $n = n/2^0$, so $\mathbb{Z} \subseteq S$. Obviously $S$ is a proper subset of $\mathbb{Q}$. If $m/2^n$ and $p/2^q$ represent fractions in lowest terms, then when $m/2^n + p/2^q$ is reduced to lowest terms, the denominator is a power of 2. Also if $m/2^n \in S$, then $-m/2^n \in S$. Therefore $S$ satisfies conditions (a) and (b) of Theorem 8.11, and hence it is a subgroup of $(\mathbb{Q}, +)$.

**46.** Define $f: H \to H \circ a$ by $f(h) = h \circ a$ for each $h \in H$. It is sufficient to show that $f$ is a one-to-one map of $H$ onto $H \circ a$. Suppose $h, k \in H$ and $f(h) = f(k)$. Then $h \circ a = k \circ a$, so $(h \circ a) \circ a^{-1} = (k \circ a) \circ a^{-1}$. Hence $h = h \circ e = h \circ (a \circ a^{-1}) = (h \circ a) \circ a^{-1} = (k \circ a) \circ a^{-1} = k \circ (a \circ a^{-1}) = k \circ e = k$. Therefore $f$ is one-to-one. Let $b \in H \circ a$. Then there exists $h \in H$ such that $b = h \circ a$. Now $f(h) = h \circ a = b$, and hence $f$ maps $H$ onto $H \circ a$.

**49.** *Hint*: Consider the mapping that sends $a \circ H$ into $H \circ a^{-1}$.

**52.** The right cosets of $H$ are $H$, $\{(12), (132)\}$, and $\{(13), (123)\}$, and the left cosets of $H$ are $H$, $\{(12), (123)\}$, and $\{(13), (132)\}$.

**55.** Since $i_S \in H$, $H \neq \varnothing$. Suppose $f, g \in H$. Then by Proposition 6.8(b), $f \circ g \in \text{Sym}(S)$. Since $f \circ g(a) = f(g(a)) = f(a) = a$, $f \circ g \in H$. Suppose $f \in H$. By Proposition 6.8(c), $f^{-1} \in \text{Sym}(S)$. Let $b = f^{-1}(a)$. Then $f(b) = a$. Since $f(a) = a$ and $f$ is one-to-one, $b = a$. Therefore $f^{-1} \in H$. So by Theorem 8.11, $(H, \circ)$ is a subgroup of $(\text{Sym}(S), \circ)$.

**60. a)**

| * | a | b | c | d |
|---|---|---|---|---|
| a | a | b | c | d |
| b | b | a | d | c |
| c | c | d | a | b |
| d | d | c | b | a |

**61.** Let $e_G$ and $e_H$ denote the identity elements of $G$ and $H$ respectively. By Theorem 8.15, $h(e_G) = e_H$. Therefore $h(a) * h(a^{-1}) = h(a \circ a^{-1}) = h(e_G) = e_H$ and hence $h(a^{-1})$ is the inverse of $h(a)$.

**64.** The function $g$ defined by $g(a) = e$, $g(b) = d$, and $g(c) = f$ is an isomorphism.

**66.** No

**68.** Let $a, b \in G$. Then $a \circ b = (a^{-1})^{-1} \circ (b^{-1})^{-1} = f(a^{-1}) \circ f(b^{-1}) = f(a^{-1} \circ b^{-1}) = (a^{-1} \circ b^{-1})^{-1} = (b^{-1})^{-1} \circ (a^{-1})^{-1} = b \circ a$. Therefore $G$ is abelian.

**71.** LEMMA. If $(G, \circ)$ and $(H, *)$ are groups and $h: G \to H$ is a homomorphism such that $\ker(h) = e_G$, then $h$ is one-to-one.

**74.** Suppose $a, b \in \ker(h)$. Then $h(a \circ b) = h(a) * h(b) = e_H * e_H = e_H$, so $a \circ b \in \ker(h)$. Suppose $a \in \ker(h)$. Then $h(a^{-1}) = (h(a))^{-1} = (e_H)^{-1} = e_H$, so $a^{-1} \in \ker(h)$. Therefore, by Theorem 8.11, $\ker(h)$ is a subgroup of $G$.

**77.** $(\{(1), (12)\}, \circ)$ is a subgroup of $(S_3, \circ)$. It is not normal because $(13) \in S_3$, $(12) \in \{(1), (12)\}$, and $(13) \circ (12) \circ (13)^{-1} = (13) \circ (12) \circ (13) = (13) \circ (132) = (23) \notin \{(1), (12)\}$.

**82.** *Hint*: Let $S = \{n \in \mathbb{N}$: if $m \in \mathbb{Z}$ then $(a^m)^n = a^{mn}\}$. The proof begins by proving that $S = \mathbb{N}$.

**85.** $\{(1), (123), (132)\}$

**89.** Since $k$ divides $n$, there exists an integer $q$ such that $n = qk$. Let $H = \langle a^q \rangle$. Clearly $(a^q)^k = a^{qk} = a^n = e$. If $p$ is a natural number less than $k$, then $qp < n$ so $(a^q)^p \neq e$. Therefore $H$ is a subgroup of $G$ of order $k$.

**91.** $\{[0]\}$, $\{[0], [9]\}$, $\{[0], [6], [12]\}$, $\{[0], [3], [6], [9], [12], [15]\}$, $\{[0], [2], [4], [6], [8], [10], [12], [14], [16]\}$, and $\mathbb{Z}_{18}$

**94.** The group $(\mathbb{Q}, +)$ is an infinite group which is not cyclic. In order to see that it is not cyclic, suppose $\mathbb{Q} = \langle a/b \rangle$, where $\gcd(a, b) = 1$. Since $a/2b \notin \langle a/b \rangle$, we have a contradiction.

**97.** Let $n=|G|$, and let $a \in G$. Then $\langle a\rangle$ is a subgroup of $G$ and by Lagrange's Theorem, $|\langle a\rangle|$ divides $n$. Thus there exists a natural number $m$ such that $m \cdot |\langle a\rangle| = n$. Therefore $a^n = a^{|\langle a\rangle| m} = (a^{|\langle a\rangle|})^m = e^m = e$.

## CHAPTER 9

**1. b)** Clearly $\max\{p-q, q-p\} = \max\{q-p, p-q\}$

**d)** Since $p \leqslant \max\{p, -p\} = |p|$ and $q \leqslant |q|$, $p+q \leqslant |p|+|q|$. Similarly $-p \leqslant |p|$ and $-q \leqslant |q|$, so $-(p+q) \leqslant |p|+|q|$. It follows that $|p+q| = \max\{(p+q), -(p+q)\} \leqslant |p|+|q|$.

**4.** $[0, 1)$

**7.** Let $\varepsilon = \frac{1}{8}$. Then $(\frac{7}{8}-\varepsilon, \frac{7}{8}+\varepsilon) \subseteq [0, 1]$.

**11. b)** Yes

**14.** *Hint*: Let $p$ and $q$ be two accumulation points of the range of $\langle x_n\rangle$ and let $\varepsilon = |p-q|/2$.

**18.** *Hint*: The sequence $\langle 1/n\rangle$ is a Cauchy sequence and so is not a counterexample, but...

**19.** *Hint*: Suppose that $x<0$ and let $\varepsilon = |x|/2$.

**21.** *Hint*: Use Theorem 9.14.

**23.** *Hint*:

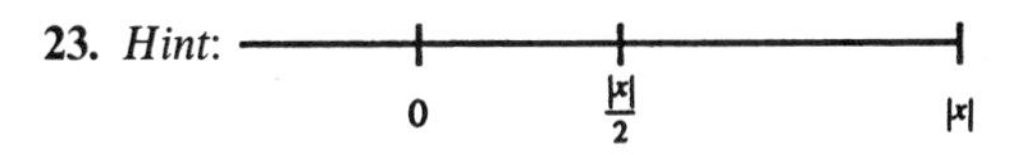


**27.** Let $g: \mathbb{N} \to \mathbb{N}$ be defined by $g(n) = 2n$. Then $\langle (-1)^{g(n)}(1-(1/g(n)))\rangle$ is a convergent subsequence.

**28.** *Hint*: Since $g$ maps into $\mathbb{N}$, $g(1) \geqslant 1$.

**29.** *Hint*: Let $\langle x_{g \circ h(n)}\rangle$ be a subsequence of $\langle x_{g(n)}\rangle$. Then $h(n+1) > h(n)$ for any $n \in \mathbb{N}$. Let $S = \{k \in \mathbb{N}: g(a+k) > g(a)$ for any $a \in \mathbb{N}\}$. By definition, $1 \in S$.

**32.** *Hint*: Use Theorem 9.15.

**34.** *Hint*: Use Theorem 9.24.

**35.** *Hint*: One way to approach this problem is to let $S$ be a nonempty set with a nonempty set $B$ of lower bounds. Argue that $B$ has a least upper bound $p$. Then argue that $p$ must be the greatest lower bound of $S$.

**36.** *Hint*: Suppose that $L_1 \neq L_2$ and let $\varepsilon = |L_1 - L_2|/2$. There are positive real numbers $\delta_1$ and $\delta_2$ such that if $0<|x-x_0|<\delta$, then $|f(x)-L_0|<\varepsilon$ and if $0<|x-x_0|<\delta_2$, then $|f(x)-L_1|<\varepsilon$. Let $\delta = \min\{\delta_1, \delta_2\}$.

**40.** *Hint*: Let $\langle x_n\rangle$ be a good sequence for $g-f$ at $x_0$.

**43.** *Hint*: Theorem 9.1 is useful.

**47.** The set $f(A)$ is nonempty and has an upper bound $f(b)$ and a lower bound $f(a)$.

**49.** *Hint*: Let $L = \text{lub}\{f(x): x \in [a, b]\}$. To see that $\lim_{x\to b} f(x) = L$, let $\varepsilon > 0$. There is an $x \in [a, b]$ such that $L-\varepsilon < f(x)$ (Why?). Let $\delta = b - x$. A similar argument applies to establish that $\lim_{x \to a} f(x) = \text{glb}\{f(x): x \in [a, b]\}$.

**52.** To see that $\lim_{x\to 1}(x^2-1)/(x-1) = 2$, let $f(x) = g(x) = x+1$, $h(x) = (x^2-1)/(x-1)$, and $\delta = 17$. Then for $x \in (1-\delta, 1) \cup (1, 1+\delta)$, $f(x) \leqslant h(x) \leqslant g(x)$; in fact, equality holds. Clearly $\lim_{x\to 1} f(x) = \lim_{x \to 1} h(x) = 2$.

**55.** Let $c$ be the least upper bound of a set $C$ of real numbers and suppose that $c \notin C$. Let $\varepsilon > 0$. There is an $x \in C$ such that $c-\varepsilon < x < c+\varepsilon$. If $(c-\varepsilon, c+\varepsilon) \cap C = \{x\}$, then $x$ is an upper bound of $C$ that is less than $c$, which is impossible. Thus every neighborhood of $c$ contains at least two members of $C$.

**56.** Yes; No

**57.** *Hint*: Use the Intermediate-Value Theorem.

# INDEX

## D



## E



## F